AF358659

Frontiers in Canine and Feline Gastrointestinal Disease

Frontiers in Canine and Feline Gastrointestinal Disease

Editors

Aarti Kathrani
Romy Heilmann

Basel • Beijing • Wuhan • Barcelona • Belgrade • Novi Sad • Cluj • Manchester

Editors

Aarti Kathrani
Clinical Science and Services
Royal Veterinary College
Hatfield
United Kingdom

Romy Heilmann
Department for Small
Animals
University of Leipzig
Leipzig
Germany

Editorial Office
MDPI
St. Alban-Anlage 66
4052 Basel, Switzerland

This is a reprint of articles from the Special Issue published online in the open access journal *Animals* (ISSN 2076-2615) (available at: www.mdpi.com/journal/animals/special_issues/canine_feline_gastrointestinal_disease).

For citation purposes, cite each article independently as indicated on the article page online and as indicated below:

Lastname, A.A.; Lastname, B.B. Article Title. *Journal Name* **Year**, *Volume Number*, Page Range.

ISBN 978-3-7258-1500-5 (Hbk)
ISBN 978-3-7258-1499-2 (PDF)
doi.org/10.3390/books978-3-7258-1499-2

Contents

About the Editors

Aarti Kathrani

Dr. Kathrani graduated from the Royal Veterinary College (RVC) in 2006, before completing her rotating small animal medicine and surgery internship at the Queen Mother Hospital for Animals, RVC, in 2007 and her Ph.D. in canine inflammatory bowel disease at the RVC in 2011. Dr. Kathrani then completed a three-year residency program in small animal internal medicine at Cornell University in 2014. She also completed a two-year small animal clinical nutrition residency program at the University of California-Davis in 2016. She is a diplomate of the American College of Veterinary Internal Medicine (Nutrition). Dr. Kathrani was appointed as a senior lecturer in Small Animal Medicine at the University of Bristol between 2016 and 2018, before returning to the RVC in 2018.

Romy Heilmann

Prof. Dr. Romy Heilmann is a professor (full) of Small Animal Internal Medicine at the University of Leipzig, Germany. She is a board-certified small animal internist (both ACVIM and ECVIM-CA), trained at the Texas A&M University Small Animal Veterinary Teaching Hospital. She holds a Dr. med. vet. (Bern, Switzerland) and a Ph.D. (Texas, USA) and is a fellow of the American Gastroenterological Association (AGA). Her research is focused on chronic enteropathies in dogs and cats, mucosal immunity, and enteroendocrinology.

Preface

Clinical signs of gastrointestinal disease are amongst the most common presenting complaints in small animal veterinary practice. Over the last decade, many scientific advances have been made regarding determining the etiopathogenesis, improving diagnostics, and optimizing the treatment of various gastrointestinal diseases in dogs and cats. Despite this, there are still large knowledge gaps in our understanding of the underlying pathogenesis and, therefore, the development of non-invasive diagnostic methods or efficacious treatments for common gastrointestinal diseases, such as chronic inflammatory enteropathy, canine protein-losing enteropathy, and feline enteric-associated T-cell lymphoma-2. Therefore, the publication of new scientific findings focused on canine and feline gastroenterology will help to advance our understanding, diagnosis, and treatment of various gastrointestinal diseases. The aim of the present Special Issue was to gather, in one publication, the most recent advances in canine and feline gastrointestinal research.

Aarti Kathrani and Romy Heilmann
Editors

Article

Chronic Inflammatory Enteropathy and Low-Grade Intestinal T-Cell Lymphoma Are Associated with Altered Microbial Tryptophan Catabolism in Cats

Patrick C. Barko [1],*, David A. Williams [1], Yu-An Wu [2], Joerg M. Steiner [2], Jan S. Suchodolski [2], Arnon Gal [3] and Sina Marsilio [4]

[1] Departments of Veterinary Clinical Medicine and Pathobiology, University of Illinois at Urbana-Champaign, Urbana, IL 61802, USA

[2] Gastrointestinal Laboratory, School of Veterinary Medicine and Biomedical Sciences, Texas A&M University, College Station, TX 77843, USA

[3] Department of Veterinary Clinical Medicine, University of Illinois at Urbana-Champaign, Urbana, IL 61802, USA

[4] Department of Veterinary Medicine and Epidemiology, UC Davis School of Veterinary Medicine, Davis, CA 95616, USA

* Correspondence: pcbarko@illinois.edu

Simple Summary: Chronic inflammatory enteropathy (CIE) and low-grade intestinal T-cell lymphoma (LGITL) are common chronic intestinal disorders in cats. Gut bacteria are implicated in the initiation and progression of chronic intestinal disorders, and changes in the composition of gut bacteria have been associated with CIE and LGITL in cats. Microbial indole catabolites of tryptophan (MICT) are chemicals produced by gut bacteria that support intestinal health. We hypothesized that, compared with healthy cats, blood concentrations of MICTs would be decreased in cats with CIE and LGITL. Using archived blood samples, we measured tryptophan and eleven of its derivatives, including eight MICTs, and compared them among cats with CIE, LGITL, and healthy controls. Consistent with our hypothesis, the concentrations of tryptophan and five different MICTs (indolepropionate, indoleacrylate, indolealdehyde, indolepyruvate, indolelactate) were decreased in cats with CIE and LGITL compared with those in healthy controls. Our findings are likely explained by changes in tryptophan metabolism related to disturbances in gut bacterial communities and intestinal inflammation and in cats with CIE and LGITL. These findings suggest that MICTs are promising biomarkers that can be used to understand how intestinal bacteria contribute to chronic intestinal disorders such as CIE and LGITL in cats.

Abstract: Chronic inflammatory enteropathy (CIE) and low-grade intestinal T-cell lymphoma (LGITL) are common chronic enteropathies (CE) in cats. Enteric microbiota dysbiosis is implicated in the pathogenesis of CE; however, the mechanisms of host–microbiome interactions are poorly understood in cats. Microbial indole catabolites of tryptophan (MICT) are gut bacterial catabolites of tryptophan that are hypothesized to regulate intestinal inflammation and mucosal barrier function. MICTs are decreased in the sera of humans with inflammatory bowel disease and previous studies identified altered tryptophan metabolism in cats with CE. We sought to determine whether MICTs were decreased in cats with CE using archived serum samples from cats with CIE ($n = 44$) or LGITL ($n = 31$) and healthy controls ($n = 26$). Quantitative LC-MS/MS was used to measure serum concentrations of tryptophan, its endogenous catabolites (kynurenine, kynurenate, serotonin) and MICTs (indolepyruvate, indolealdehyde, indoleacrylate, indoleacetamide, indoleacetate, indolelactate, indolepropionate, tryptamine). Serum concentrations of tryptophan, indolepropionate, indoleacrylate, indolealdehyde, indolepyruvate, indolelactate were significantly decreased in the CIE and LGITL groups compared to those in healthy controls. Indolelactate concentrations were significantly lower in cats with LGITL compared to CIE ($p = 0.006$). Significant correlations were detected among serum MICTs and cobalamin, folate, fPLI, and fTLI. Our findings suggest that MICTs are promising biomarkers to investigate the role of gut bacteria in the pathobiology of chronic enteropathies in cats.

Citation: Barko, P.C.; Williams, D.A.; Wu, Y.-A.; Steiner, J.M.; Suchodolski, J.S.; Gal, A.; Marsilio, S. Chronic Inflammatory Enteropathy and Low-Grade Intestinal T-Cell Lymphoma Are Associated with Altered Microbial Tryptophan Catabolism in Cats. *Animals* **2024**, *14*, 67. https://doi.org/10.3390/ani14010067

Academic Editor: Sylvia García-Belenguer

Received: 23 October 2023
Revised: 5 December 2023
Accepted: 12 December 2023
Published: 23 December 2023

Keywords: microbial indole catabolites of tryptophan; indole; chronic enteropathy; inflammatory bowel disease; alimentary small cell lymphoma

1. Introduction

Chronic inflammatory enteropathy (CIE), also called inflammatory bowel disease (IBD), and low-grade intestinal T-cell lymphoma (LGITL) are two common forms of chronic enteropathy (CE) in cats [1–3]. The coexistence of inflammatory and neoplastic lesions in cats with CE and the progression of CIE to LGITL in some affected cats suggest that these two conditions share a common etiopathogenesis [1,2]. Complex and reciprocal interactions among commensal microbes and host tissues are critical for regulating inflammatory responses and cell proliferation in the enteric mucosa, and these interactions are largely mediated by bacterial metabolites, including short chain fatty acids, secondary bile acids, and tryptophan derivatives [3,4]. Conversely, enteric microbiota dysbiosis (EMD), defined by an imbalance between beneficial commensal microbes and opportunistic pathobionts, contributes to the pathogeneses of inflammatory and neoplastic disorders of the intestinal mucosa [5]. EMD has been identified in cats with CIE and LGITL, but there is a paucity of data regarding the mechanisms through which enteric microbiota influence mucosal health and contribute to the pathogenesis of CE in cats [6,7].

Tryptophan is an essential amino acid and emerging evidence from research in humans and rodents suggests that catabolic derivatives of tryptophan participate in regulating intestinal mucosal health and homeostasis. Tryptophan is catabolized through endogenous and exogenous (microbial) pathways (Figure 1A) [8–12]. Approximately 90–95% of tryptophan is catabolized through the endogenous kynurenine pathway (KP), which is controlled by the rate-limiting enzymes tryptophan 2,3-dioxygenase (TDO) in the liver and indoleamine 2,3-dioxygenase-1 (IDO1) in extrahepatic tissues, including leukocytes and intestinal epithelial cells [8]. The TDO-dependent hepatic KP is primarily regulated by endocrine factors (e.g., glucocorticoids, glucagon), whereas the IDO-dependent extrahepatic KP is stimulated by inflammatory cytokines [13]. Approximately 5% of ingested tryptophan escapes absorption in the small intestine and becomes a substrate for metabolism by gut microbiota, resulting in the generation of microbial indole catabolites of tryptophan (MICT), which are readily absorbed by the jejunal, ileal, and colonic mucosa [12,14,15]. MICTs, including indolepropionate, indolealdehyde, indolelactate, and indoleacrylate, are exclusive products of gut microbial metabolism [15–17]. Smaller fractions of ingested tryptophan (1–2%) are used for the synthesis of serotonin by enterochromaffin cells, and tryptamine, which can be produced by both bacterial and mammalian cells [18–20]. Emerging evidence suggests that disruptions in tryptophan catabolism are associated with inflammatory bowel diseases and that MICTs play a role in regulating gastrointestinal immunity and the mucosal barrier function [12,15,21].

Recent studies have revealed altered tryptophan catabolism in humans with IBD, characterized by decreased serum concentrations of tryptophan and indolepropionate, decreased fecal concentrations of indoleacetate, and increased serum concentrations of kynurenine pathway catabolites [22–26]. Serum tryptophan concentrations were inversely correlated with disease activity and serum C-reactive protein, whereas increasing ratios of kynurenine–tryptophan were associated with increased odds of observing endoscopic inflammatory lesions and worse clinical outcomes in humans with IBD [23,24]. Mechanistic investigations in rodents and in vitro with human intestinal epithelial cell cultures implicate MICTs in regulating intestinal mucosal inflammation and mucosal barrier function (Figure 1B). MICTs induce aryl hydrocarbon (AhR) and pregnane-X (PXR) receptor-mediated expression of tight junction proteins (TJP), mucin, anti-microbial peptides (AMP), and anti-inflammatory cytokines (e.g., IL-10) while decreasing the expression of pro-inflammatory cytokines (e.g., TNFα) in the enteric mucosa. [12,15,20,27–31]. A variety of bacteria are known to catabolize tryptophan into MICTs, including various species of *Lactobacillus*, *Bacteroides*, *Bifidobacterium*, *Peptostreprococcus*, and *Clostridium*, which are often decreased in EMD associated with chronic enteropathies in humans and other animal species, including cats [15,30]. Not only are MICTs potential biomarkers for disruptions in gut microbial metabolism, but they are also promising therapeutic targets. For example, the administration of probiotics containing MICT-generating species and/or dietary MICT supplementation ameliorates mucosal inflammation and EMD in animal models of IBD [30,32,33].

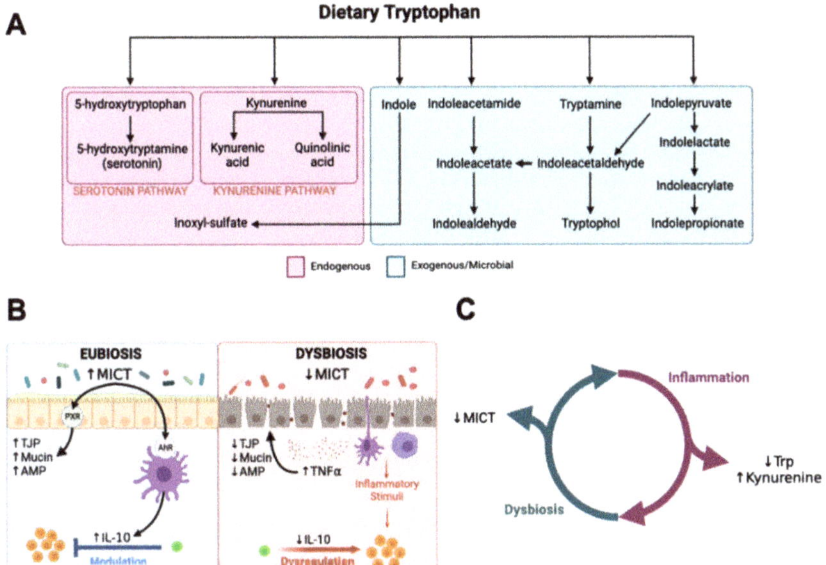

Figure 1. Microbial indole catabolites of tryptophan. (**A**) Arrows represent hypothetical pathways for tryptophan catabolism summarized from previous studies. Endogenous (mammalian) catabolic pathways are presented in the pink box. Gut microbial pathways are presented in the green box. (**B**) Proposed mechanisms through which microbial indole catabolites of tryptophan impact intestinal mucosal health from research in rodents and in vitro in human intestinal epithelial cell cultures. MICTs support intestinal mucosal homeostasis via interactions with aryl hydrocarbon (AhR) and pregnane-X (PXR) receptors to increase expression of tight junction proteins (TJP), mucin, antimicrobial peptides (AMP), and IL-10. Synthesis of MICTs is reduced in dysbiosis associated with IBD and this can perpetuate mucosal inflammation and increase intestinal permeability due to decreased IL-10 and increased expression of pro-inflammatory cytokines (TNFα). (**C**) Graphical hypothesis. Mucosal inflammation in CE will increase catabolism of tryptophan via the kynurenine pathway and CE-associated dysbiosis will decrease synthesis of MICTs by gut microbes in cats with CE. Figures created with BioRender.com. (**B**) was adapted from "Immune Response in Inflammatory Bowel Disease", by BioRender.com (2023), retrieved from https://app.biorender.com/biorender-templates, accessed on 14 December 2023.

A previous untargeted metabolomics study of feces from cats with CE identified evidence of altered tryptophan metabolism in cats with CE, and a pilot study revealed decreased concentrations of tryptophan and indole-3-lactic acid in the sera of cats with CIE and LGITL compared with healthy controls [34,35]. We hypothesized that CIE and LGITL would be associated with altered tryptophan catabolism in cats, characterized by decreased serum concentrations of tryptophan and MICTs, and increased serum concentrations of kynurenine (Figure 1C). Here, we report the results of a comprehensive survey of tryptophan catabolites in the sera of cats with CE and healthy controls using quantitative LC-MS/MS. By investigating associations between microbial tryptophan catabolites and

CIE and LGITL, we aim to generate novel insights that can be used to understand the role of gut microbiota in the pathogenesis and pathophysiology of CE in cats.

2. Materials and Methods

2.1. Patient Population and Sample Acquisition

This was a case–control investigation using archived serum samples from client-owned cats with spontaneously occurring CIE and LGITL. Samples from cats with CE were collected between 2015 and 2019 at three different institutions for either routine diagnostics or from participants in previous investigations unrelated to this study. These cats had undergone comprehensive diagnostic evaluation by their primary veterinarians and/or the clinical investigators. The medical records and experimental metadata for these cats were reviewed by the authors to extract data related to clinical signs, physical examination findings, clinicopathologic results (including serum concentrations of cobalamin, folate, fPLI, and fTLI), results of histopathology, medication history, and other concurrent diagnoses. In cases with missing values for serum concentrations of cobalamin, folate, fPLI, and fTLI, these assays were performed on residual serum, when available. Though diagnostic evaluations varied among institutions, we applied uniform inclusion and exclusion criteria. To be eligible for inclusion, cats had to have documentation of active clinical signs of gastrointestinal dysfunction, including vomiting, diarrhea, weight loss, anorexia, or some combination thereof, that had persisted for more than three weeks. Additionally, all cats with clinical signs consistent with CE had to have histopathologic findings consistent with CIE or LGITL based on examination of endoscopic or full-thickness, surgical biopsies by a board-certified veterinary pathologist. For cases in which the pathologist could not differentiate between CIE and LGITL based on routine histopathology, ancillary testing with immunohistochemistry and PCR for antigen receptor rearrangement (PARR) was performed for diagnostic confirmation. A final diagnosis of CIE or LGITL was reached upon integration of results from histopathology with immunohistochemistry and PARR when available, consistent with previous studies in cats and current EuroClonality/BIOMED-2 guidelines and a recent consensus statement from the American College of Veterinary Internal Medicine (ACVIM) [36–42]. Cats were excluded if they were documented to have received antibiotics or immunomodulatory drugs within 4 weeks of sample collection. Cats with documented hyperthyroidism were excluded unless clinical signs of gastrointestinal dysfunction persisted following the treatment of hyperthyroidism. Cats with documented diagnoses of other neoplasms (aside from LGITL) or exocrine pancreatic insufficiency (serum fTLI concentrations ≤ 12.0 µg/L) were excluded.

Archived serum samples from healthy cats collected for previous, unrelated investigations between 2015 and 2021 were used as healthy controls. The health status of cats in this group was verified by medical histories covering the following areas: attitude/activity, appetite, drinking, urination, chronic illnesses, weight loss, vomiting, diarrhea, and treatment with antibiotics or immunomodulatory drugs. Physical examinations were performed by board-certified small animal internal medicine specialists. The body condition scores were assessed using a previously established nine-point scoring system [43]. Blood was collected from a peripheral vein or the jugular vein, and the following tests were performed: complete blood count, serum chemistry profile, and serum concentrations of total T4, cobalamin, folate, feline pancreatic lipase immunoreactivity (Idexx Spec fPL), and feline trypsin-like immunoreactivity (fTLI; TAMU). Cats with gastrointestinal signs (weight loss, hyporexia, vomiting > 2x/month, diarrhea) within 6 months prior to enrollment were excluded. Cats with systemic diseases, chronic illnesses, or clinically significant laboratory abnormalities were also excluded from the study. Cats with serum concentrations of cobalamin < 290 ng/L, folate < 9.7 µg/L or > 21.6 µg/L, fPLI > 3.5 µg/L, or fTLI ≤ 12.0 µg/L were excluded. Finally, cats that had received any antibiotics or immunomodulatory drugs within 6 months prior to sample collection were excluded.

Archived sera from the cats with CE and healthy controls had been stored below $-70\ ^{\circ}$C at participating institutions and mailed overnight on dry ice to the investigators for use in the present study. Following receipt, they were stored at $-80\,^{\circ}\mathrm{C}$ and transported on dry ice for measurement of serum tryptophan catabolites.

2.2. Quantification of Tryptophan Catabolites

Quantitative (targeted) liquid chromatography–mass spectrometry (LC-MS) was used to measure serum concentrations of tryptophan (Sigma-Aldrich Cat. PHR1176), seven different MICTs, and other tryptophan catabolites, including kynurenine, kynurenic acid, tryptamine, and serotonin against known dilutions of analytic standards. All analytic standards were sourced from the same chemical supplier (Sigma-Aldrich, St. Louis, MO, USA) at the highest purity available: tryptophan (Cat. PHR1176), indolepyruvate acid (Cat. I7017), indolealdehyde (Cat. 129445), indoleacrylate (Cat. I2273), indoleacetamide (Cat. 286281), indoleacetate (Cat. I3750), indolelactate (Cat. I5508), indolepropionate (Cat. 220027), kynurenine (Cat. 67653), kynurenic acid (Cat. 67667), tryptamine (Cat. 76706), and serotonin (Cat. 14927). The analytic standards were diluted in 70% methanol and the dilution range for the calibration curve was 0.25–5000 ng/mL. LC-MS/MS was performed at the Carver Metabolomics Core in the Roy J. Carver Biotechnology Center (University of Illinois, Urbana, IL, USA). Serum samples (30 μL) were spiked with 10 μL of internal standards (1 μg/mL), deproteinized with methanol (70 μL), centrifuged, and 2 μL of the supernatant was injected into the LC-MS. Chromatography was performed using the Vanquish system (Thermo Scientific, Waltham, MA, USA), with a Waters Acquity UPLC BEH C18, (2.1 × 150 mm; 1.7 μm) column with a flow rate of 300 μL/min and two mobile phases (0.1% formic acid in water; 0.1% formic acid in acetonitrile) with a column chamber temperature of $40\,^{\circ}\mathrm{C}$. Mass spectrometry utilized a TSQ Altis LC-MS/MS system (Thermo Scientific). Data were acquired in both positive and negative SRM modes at 3500 V and 5000 V, respectively. Peak integration and quantitation were performed with Thermo TraceFinder software (version 4.1).

2.3. Statistical Analysis

Statistical analyses were performed using the R language for statistical computing (v. 4.2.1) [44]. Reproductive status was compared among the groups using the Chi-square (χ^2) test. The distribution of continuous numerical variables was assessed by examining histograms and Shapiro–Wilk tests. Owing to violations of normality for nearly all analytes, non-parametric methods were utilized for inferential statistics and the results are described using the median and interquartile range (IQR). Kruskal–Wallis tests were used to compare numerical variables among the groups. Variables that varied significantly in the overall test were compared group-wise using post hoc Dunn's tests. To detect correlations among gastrointestinal biomarkers and serum concentrations of tryptophan catabolites, Spearman's rank correlation coefficients and corresponding p-values were calculated for all pairwise combination of variables. As the Kruskal–Wallis and Spearman rank correlation tests utilized multiple comparisons, p-values were adjusted (p_{adj}) using the Benjamini–Hochberg method to control for false discovery [45]. Features were considered significantly different using a two-tailed significance threshold of $\alpha < 0.05$.

3. Results

3.1. Patient Population

Archived sera from 101 cats were obtained, including 44 cats with CIE, 31 cats with LGITL, and 26 healthy controls. Demographic and clinical characteristics of this patient cohort are summarized in Table 1. There were significant differences in age ($p = 0.01$) and body condition score ($p < 0.001$) among the groups. Cats in the LGITL group (median 12 years, IQR = 8.0–13.0; $p < 0.001$) were significantly older than healthy controls (median 10 years, IQR = 8–11), but there were no other statistically significant differences in age among the groups. Compared with healthy controls (median 6, IQR = 5–8), cats in the CIE (median 5, IQR = 4–5; $p < 0.001$) and LGITL groups (median 4, IQR = 4–5; $p < 0.001$) had significantly lower body condition scores (1–9 scale), but there were no significant differences between the CIE and LGITL groups (p = 0.39). All cats were either spayed females or neutered males and there were no statistically significant differences in the proportions of spayed females or neutered male cats among the groups (p = 0.853). Serum concentrations of cobalamin (p < 0.001) and fTLI (p < 0.001) were significantly different among the groups and were significantly lower in cats with CIE (median 743.0 ng/L, IQR = 378.8–891.3; p < 0.001) and LGITL (median 615 ng/L, IQR = 209.0–906.5; p < 0.001) compared with healthy controls (median 1000 ng/L, IQR = 826.3–1000.0). There was no significant difference in serum cobalamin concentrations between the CIE and LGITL groups (p = 0.50). Serum fTLI concentrations were significantly higher in cats with LGITL (median 56.8 µg/L, IQR = 38.5–84.4) than the CIE (median 32.2 µg/L, IQR = 27.8–47.5; p = 0.003) and healthy control groups (median 32.6 µg/L, IQR = 23.1–39.0; p < 0.001), but there were no other significant differences between the CIE and LGITL groups. Serum fPLI concentrations were significantly different among the groups (p = 0.045) in the Kruskal–Wallis test and were higher in the LGITL group compared with the CIE and healthy control groups, but the differences were not statistically significant in the pairwise Dunn's tests. There were no significant differences in serum folate concentrations among the groups (p = 0.66). Extra-intestinal comorbidities in the CE groups included chronic renal disease (n = 17) and hyperthyroidism (n = 6), with two of these cats having both chronic renal disease and hyperthyroidism. Histopathologic findings from the livers and pancreata of 57 cats with CE were available for review in the medical records. Inflammatory infiltrates were identified in the livers of 11 cats and in the pancreata of 3 cats, and all 3 cats with pancreatitis also had inflammatory infiltrates present in the liver.

Table 1. Demographic and clinical data. The p-values were generated from Kruskal–Wallis tests (age, BCS, cobalamin, folate, fPLI, fTLI) or Chi-Square tests (sex). Post hoc Dunn's tests were performed for numeric variables that were significantly different (p < 0.05) among the groups in the Kruskal–Wallis test. Groups with the same superscript did not differ significantly, whereas those with different superscripts were significantly different in post hoc Dunn's tests (p_{adj} < 0.05). CIE, chronic inflammatory enteropathy; LGITL, low-grade intestinal T-cell lymphoma. The upper and lower limits of detection for serum cobalamin were 150 ng/L and 1000 ng/L, respectively.

	Healthy (n = 26)	CIE (n = 44)	LGITL (n = 31)	p-Value
Age (years)				0.010
Median	10.0 [a]	10.5 [a,b]	12.0 [b]	
IQR	8.0–11.0	7.0–13.0	8.0–13.0	
Min–Max	3–14	2–17	7–15	
Sex				0.85
Spayed Female	15 (57.7%)	23 (52.3%)	18 (58.1%)	
Neutered Male	11 (42.3%)	21 (47.7%)	13 (41.9%)	
BCS (1–9)				<0.001
Median	6 [a]	5 [b]	4 [b]	
IQR	5–8	4–5	4–5	
Min–Max	5–9	3–7	1–9	
Cobalamin (ng/L)				<0.001
Median	1000.0 [a]	743.0 [b]	615.0 [b]	
IQR	826.3–1000.0	378.8–891.3	209.0–906.5	
Min–Max	311.0–1000.0	150.0–1000.0	150.0–1000.0	
Folate (µg/L)				0.66
Median	15.8	16.1	16.0	
IQR	14.1–17.9	10.7–19.7	13.2–21.2	
Min–Max	11.0–21.5	5.2–38.0	9.2–48.0	

Table 1. *Cont.*

	Healthy (*n* = 26)	**CIE (*n* = 44)**	**LGITL (*n* = 31)**	***p*-Value**
fPLI (µg/L)				0.045
Median	2.0 [a]	2.0 [a]	2.7 [a]	
IQR	1.4–2.5	1.3–2.98	1.5–5.98	
Min–Max	0.9–3.2	0.5–21.6	1.0–51.0	
fTLI (µg/L)				<0.001
Median	32.6 [a]	32.2 [a]	56.8 [b]	
IQR	23.1–39.0	27.8–47.5	38.5–84.4	
Min–Max	15.9–63.7	14.3–231.0	15.4–243.7	

3.2. Serum Concentrations of Tryptophan Catabolites

Serum concentrations of tryptophan ($p < 0.001$) varied significantly among the groups (Table 2; Figure 2) and were lower in the CIE (median 10.8 µg/mL, IQR = 9.0–13.3; µg/mL; $p < 0.001$) and LGITL (median 9.3 µg/mL, IQR = 8.4–11.7; µg/mL; $p < 0.001$) groups compared with those in the healthy controls (median 16.9 µg/mL, IQR = 15.5–18.7). There were no significant differences in serum tryptophan concentrations between the CIE and LGITL groups ($p = 0.17$). Serum concentrations of kynurenine ($p = 0.23$) and kynurenate ($p = 0.63$) did not vary significantly among the groups, despite being higher in the CIE and LGITL groups compared with those in the healthy controls. There were no significant differences in serum serotonin concentrations ($p = 0.15$) among the groups.

Table 2. Serum concentrations of tryptophan catabolites. The *p*-values were generated from Kruskal–Wallis tests and post hoc Dunn's tests were performed for variables that were significantly different ($p < 0.05$) among the groups. In rows containing metabolites that were significantly different, groups with the same superscript indicate variables that did not differ significantly, whereas those with different superscripts indicate variables that were significantly different in post hoc Dunn's tests ($p_{adj} < 0.05$). CIE, chronic inflammatory enteropathy; LGITL, low-grade intestinal T-cell lymphoma.

	Healthy (*n* = 26)	**CIE (*n* = 44)**	**LGITL (*n* = 31)**	**Kruskal–Wallis *p*-Value**
Tryptophan (µg/mL)				<0.001
Median	16.9 [a]	10.8 [b]	9.3 [b]	
IQR	15.5–18.7	9.0–13.3	8.4–11.7	
Min–Max	12.2–25.0	3.2–21.5	4.1–18.9	
Kynurenine (µg/mL)				0.23
Median	1.04	1.14	1.26	
IQR	0.888–1.237	0.843–1.57	1.01–1.53	
Min–Max	085–1.94	0.513–9.81	0.601–2.52	
Kynurenate (ng/mL)				0.63
Median	8.9	10.3	9.0	
IQR	7.9–10.9	6.5–14.9	6.6–13.8	
Min–Max	4.6–22.7	3.7–136.2	4.6–37.3	
Serotonin (ng/mL)				0.15
Median	888.0	760.9	954.2	
IQR	677.8–1160.0	516.0–1038.8	759.5–1235.9	
Min–Max	293.9–1439.3	12.8–3214.1	13.4–2158.5	
Indoleacetate (ng/mL)				0.21
Median	182.8	267.3	210.0	
IQR	139.9–300.4	172.9–420.0	95.7–418.1	
Min–Max	62.7–980.1	22.1–3219.3	51.8–1946.7	

Table 2. *Cont.*

	Healthy (*n* = 26)	CIE (*n* = 44)	LGITL (*n* = 31)	Kruskal–Wallis *p*-Value
Indolepropionate (ng/mL)				<0.001
Median	55.4 [a]	28.2 [b]	26.1 [b]	
IQR	43.7–71.5	18.0–48.0	16.1–42.7	
Min–Max	7.2–127.0	3.9–416.5	3.6–601.7	
Indoleacetamide (ng/mL)				0.034
Median	1.4 [a]	1.4 [a]	1.4 [a]	
IQR	1.4–1.4	1.4–1.4	1.4–1.4	
Min–Max	1.3–2.0	1.3–2.1	1.3–1.5	
Indoleacrylate (ng/mL)				<0.001
Median	914.2 [a]	606.0 [b]	488.3 [b]	
IQR	834.5–1006.2	478.3–687.8	445.5–615.3	
Min–Max	657.5–1448.8	165.0–1239.3	202.7–1018.4	
Indolelactate (ng/mL)				<0.001
Median	252.7 [a]	165.5 [b]	105.0 [c]	
IQR	187.8–343.8	102.9–271.8	79.0–145.2	
Min–Max	93.8–713.1	56.0–691.7	42.1–355.6	
Indolepyruvate (ng/mL)				0.001
Median	1154.7 [a]	1127.0 [b]	1119.1 [b]	
IQR	1127.1–1177.5	1109.4–1142.7	1108.7–1130.6	
Min–Max	1105.4–1252.6	1090.9–1387.2	1096.3–1207.7	
Indolealdehyde (ng/mL)				<0.001
Median	39.2 [a]	25.9 [b]	22.2 [b]	
IQR	37.4–44.9	20.2–32.3	19.3–26.7	
Min–Max	27.9–60.5	6.4–53.2	8.5–45.1	
Tryptamine (ng/mL)				0.160
Median	3.5	3.4	3.6	
IQR	3.2–4.0	3.0–3.7	3.3–3.9	
Min–Max	2.6–4.2	0.0–6.3	2.4–4.8	

Concentrations of several serum MICTs varied significantly among the groups (Table 2; Figure 2): indolepropionate ($p < 0.001$), indoleacrylate ($p < 0.001$), indolelactate ($p < 0.001$), indolepyruvate ($p = 0.001$), and indolealdehyde ($p < 0.001$). Indolacetamide concentrations were found to be significantly different ($p = 0.034$) among the groups in the overall test; however, the medians and IQRs did not differ among the groups and there were several extreme outliers. Thus, indolacetamide was not considered significantly different and was not subjected to post hoc testing. Post hoc pairwise comparisons between groups revealed that serum concentrations of indolepropionate, indoleacrylate, indolepyruvate, and indolealdehyde were significantly decreased in the CIE and LGITL groups compared with healthy controls (Table 3). Serum indolelactate concentrations were significantly lower ($p_{adj} = 0.008$) in cats with LGITL (median 105.0 ng/mL, IQR = 79.0–145.2) compared with IBD (median 165.5 ng/mL, IQR = 102.9–271.8), but there were no significant differences in the serum concentrations of other MICTs between the CIE and LGITL groups. Serum concentrations of tryptamine were not significantly different among the groups ($p = 0.16$). Graphical outputs (boxplots) for all analytes are shown in the Supplementary Materials (File S1).

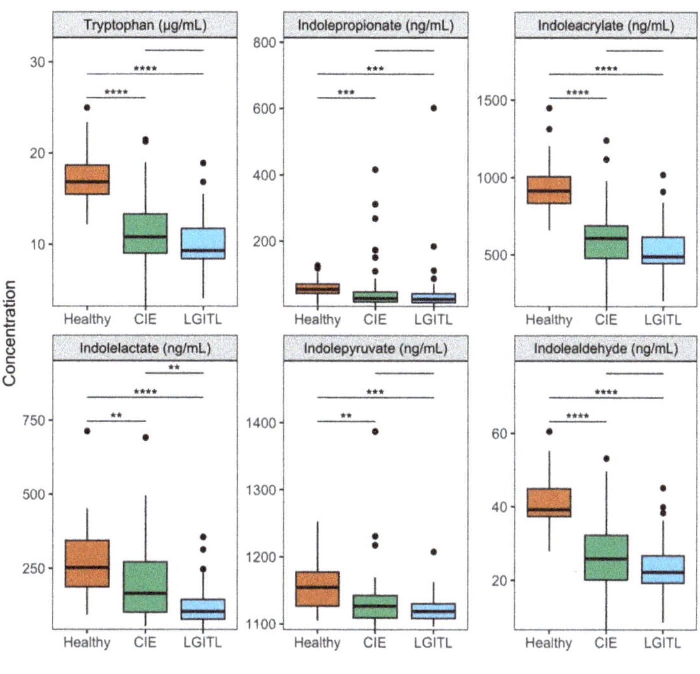

Figure 2. Serum concentrations of significantly variable tryptophan catabolites. Boxplots were drawn for each variable that differed significantly among the groups in the overall Kruskal–Wallis tests. The upper and lower boundaries of the box represent the 25th and 75th percentiles and the horizontal line represents the median. The whiskers represent the maximum and minimum values below and above the upper (75th percentile + IQR) and lower (and 25th percentile—IQR) fences, respectively. Bars denote statistically significant group-wise comparisons from post hoc Dunn's tests and the asterisks correspond to the resulting adjusted p-values (p_{adj}). **, $p_{adj} < 0.01$; ***, $p_{adj} < 0.001$, ****, $p_{adj} < 0.0001$. CIE, chronic inflammatory enteropathy; LGITL, low-grade intestinal T-cell lymphoma.

Table 3. Results of post hoc Dunn's tests. The Z-statistics and adjusted p-values (p_{adj}) from Dunn's tests are shown for group-wise contrasts of every variable that differed significantly among the groups in the overall Kruskal–Wallis tests. CIE, chronic inflammatory enteropathy; LGITL, low-grade intestinal T-cell lymphoma.

Variable	Contrast	Dunn's Z Statistic	p_{adj} Value
Tryptophan	Healthy vs. CIE	−4.96	<0.001
	Healthy vs. LGITL	−5.83	<0.001
	CIE vs. LGITL	−1.37	0.17
Indolepropionate	Healthy vs. CIE	−3.37	0.001
	Healthy vs. LGITL	−3.60	<0.001
	CIE vs. LGITL	−0.53	0.60
Indolepyruvate	Healthy vs. CIE	−3.14	0.003
	Healthy vs. LGITL	−3.43	0.002
	CIE vs. LGITL	−0.38	0.56

Table 3. *Cont.*

Variable	Contrast	Dunn's Z Statistic	p_{adj} Value
Indolelactate	Healthy vs. CIE	−2.60	0.009
	Healthy vs. LGITL	−4.87	<0.001
	CIE vs. LGITL	−2.77	0.008
Indolealdehyde	Healthy vs. CIE	−4.88	<0.001
	Healthy vs. LGITL	−5.74	<0.001
	CIE vs. LGITL	−1.36	0.17
Indoleacrylate	Healthy vs. CIE	−5.08	<0.001
	Healthy vs. LGITL	−5.93	<0.001
	CIE vs. LGITL	−1.36	0.17

3.3. Correlation Analysis

Several serum tryptophan catabolite concentrations were significantly correlated with serum cobalamin, folate, fTLI and fPLI concentrations (Table 4). Serum cobalamin concentrations were positively correlated with indolealdehyde (r = 0.39; p_{adj} < 0.001), tryptophan (r = 0.39; p_{adj} = 0.001), indoleacrylate (r = 0.37; p_{adj} = 0.001), indolepyruvate (r = 0.31; p_{adj} = 0.007), and indolelactate (r = 0.27; p_{adj} = 0.02); and negatively correlated with kynurenine (r = −0.25; p_{adj} = 0.03). Serum fPLI concentrations were correlated positively with kynurenine (r = 0.28; p_{adj} = 0.016) and negatively with tryptophan (r = −0.28; p_{adj} = 0.016), indoleacrylate (r = −0.29; p_{adj} = 0.016), and indolealdehyde (r = −0.29; p_{adj} = 0.016). Similarly, serum fTLI concentrations were correlated positively with kynurenine (r = 0.29; p_{adj} = 0.013) and negatively with tryptophan (r = −0.38; p_{adj} = 0.001), indolealdehyde (r = −0.38; p_{adj} = 0.001), indoleacrylate (r = −0.38; p_{adj} = 0.001), and indolelactate (r = −0.28; p_{adj} = 0.013). There were numerous statistically significant correlations among tryptophan catabolites, which are summarized (Table 5, Figure 3). The complete results of the correlation analysis are shown in the Supplementary Materials (File S2).

Table 4. Correlations among tryptophan catabolites and established gastrointestinal and pancreatic biomarkers in serum. Spearman's rank correlation coefficients (rho) were calculated for combinations of tryptophan catabolites and clinical gastrointestinal biomarkers. *p*-values are adjusted for multiple comparisons (p_{adj}) using the Benjamini–Hochberg method to control for false discovery. Shown here are statistically significant correlations (p_{adj} < 0.05).

Clinical Variable	Tryptophan Catabolite	Spearman's Rho	p_{adj}
Cobalamin	Indolealdehyde	0.39	<0.001
	Tryptophan	0.39	0.001
	Indoleacrylate	0.37	0.001
	Indolepyruvate	0.31	0.007
	Indolelactate	0.27	0.02
	Kynurenine	−0.25	0.03
fPLI	Kynurenine	0.28	0.016
	Tryptophan	−0.28	0.016
	Indoleacrylate	−0.29	0.016
	Indolealdehyde	−0.29	0.016
fTLI	Kynurenine	0.29	0.013
	Indolelactate	−0.28	0.013
	Indoleacrylate	−0.38	0.001
	Indolealdehyde	−0.38	0.001
	Tryptophan	−0.38	0.001

Table 5. Correlations among tryptophan catabolites. Spearman's rank correlation coefficients (rho) were calculated for pairwise combinations of tryptophan catabolites. *p*-values are adjusted for multiple comparisons (p_{adj}) using the Benjamini–Hochberg method to control for false discovery. Shown here are statistically significant correlations ($p_{adj} < 0.05$).

		Spearman's Rho	p_{adj}
Tryptamine	Serotonin	0.98	<0.001
Tryptophan	Indolealdehyde	0.99	<0.001
	Indoleacrylate	0.99	<0.001
	Indolelactate	0.64	<0.001
	Indolepyruvate	0.42	<0.001
Serotonin	Indoleacetate	0.26	0.034
Kynurenate	Kynurenine	0.72	<0.001
Indolepropionate	Indoleacetate	0.27	0.026
Indolepyruvate	Indoleacrylate	0.41	<0.001
	Indolelactate	0.34	0.003
Indolelactate	Indoleacrylate	0.62	<0.001
	Indolepropionate	0.34	0.0023
	Indoleacetamide	0.28	0.02
Indolealdehyde	Indoleacrylate	0.99	<0.001
	Indolelactate	0.65	<0.001
	Indolepyruvate	0.41	<0.001
Indoleacetamide	Indoleacetate	0.3	0.01
	Indolepropionate	0.27	0.027

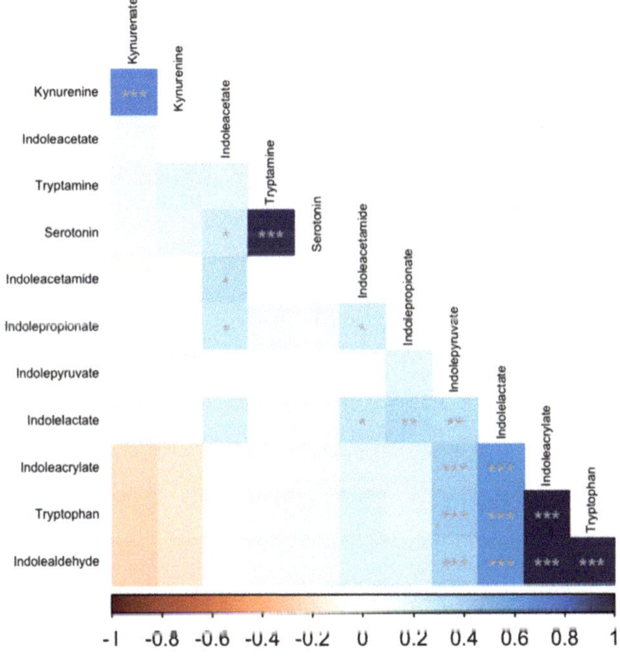

Figure 3. Correlation matrix of serum tryptophan catabolites. Spearman's rank correlation coefficients were calculated for all pairwise combinations. The intensity of the colored cells is proportional

to Spearman's rho, with blue cells indicating a positive correlation and red cells indicating a negative correlation. The matrix is ordered by the first principal component of the correlation matrix. Each cell is annotated with asterisks corresponding to the *p*-value for a given correlation: * $p < 0.05$; ** $p < 0.01$; *** $p < 0.001$.

4. Discussion

We sought to determine whether serum concentrations of tryptophan and its endogenous and exogenous (microbial) catabolites varied among cats with CIE, LGITL, and healthy controls. We hypothesized that serum concentrations of tryptophan and MICTs would be decreased and those of kynurenine would be increased in cats with CIE and LGITL compared with healthy controls. Consistent with our hypotheses, serum concentrations of tryptophan and several MICTs (indolepropionate, indolepyruvate, indolelactate, indolealdehyde, indoleacrylate) were significantly lower in cats with CIE and LGITL compared with healthy controls. Median serum concentrations of kynurenine were higher in the sera of cats with CIE and LGITL than in healthy controls, but the differences were not statistically significant. There were statistically significant correlations among serum MICTs and established serum biomarkers of gastrointestinal and pancreatic health, including cobalamin, fPLI, and fTLI. Finally, there were also numerous statistically significant correlations among tryptophan, MICTs, and endogenous tryptophan catabolites (kynurenine, kynurenate, serotonin). These findings have established strong evidence for altered tryptophan catabolism in cats with CIE and LGITL.

Our findings are consistent with previous studies that have documented decreased tryptophan and increased kynurenine in humans with IBD [23,26]. Similarly, another investigation identified decreased plasma concentrations of tryptophan in cats with CE and an inverse association between plasma tryptophan concentrations and the severity of clinical signs [46]. It is plausible that, like humans with IBD, decreased serum concentrations of tryptophan were associated with increased catabolism of tryptophan through the kynurenine pathway (KP) due to intestinal inflammation. Inflammatory cytokines including IFN-γ, IFN-α, and IL-6 induce upregulation of IDO1 activity, resulting in increased catabolism of tryptophan through the IDO1-dependent extrahepatic KP in intestinal epithelial cells and leukocytes during pro-inflammatory conditions [8,11]. In humans and rodents, the IDO1-mediated KP controls the systemic balance of kynurenine and tryptophan and its activation is known to result in decreased concentrations of tryptophan and increased concentrations of kynurenine in serum [8]. Kynurenine is an endogenous aryl hydrocarbon receptor (AhR) agonist, which regulates both innate and adaptive immune responses. Activation of the IDO1-dependent KP is associated with increased secretion of immunomodulatory cytokines IL-10 and IL-22, development of Foxp3[+] T-regulatory cells, lymphocyte apoptosis, and T-cell suppression [11]. Thus, activation of the IDO1-dependent KP may be a counterregulatory mechanism to modulate inflammatory responses in diseased intestinal mucosal tissues [25,28]. Intestinal malabsorption due to mucosal infiltrative lesions is another likely contributor to decreased serum tryptophan concentrations in cats with CE as we have observed increased fecal concentrations of tryptophan in cats with CE (unpublished preliminary data). Follow-up investigations are needed to understand the mechanisms of altered tryptophan homeostasis, and to determine whether the serum tryptophan is a useful biomarker in cats with CE.

MICTs are considered exclusive products of microbial tryptophan catabolism and previous studies in cats with CE have identified decreased abundances of bacteria known to catabolize tryptophan into MICTs in humans and rodents, including *Bacteroides* and *Bifidobacterium* [6,7,12,17]. Thus, decreased serum concentrations of MICTs observed in cats with CIE and LGITL may be influenced by altered microbial tryptophan catabolism associated with enteric microbiota dysbiosis. MICTs modulate intestinal mucosal homeostasis by regulating inflammation and mucosal barrier function via aryl hydrocarbon (AhR) and pregnane-X (PXR) receptor-mediated expression of tight junction proteins

(TJP), mucin, anti-microbial peptides (AMD), and anti-inflammatory cytokines (IL-10, IL-22) and decreased expression of pro-inflammatory cytokines (TNFα) in the enteric mucosa [20,21,25,28,29,31,47,48]. MICTs are not only potential biomarkers for EMD-associated changes in host–microbiome signaling and associated disruptions in intestinal homeostasis but are also promising therapeutic targets. Administration of probiotics containing MICT-generating species and/or dietary MICT supplementation may ameliorate mucosal inflammation and EMD in animal models of IBD [30,32,33,49]. Given the putative roles MICTs play in regulating intestinal health, it is plausible that altered tryptophan catabolism is a mechanism through which dysbiosis can contribute to the pathogenesis and/or pathophysiology of CIE and LGITL in cats. Future investigations are needed to determine whether enteric microbiota dysbiosis in CIE and LGITL is associated with altered microbial tryptophan catabolism and identify commensal microbiota that can catabolize tryptophan into MICTs in cats.

Other catabolites of tryptophan, including serotonin and tryptamine, did not differ significantly among the groups. However, serum concentrations of serotonin and tryptamine were strongly and positively correlated, and serotonin was also positively correlated with indoleacetate. Like other MICTs, indoleacetate is presumed to be an exclusive product of microbial tryptophan catabolism, whereas tryptamine and folate can also be synthesized by gut microbes. These associations suggest that enteric microbiota may influence serotonin metabolism in cats. This interpretation is supported by previous studies that have established a mechanistic role for enteric microbiota in regulating intestinal synthesis of serotonin in rodents [19,50]. Serotonin concentrations are decreased in sera, feces, and the colonic mucosa of germ-free mice (GF), and the expression of colonic tryptophan hydroxylase-1 (Tph1), the rate-limiting enzyme in serotonin synthesis, is downregulated compared with specific-pathogen-free mice (SPF) [17,19]. Conventionalization of GF mice with microbiota from SPF mice results in the restoration of serotonin concentrations in the serum and colonic mucosa, and an increased expression of Tph1, whereas antibiotic treatment recapitulates the GF state of serotonin depletion [19]. Both serotonin and tryptamine can influence GI motility and secretions via interactions with serotonin receptors [18,51]. The roles, if any, of serotonin and tryptamine in GI homeostasis and the pathophysiology of chronic enteropathies in cats are undetermined and should be targets of future investigations.

Our findings should be interpreted considering several limitations. Archived samples collected from cats with CIE and LGITL were used for this investigation. These samples were collected and stored in different locations for variable periods of time and may have been exposed to freeze–thaw cycles. The long-term stabilities of MICTs in frozen feline serum have not been specifically studied and we cannot exclude the possibility that these pre-analytic conditions may have affected the concentrations of tryptophan or its catabolites in the serum samples. Though we measured tryptophan and its catabolic derivatives against known concentrations of analytic standards (absolute quantification), the LC-MS/MS panel used has not been validated for feline serum samples. Investigations to assess the stability of analytes and to validate the assays for feline serum are currently underway. Though the diagnostic investigation of cats with CE was very similar among institutions, it was not identical. Different procedures were used to collect intestinal biopsies, including exploratory laparotomy for full-thickness biopsies and endoscopy for partial-thickness mucosal biopsies. It is possible that different histopathologic diagnoses could have been obtained if full-thickness surgical biopsies had been analyzed for all patients. Also, despite extensive testing and exclusion of ambiguous cases, it remains possible that some cats were misclassified as either CIE or LGITL. There are currently no validated criteria for the definitive differentiation of CIE and LGITL. Based on BIOMED2 guidelines in humans and a recent consensus statement from the American College of Veterinary Internal Medicine, all data, including clinical, laboratory, histopathological, immunohistochemical and clonality data, should be used to determine a final diagnosis [37,42]. This approach has been adopted in this study. It is possible these factors could have impacted the comparisons among CIE and LGITL cats; however, our findings of decreased tryptophan and MICTs were nearly identical in these groups and significantly different from healthy control cats in a manner consistent with IBD in humans and rodent models. Thus, we conclude that altered microbial tryptophan catabolism is present in both cats with CIE and LGITL, regardless of the histopathologic distinction between these disorders. Intestinal biopsies were not collected from the healthy cats, and we cannot conclude that they lacked mucosal infiltrates, which could also have impacted our results. A previous investigation identified lesions consistent with CIE and LGITL in cats lacking clinical signs of gastrointestinal dysfunction [52]. It is possible that some cats in the healthy control group had subclinical CIE or LGITL. Nonetheless, these were cats lacking clinical signs or clinicopathologic evidence of CE and we were able to demonstrate statistically significant differences in tryptophan and its catabolites when apparently healthy cats were compared with cats with active clinical signs and histopathologic lesions consistent with CIE or LGITL. Using archived samples, the authors were not able to control for the diets of the cats involved in this study. A previous study identified diet-associated differences in fecal indole concentrations, and it is possible our results could have been affected by differences in diet among cats [53]. Some cats included in this investigation had extra-intestinal comorbidities. Though we cannot exclude the possibility that the presence of these disorders affected our results, a sub-analysis did not reveal any statistically significant differences among cats with CE with respect to different extra-intestinal comorbidities (File S3). As chronic renal disease and hyperthyroidism are common in geriatric cats, and inflammatory disorders of the liver and pancreas are common in cats with CE, it will be important to establish the impact of these comorbidities on tryptophan catabolism in cats before their potential as biomarkers in feline CE can be determined. Owing to the use of archived samples and retrospective patient data, information about cats' demographic data and clinical characteristics was unavailable or collected inconsistently (e.g., breed, diet, results of routine clinicopathologic tests). This prevented a complete demographic and clinical description of the cats included in this study. Finally, it must be noted that this study was not intended to investigate the clinical aspect of CIE or LGITL in cats and the clinical significance of our findings is unknown. Prospective and mechanistic studies that address these limitations are needed to confirm our findings and determine their pathophysiologic and clinical significance.

5. Conclusions

For the first time, and consistent with studies in humans and animal models of IBD, we have identified decreased concentrations of microbial indole catabolites of tryptophan (i.e., indolepropionate, indolepyruvate, indolelactate, indolealdehyde, and indoleacrylate) in cats with CIE and LGITL. It is plausible that decreased microbial indole catabolites of tryptophan contribute to the pathophysiology of CIE and LGITL by promoting intestinal inflammation and decreasing mucosal barrier integrity. The discovery and characterization of novel biomarkers of microbial metabolism are important to identify sub-populations of cats with chronic enteropathies in which different pathomechanisms of dysbiosis are active. In the future, these insights could be exploited to develop novel diagnostics and therapies targeting metabolic pathways in enteric microbiota that regulate intestinal mucosal health.

Supplementary Materials: The following supporting information can be downloaded at: https://www.mdpi.com/article/10.3390/ani14010067/s1, File S1: Boxplots comparing all tryptophan catabolites among cats with chronic inflammatory enteropathy, low-grade intestinal T-cell lymphoma, and healthy controls. The upper and lower boundaries of the box represent the 25th and 75th percentiles and the horizontal line represents the median. The whiskers represent the maximum and minimum values below and above the upper (75th percentile + IQR) and lower (and 25th percentile—IQR) fences, respectively. Plots are annotated with the *p*-values from Kruskal–Wallis tests (KW). CIE, chronic inflammatory enteropathy; LGITL, low-grade intestinal T-cell lymphoma. File S2. Complete results for correlation analysis among tryptophan catabolites. Spearman's rank correlation coefficients (rho) were calculated for pairwise combinations of tryptophan catabolites. *p*-values are adjusted for multiple comparisons (p_{adj}) using the Benjamini–Hochberg method to control for false discovery. File S3. Comparison of tryptophan catabolites among cats with chronic enteropathy and extra-intestinal comorbidities. *p*-values were derived from Kruskal–Wallis tests and values are presented as medians, interquartile ranges (IQR), maximum (max), and minimum (min). Supplemental Table S1: Serum concentrations of tryptophan catabolites from cats with chronic inflammatory enteropathy and low-grade intestinal T-cell lymphoma were compared among cats with and without (none) histopathologic evidence of pancreatitis, inflammatory liver infiltrates, and combinations thereof. Data are derived from the subset of cats in which histopathologic analysis of the small intestines, pancreata, and livers was available. Supplemental Table S2: Serum concentrations of tryptophan catabolites from cats with chronic inflammatory enteropathy and low-grade intestinal T-cell lymphoma were compared among cats with and without (none) previous diagnoses of chronic kidney disease (CKD), hyperthyroidism, and combinations thereof. Data are derived from the subset of cats in which medical records contained sufficient information to assess these previous diagnoses.

Author Contributions: Conceptualization, P.C.B., D.A.W. and S.M.; Methodology, P.C.B.; Formal Analysis, P.C.B.; Investigation, P.C.B.; Resources, P.C.B., D.A.W., S.M., Y.-A.W., J.M.S., J.S.S. and A.G.; Data Curation, P.C.B.; Writing—Original Draft Preparation, P.C.B.; Writing—Review and Editing, P.C.B., D.A.W., S.M., Y.-A.W., J.M.S., J.S.S. and A.G.; Visualization, P.C.B.; Supervision, D.A.W.; Project Administration, D.A.W.; Funding Acquisition, P.C.B., D.A.W., S.M. and J.M.S. All authors have read and agreed to the published version of the manuscript.

Funding: This project was supported by a Miller Trust grant from the Every Cat Health Foundation (grant number MT21-003). The contents of this publication are solely the responsibility of the authors and do not necessarily represent the views of EveryCat.

Institutional Review Board Statement: All samples used in this investigation were from archived samples left over from routine diagnostic or previous, unrelated investigations. The original studies were approved by the Institutional Animal Care and Use Committees at the University of Illinois (protocol ID# 20141) and University of California at Davis (protocol ID# 2014-0369 CA).

Informed Consent Statement: For all samples collected by veterinarians for routine diagnostics in a clinical setting, the client gave informed consent as part of a valid veterinary–client relationship. All samples collected by primary care veterinarians from client-owned cats in a veterinary clinical setting provided consent as a part of a valid veterinary–client relationship. For samples from cats collected from previous, unrelated investigations, written consent was provided.

Data Availability Statement: All data and R code necessary to replicate this analysis are located in a GitHub repository (https://github.com/pcbarko/FCE_Microbial_Indole_Catabolites; accessed on 14 December 2023).

Acknowledgments: The authors thank Alexander Ulanov and Michael La Frano at the Carver Metabolomics Core in the Roy J. Carver Biotechnology Center for conducting LC-MS analysis.

Conflicts of Interest: The authors declare no conflicts of interest.

References

1. Moore, P.F.; Rodriguez-Bertos, A.; Kass, P.H. Feline Gastrointestinal Lymphoma: Mucosal Architecture, Immunophenotype, and Molecular Clonality. *Vet. Pathol.* **2012**, *49*, 658–668. [CrossRef]
2. Moore, P.F.; Woo, J.C.; Vernau, W.; Kosten, S.; Graham, P.S. Characterization of feline T cell receptor gamma (TCRG) variable region genes for the molecular diagnosis of feline intestinal T cell lymphoma. *Vet. Immunol. Immunopathol.* **2005**, *106*, 167–178. [CrossRef]
3. Yu, L.C.H. Microbiota dysbiosis and barrier dysfunction in inflammatory bowel disease and colorectal cancers: Exploring a common ground hypothesis. *J. Biomed. Sci.* **2018**, *25*, 79. [CrossRef]
4. Hertli, S.; Zimmermann, P. Molecular interactions between the intestinal microbiota and the host. *Mol. Microbiol.* **2022**, *117*, 1297. [CrossRef]
5. Quaglio, A.E.V.; Grillo, T.G.; De Oliveira, E.C.S.; Di Stasi, L.C.; Sassaki, L.Y. Gut microbiota, inflammatory bowel disease and colorectal cancer. *World J. Gastroenterol.* **2022**, *28*, 4053–4060. [CrossRef]
6. Marsilio, S.; Pilla, R.; Sarawichitr, B.; Chow, B.; Hill, S.L.; Ackermann, M.R.; Estep, J.S.; Lidbury, J.A.; Steiner, J.M.; Suchodolski, J.S. Characterization of the fecal microbiome in cats with inflammatory bowel disease or alimentary small cell lymphoma. *Sci. Rep.* **2019**, *9*, 19208. [CrossRef]
7. Sung, C.-H.; Marsilio, S.; Chow, B.; Zornow, K.A.; Slovak, J.E.; Pilla, R.; Lidbury, J.A.; Steiner, J.M.; Park, S.Y.; Hong, M.-P.; et al. Dysbiosis index to evaluate the fecal microbiota in healthy cats and cats with chronic enteropathies. *J. Feline Med. Surg.* **2022**, *24*, e1–e12. [CrossRef]

8.   Badawy, A.A.B. Kynurenine Pathway of Tryptophan Metabolism: Regulatory and Functional Aspects. *Int. J. Tryptophan Res.* **2017**, *10*, 1178646917691938. [CrossRef]
9.   Dehhaghi, M.; Kazemi Shariat Panahi, H.; Guillemin, G.J. Microorganisms, Tryptophan Metabolism, and Kynurenine Pathway: A Complex Interconnected Loop Influencing Human Health Status. *Int. J. Tryptophan Res.* **2019**, *12*, 1178646919852996. [CrossRef] [PubMed]
10. Murakami, Y.; Saito, K. Species and cell types difference in tryptophan metabolism. *Int. J. Tryptophan Res.* **2013**, *6*, 47–54. [CrossRef] [PubMed]
11. Haq, S.; Grondin, J.A.; Khan, W.I. Tryptophan-derived serotonin-kynurenine balance in immune activation and intestinal inflammation. *FASEB J.* **2021**, *35*, e21888. [CrossRef]
12. Roager, H.M.; Licht, T.R. Microbial tryptophan catabolites in health and disease. *Nat. Commun.* **2018**, *9*, 3294. [CrossRef]
13. Li, X.; Zhang, Z.H.; Zabed, H.M.; Yun, J.; Zhang, G.; Qi, X. An Insight into the Roles of Dietary Tryptophan and Its Metabolites in Intestinal Inflammation and Inflammatory Bowel Disease. *Mol. Nutr. Food Res.* **2021**, *65*, 2000461. [CrossRef]
14. Fordtran, J.S.; Scroggie, W.B.; Polter, D.E. Colonic absorption of tryptophan metabolites in man. *J. Lab. Clin. Med.* **1964**, *64*, 125–132.
15. Agus, A.; Planchais, J.; Sokol, H. Cell Host & Microbe Review Gut Microbiota Regulation of Tryptophan Metabolism in Health and Disease. *Cell Host Microbe* **2018**, *23*, 716–724. [CrossRef]
16. Jin, U.-H.; Lee, S.-O.; Sridharan, G.; Lee, K.; Davidson, L.A.; Jayaraman, A.; Chapkin, R.S.; Alaniz, R.; Safe, S. Microbiome-derived tryptophan metabolites and their aryl hydrocarbon receptor-dependent agonist and antagonist activities. *Mol. Pharmacol.* **2014**, *85*, 777–788. [CrossRef]
17. Wikoff, W.R.; Anfora, A.T.; Liu, J.; Schultz, P.G.; Lesley, S.A.; Peters, E.C.; Siuzdak, G. Metabolomics analysis reveals large effects of gut microflora on mammalian blood metabolites. *Proc. Natl. Acad. Sci. USA* **2009**, *106*, 3698. [CrossRef]
18. Bhattarai, Y.; Williams, B.B.; Battaglioli, E.J.; Whitaker, W.R.; Till, L.; Grover, M.; Linden, D.R.; Akiba, Y.; Kandimalla, K.K.; Zachos, N.C.; et al. Gut Microbiota-Produced Tryptamine Activates an Epithelial G-Protein-Coupled Receptor to Increase Colonic Secretion. *Cell Host Microbe* **2018**, *23*, 775–785.e5. [CrossRef]
19. Yano, J.M.; Yu, K.; Donaldson, G.P.; Shastri, G.G.; Ann, P.; Ma, L.; Nagler, C.R.; Ismagilov, R.F.; Mazmanian, S.K.; Hsiao, E.Y. Indigenous bacteria from the gut microbiota regulate host serotonin biosynthesis. *Cell* **2015**, *161*, 264–276. [CrossRef]
20. Scott, S.A.; Fu, J.; Chang, P.V. Microbial tryptophan metabolites regulate gut barrier function via the aryl hydrocarbon receptor. *Proc. Natl. Acad. Sci. USA* **2020**, *117*, 19376–19387. [CrossRef]
21. Gupta, N.K.; Thaker, A.I.; Kanuri, N.; Riehl, T.E.; Rowley, C.W.; Stenson, W.F.; Ciorba, M.A. Serum analysis of tryptophan catabolism pathway: Correlation with Crohn's disease activity. *Inflamm. Bowel Dis.* **2012**, *18*, 1214–1220. [CrossRef]
22. Nikolaus, S.; Schulte, B.; Al-Massad, N.; Thieme, F.; Schulte, D.M.; Bethge, J.; Rehman, A.; Tran, F.; Aden, K.; Häsler, R.; et al. Increased Tryptophan Metabolism Is Associated With Activity of Inflammatory Bowel Diseases. *Gastroenterology* **2017**, *153*, 1504–1516.e2. [CrossRef]
23. Sofia, M.A.; Ciorba, M.A.; Meckel, K.; Lim, C.K.; Guillemin, G.J.; Weber, C.R.; Bissonnette, M.; Pekow, J.R. Tryptophan Metabolism through the Kynurenine Pathway is Associated with Endoscopic Inflammation in Ulcerative Colitis. *Inflamm. Bowel Dis.* **2018**, *24*, 1471–1480. [CrossRef]
24. Alexeev, E.E.; Lanis, J.M.; Kao, D.J.; Campbell, E.L.; Kelly, C.J.; Battista, K.D.; Gerich, M.E.; Jenkins, B.R.; Walk, S.T.; Kominsky, D.J.; et al. Microbiota-Derived Indole Metabolites Promote Human and Murine Intestinal Homeostasis through Regulation of Interleukin-10 Receptor. *Am. J. Pathol.* **2018**, *188*, 1183–1194. [CrossRef]
25. Lamas, B.; Richard, M.L.; Leducq, V.; Pham, H.-P.; Michel, M.-L.; Da Costa, G.; Bridonneau, C.; Jegou, S.; Hoffmann, T.W.; Natividad, J.M.; et al. CARD9 impacts colitis by altering gut microbiota metabolism of tryptophan into aryl hydrocarbon receptor ligands. *Nat. Med.* **2016**, *22*, 598–605. [CrossRef]
26. Vyhlídalová, B.; Krasulová, K.; Pečinková, P.; Marcalíková, A.; Vrzal, R.; Zemánková, L.; Vančo, J.; Trávníček, Z.; Vondráček, J.; Karasová, M.; et al. Gut microbial catabolites of tryptophan are ligands and agonists of the aryl hydrocarbon receptor: A detailed characterization. *Int. J. Mol. Sci.* **2020**, *21*, 2614. [CrossRef]
27. Bansal, T.; Alaniz, R.C.; Wood, T.K.; Jayaraman, A. The bacterial signal indole increases epithelial-cell tight-junction resistance and attenuates indicators of inflammation. *Proc. Natl. Acad. Sci. USA* **2010**, *107*, 228–233. [CrossRef]
28. Wlodarska, M.; Luo, C.; Kolde, R.; D'hennezel, E.; Annand, J.W.; Heim, C.E.; Krastel, P.; Schmitt, E.K.; Omar, A.S.; Creasey, E.A.; et al. Indoleacrylic Acid Produced by Commensal Peptostreptococcus Species Suppresses Inflammation. *Cell Host Microbe* **2017**, *22*, 25–37.e6. [CrossRef]
29. Ehrlich, A.M.; Pacheco, A.R.; Henrick, B.M.; Taft, D.; Xu, G.; Huda, M.N.; Mishchuk, D.; Goodson, M.L.; Slupsky, C.; Barile, D.; et al. Indole-3-lactic acid associated with *Bifidobacterium*-dominated microbiota significantly decreases inflammation in intestinal epithelial cells. *BMC Microbiol.* **2020**, *20*, 357. [CrossRef]

30. Vyhlídalová, B.; Bartoňková, I.; Jiskrová, E.; Li, H.; Mani, S.; Dvořák, Z. Differential activation of human pregnane X receptor PXR by isomeric mono-methylated indoles in intestinal and hepatic in vitro models. *Toxicol. Lett.* **2020**, *324*, 104–110. [CrossRef]
31. Descamps, H.C.; Herrmann, B.; Wiredu, D.; Thaiss, C.A. The path toward using microbial metabolites as therapies. *EBioMedicine* **2019**, *44*, 747–754. [CrossRef]
32. Pernomian, L.; Duarte-Silva, M.; de Barros Cardoso, C.R. The Aryl Hydrocarbon Receptor (AHR) as a Potential Target for the Control of Intestinal Inflammation: Insights from an Immune and Bacteria Sensor Receptor. *Clin. Rev. Allergy Immunol.* **2020**, *59*, 382–390. [CrossRef]
33. Zhang, J.; Zhu, S.; Ma, N.; Johnston, L.J.; Wu, C.; Ma, X. Metabolites of microbiota response to tryptophan and intestinal mucosal immunity: A therapeutic target to control intestinal inflammation. *Med. Res. Rev.* **2021**, *41*, 1061–1088. [CrossRef]
34. Barko, P.; Wu, Y.-A.; Steiner, J.; Norsworthy, G.; Gal, A.; Williams, D.; Estep, S. Altered Tryptophan Metabolism in Cats with Chronic Inflammatory Enteropathy or Alimentary Small Cell Lymphoma. Research Communications of the 32nd ECVIM-CA Online Congress. *J. Vet. Intern. Med.* **2022**, *36*, 2455–2551. [CrossRef]
35. Briscoe, K.; Krockenberger, M.; Beatty, J.; Crowley, A.; Dennis, M.; Canfield, P.; Dhand, N.; Lingard, A.; Barrs, V. Histopathological and Immunohistochemical Evaluation of 53 Cases of Feline Lymphoplasmacytic Enteritis and Low-Grade Alimentary Lymphoma. *J. Comp. Pathol.* **2011**, *145*, 187–198. [CrossRef]
36. Langerak, A.W.; Groenen, P.J.T.A.; Brüggemann, M.; Beldjord, K.; Bellan, C.; Bonello, L.; Boone, E.; I Carter, G.; Catherwood, M.; Davi, F.; et al. EuroClonality/BIOMED-2 guidelines for interpretation and reporting of Ig/TCR clonality testing in suspected lymphoproliferations. *Leukemia* **2012**, *26*, 2159–2171. [CrossRef]
37. Paulin, M.V.; Couronné, L.; Beguin, J.; Le Poder, S.; Delverdier, M.; Semin, M.-O.; Bruneau, J.; Cerf-Bensussan, N.; Malamut, G.; Cellier, C.; et al. Feline low-grade alimentary lymphoma: An emerging entity and a potential animal model for human disease. *BMC Vet. Res.* **2018**, *14*, 306. [CrossRef]
38. Marsilio, S. Differentiating Inflammatory Bowel Disease from Alimentary Lymphoma in Cats: Does It Matter? *Vet. Clin. North. Am. Small Anim. Pract.* **2021**, *51*, 93–109. [CrossRef] [PubMed]
39. Marsilio, S.; Newman, S.J.; Estep, J.S.; Giaretta, P.R.; Lidbury, J.A.; Warry, E.; Flory, A.; Morley, P.S.; Smoot, K.; Seeley, E.H.; et al. Differentiation of lymphocytic-plasmacytic enteropathy and small cell lymphoma in cats using histology-guided mass spectrometry. *J. Vet. Intern. Med.* **2020**, *34*, 669–677. [CrossRef] [PubMed]
40. Freiche, V.; Paulin, M.V.; Cordonnier, N.; Huet, H.; Turba, M.; Macintyre, E.; Molina, T.; Hermine, O.; Couronné, L.; Bruneau, J. Histopathologic, phenotypic, and molecular criteria to discriminate low-grade intestinal T-cell lymphoma in cats from lymphoplasmacytic enteritis. *J. Vet. Intern. Med.* **2021**, *35*, 2673–2684. [CrossRef] [PubMed]
41. Marsilio, S.; Freiche, V.; Johnson, E.; Leo, C.; Langerak, A.W.; Peters, I.; Ackermann, M.R. ACVIM consensus statement guidelines on diagnosing and distinguishing low-grade neoplastic from inflammatory lymphocytic chronic enteropathies in cats. *J. Vet. Intern. Med.* **2023**, *37*, 794–816. [CrossRef] [PubMed]
42. Laflamme, D. Development and validation of a body condition score system for cats: A clinical tool. *Feline Pract.* **1997**, *5*, 13–18.
43. R Core Team. *R: A Language and Environment for Statistical Computing*; R Foundation for Statistical Computing: Vienna, Austria, 2022.
44. Benjamini, Y.; Hochberg, Y. Controlling the False Discovery Rate: A Practical and Powerful Approach to Multiple Testing. *J. R. Stat. Soc. Ser. B (Methodol.)* **1995**, *57*, 289–300. [CrossRef]
45. Sakai, K.; Maeda, S.; Yonezawa, T.; Matsuki, N. Decreased plasma amino acid concentrations in cats with chronic gastrointestinal diseases and their possible contribution in the inflammatory response. *Vet. Immunol. Immunopathol.* **2018**, *195*, 1–6. [CrossRef] [PubMed]
46. Monteleone, I.; Rizzo, A.; Sarra, M.; Sica, G.; Sileri, P.; Biancone, L.; Macdonald, T.T.; Pallone, F.; Monteleone, G. Aryl hydrocarbon receptor-induced signals up-regulate IL-22 production and inhibit inflammation in the gastrointestinal tract. *Gastroenterology* **2011**, *141*, 237–248.e1. [CrossRef] [PubMed]
47. Dong, F.; Perdew, G.H. The aryl hydrocarbon receptor as a mediator of host-microbiota interplay. *Gut Microbes* **2020**, *12*, 1859812. [CrossRef] [PubMed]
48. Reigstad, C.S.; Salmonson, C.E.; Rainey, J.F., III; Szurszewski, J.H.; Linden, D.R.; Sonnenburg, J.L.; Farrugia, G.; Kashyap, P.C. Gut microbes promote colonic serotonin production through an effect of short-chain fatty acids on enterochromaffin cells. *FASEB J.* **2015**, *29*, 1395–1403. [CrossRef] [PubMed]
49. Marsilio, S.; Chow, B.; Hill, S.L.; Ackermann, M.R.; Estep, J.S.; Sarawichitr, B.; Pilla, R.; Lidbury, J.A.; Steiner, J.M.; Suchodolski, J.S. Untargeted metabolomic analysis in cats with naturally occurring inflammatory bowel disease and alimentary small cell lymphoma. *Sci. Rep.* **2021**, *11*, 9198. [CrossRef]
50. Blake, A.B.; Guard, B.C.; Honneffer, J.B.; Lidbury, J.A.; Steiner, J.M.; Suchodolski, J.S. Altered microbiota, fecal lactate, and fecal bile acids in dogs with gastrointestinal disease. *PLoS ONE* **2019**, *14*, e0224454. [CrossRef]
51. Mawe, G.M.; Hoffman, J.M. Serotonin Signaling in the Gastrointestinal Tract: Functions, dysfunctions, and therapeutic targets. *Nat. Rev. Gastroenterol. Hepatol.* **2013**, *10*, 473–486. [CrossRef]

52. Marsilio, S.; Ackermann, M.R.; Lidbury, J.A.; Suchodolski, J.S.; Steiner, J.M. Results of histopathology, immunohistochemistry, and molecular clonality testing of small intestinal biopsy specimens from clinically healthy client-owned cats. *J. Vet. Intern. Med.* **2019**, *33*, 551–558. [CrossRef] [PubMed]
53. Lee, A.H.; Jha, A.R.; Do, S.; Scarsella, E.; Shmalberg, J.; Schauwecker, A.; Steelman, A.J.; Honaker, R.W.; Swanson, K.S. Dietary enrichment of resistant starches or fibers differentially alter the feline fecal microbiome and metabolite profile. *Anim. Microbiome* **2022**, *4*, 61. [CrossRef] [PubMed]

Article

Untargeted Analysis of Serum Metabolomes in Dogs with Exocrine Pancreatic Insufficiency

Patrick C. Barko [1,2,*], Stanley I. Rubin [3], Kelly S. Swanson [4], Maureen A. McMichael [5], Marcella D. Ridgway [1] and David A. Williams [1,2]

[1] Department of Veterinary Clinical Medicine, University of Illinois at Urbana-Champaign, Urbana, IL 61802, USA
[2] Department of Pathobiology, University of Illinois at Urbana-Champaign, Urbana, IL 61802, USA
[3] VCA Animal Hospitals, Los Angeles, CA 90064, USA
[4] Department of Animal Sciences and Division of Nutritional Sciences, College of Agricultural, Consumer and Environmental Sciences, University of Illinois at Urbana-Champaign, Urbana, IL 61801, USA
[5] Department of Clinical Sciences, Auburn University, Auburn, AL 36849, USA
* Correspondence: pcbarko@illinois.edu

Simple Summary: Exocrine pancreatic insufficiency (EPI) is a digestive disorder in dogs resulting from insufficient secretion of digestive enzymes from the exocrine pancreas. EPI is treated with oral pancreatic enzyme replacement therapy (PERT), but the persistence of clinical signs, especially diarrhea, is common after treatment. We sought to develop new insights into EPI using untargeted metabolomics analysis, a method that can measure hundreds of biochemicals (metabolites) in a sample. We analyzed 759 serum metabolites and identified 114 that varied significantly between dogs with EPI and healthy controls. Differences in fatty acid and amino acid metabolites were consistent with a state of malnourishment, and decreased vitamin B and C metabolites were suggestive of micronutrient deficiencies in dogs with EPI. Disturbances in gut microbial metabolites indicated the altered composition of the intestinal microbiome in dogs with EPI. Increased kynurenine, a tryptophan metabolite, in dogs with EPI may be associated with intestinal inflammation. As an exploratory study, causation cannot be determined from these results, but our findings have generated new data that can be used to inform future investigations of gastrointestinal and metabolic disturbances underlying the persistence of clinical signs in dogs with EPI treated with PERT.

Abstract: Exocrine pancreatic insufficiency (EPI) is a malabsorptive syndrome resulting from insufficient secretion of pancreatic digestive enzymes. EPI is treated with pancreatic enzyme replacement therapy (PERT), but the persistence of clinical signs, especially diarrhea, is common after treatment. We used untargeted metabolomics of serum to identify metabolic disturbances associated with EPI and generate novel hypotheses related to its pathophysiology. Fasted serum samples were collected from dogs with EPI ($n = 20$) and healthy controls ($n = 10$), all receiving PERT. Serum metabolomes were generated using UPLC-MS/MS, and differences in relative metabolite abundances were compared between the groups. Of the 759 serum metabolites detected, 114 varied significantly ($p < 0.05$, $q < 0.2$) between dogs with EPI and healthy controls. Differences in amino acids (arginate, homoarginine, 2-oxoarginine, N-acetyl-cadaverine, and α-ketoglutaramate) and lipids (free fatty acids and docosahexaenoylcarnitine) were consistent with increased proteolysis and lipolysis, indicating a persistent catabolic state in dogs with EPI. Relative abundances of gut microbial metabolites (phenyllactate, 4-hydroxyphenylacetate, phenylacetyl-amino acids, catechol sulfates, and o-cresol-sulfate) were altered in dogs with EPI, consistent with disruptions in gut microbial communities. Increased kynurenine is consistent with the presence of intestinal inflammation in dogs with EPI. Whether these metabolic disturbances participate in the pathophysiology of EPI or contribute to the persistence of clinical signs after treatment is unknown, but they are targets for future investigations.

Keywords: EPI; pancreas; dogs; canine; metabolomics

Citation: Barko, P.C.; Rubin, S.I.; Swanson, K.S.; McMichael, M.A.; Ridgway, M.D.; Williams, D.A. Untargeted Analysis of Serum Metabolomes in Dogs with Exocrine Pancreatic Insufficiency. *Animals* **2023**, *13*, 2313. https://doi.org/10.3390/ani13142313

Academic Editors: Aarti Kathrani and Romy M. Heilmann

Received: 18 April 2023
Revised: 3 July 2023
Accepted: 4 July 2023
Published: 14 July 2023

1. Introduction

Canine exocrine pancreatic insufficiency (EPI) is a malabsorptive syndrome caused by insufficient secretion of pancreatic digestive enzymes [1,2]. Characteristic clinical signs of EPI include diarrhea, weight loss, and polyphagia. Clinical signs result from failure to digest dietary macromolecules (i.e., fats, carbohydrates, and proteins) resulting in malabsorption, negative energy balance, and altered composition of the gut microbiome (small intestinal bacterial overgrowth) [2]. Due to malabsorption, EPI is also associated with deficiencies in specific micronutrients, such as cobalamin and lipid-soluble vitamins [3–5]. The most common cause of canine EPI is end-stage pancreatic acinar atrophy (PAA), and clinical signs of EPI develop following an approximate 90% reduction in pancreatic acinar mass [6–9]. Serum concentrations of canine trypsin-like immunoreactivity (cTLI) less than 2.5 µg/L are diagnostic for EPI [10,11]. Dogs with EPI require life-long pancreatic enzyme replacement therapy (PERT), typically via the addition of enzyme extracts to each meal, and most require supplementation with cobalamin. The long-term prognosis for EPI is good, but responses to enzyme replacement therapy are variable, and a substantial proportion (~40%) of dogs with EPI have persistent clinical signs after the initiation of PERT, most commonly diarrhea and weight loss [12–14]. Thus, many dogs require adjunctive therapies, including therapeutic diets, antibiotics, probiotics, or some combination thereof, to control their clinical signs.

Little is known about the pathogenesis of PAA or its progression to EPI. Immunohistochemical and serologic studies have identified lymphocytic pancreatitis and circulating antibodies against pancreatic acini in dogs with subclinical PAA; however, in end-stage PAA associated with EPI, there is the replacement of acini with adipose tissue with minimal inflammation or fibrosis [9,15,16]. Over-representation of certain breeds, especially German Shepherds, among dogs with EPI has prompted investigations to identify a genetic basis for the disorder. The heritability of EPI is complex and does not follow a Mendelian pattern [17,18]. Recent genomics studies have identified a common risk allele in the dog leukocyte antigen complex in German Shepherd dogs and Pembroke Welsh Corgis with EPI [19,20]. However, the risk allele was not present in all dogs with EPI and was identified in some healthy dogs. Collectively, these findings suggest that there is an immune-mediated component to the pathogenesis of PAA and that one or more genetic or environmental risk factors may contribute to its pathogenesis.

Given the incomplete understanding of its etiopathogenesis, there is a need to generate new insights into the pathogenesis of EPI and its pathophysiologic consequences. There is also a need to identify pathophysiologic disturbances present in dogs with EPI after the initiation of PERT to develop novel therapeutic approaches for dogs with persistent signs of gastrointestinal dysfunction. Untargeted metabolomics analysis is an investigative tool that can provide new insights into poorly understood disease processes, especially those associated with nutritional and metabolic disturbances, such as EPI. The metabolome is the sum of all small molecules (<1500 kDa) present in an organism or biologic sample that are precursors, intermediates, and end-products of metabolic processes. Metabolites of endogenous and exogenous (diet, environment, and microbial) origin contribute to the serum metabolome of mammals [21]. The quantification of metabolites in a biologic sample can yield valuable insights into the metabolic phenotypes (i.e., metabotypes) of disease states by elucidating pathophysiologic mechanisms of disease and facilitating the discovery of novel biomarkers. The objectives of this investigation were to (1) detect differences in serum metabolite profiles between dogs with EPI and healthy controls and (2) to identify novel metabolic signatures relevant to the pathophysiology of EPI. While this was an untargeted, hypothesis-generating investigation, we hypothesized that metabolomics profiling would identify novel serum biomarkers plausibly associated with persistent gastrointestinal dysfunction in dogs with EPI following treatment with PERT.

2. Materials and Methods

2.1. Patient Population and Sample Collection

Adult ($\geq$1 year of age) dogs with previously diagnosed EPI were recruited from an online patient registry (epi4dogs.com/epi-registry; registry data accessed 5 February 2016). Dogs diagnosed with EPI were identified from the registry data, and their owners and primary care veterinarians were contacted to collect patient histories and medical records. Dogs were eligible for inclusion if a review of their medical records identified subnormal serum cTLI concentrations (cTLI < 2.5 µg/L; reference interval: 5.7–45.2 µg/L) from two or more consecutive measurements. Additionally, dogs with EPI had to have been receiving oral PERT with a commercially available enzyme extract of porcine origin for at least 30 days prior to enrollment. The diets of participating dogs were not standardized, and they were fed their current diet under the supervision of their owners and primary veterinarians. Information regarding the diets and PERT of dogs with EPI included in this study are presented in Supplementary File S1. Dogs with EPI were excluded if they had been diagnosed with any other gastrointestinal, metabolic, endocrine, systemic, or neoplastic diseases by their primary care veterinarian. After an overnight fast, samples of whole blood were collected by the dogs' primary care veterinarians via jugular or lateral saphenous venipuncture using sterile 20 g hypodermic needles and syringes. Whole blood samples were collected into empty, sterile tubes, allowed to clot at room temperature for 30 min, and centrifuged on site at 2500–3500 rpm for 10 min. The resulting serum was aliquoted into empty, sterile polypropylene tubes and immediately frozen on site in clinical freezers (approximately $-20\ ^\circ$C). To standardize sample collection, all required materials (hypodermic needles, syringes, transfer pipets, sample tubes) and detailed instructions were provided to participating veterinarians by the investigators. All samples were mailed overnight on dry ice to the investigators, where they were stored at $-80\ ^\circ$C following acquisition. Samples were collected and received by the investigators between September, 2016 and February, 2017. Aliquots of serum were submitted for measurement of serum cTLI, cobalamin, and folate using solid-phase chemiluminescent assays run on the IMMULITE 2000 PXI immunoassay system (Siemens Inc., Munich, Germany) at the Texas A&M Gastrointestinal Laboratory (College Station, TX, USA). Dogs with serum cTLI concentrations <2.5 µg/L were assigned to the EPI group, and those with serum cTLI concentrations $\geq$2.5 µg/L were excluded.

Healthy, adult ($\geq$1 year of age), client-owned dogs were recruited as a control group. Owners completed a survey to gather information about their dogs' medical history, while their medical records were reviewed by the investigators. To screen for confounding diseases, each healthy dog received a physical exam, and the following diagnostic screening tests were performed: complete blood count, serum biochemistry panel, urinalysis, serum total T4, serum cTLI, serum cobalamin, serum folate, and fecal flotation. Complete blood count, serum chemistries, urinalysis, total T4, and fecal flotation assays were performed at the University of Illinois Veterinary Diagnostic Laboratory (Urbana, IL, USA). Dogs were excluded from this group if they had any historical clinical signs of gastrointestinal disease (vomiting, diarrhea, weight loss, anorexia, polyphagia) within 6 months of recruitment; serum cTLI concentrations <5.7 µg/L; clinical signs or laboratory abnormalities consistent with any systemic, metabolic, gastrointestinal, or pancreatic diseases; or received antibiotics, corticosteroids, or immunomodulatory drugs within 6 months of recruitment. To account for the potential impact of pancreatic enzyme replacement therapy on serum metabolite profiles, a powdered pancreatic enzyme extract (Pancreatin 6X; American Laboratories Inc., Omaha, NE, USA) was administered orally (1 tsp/cup of food) to the healthy dogs for 14 days prior to sample collection. Following the supplementation period, serum samples were collected after an overnight fast as described above and immediately frozen at $-80\ ^\circ$C.

2.2. Untargeted Serum Metabolomics

Serum metabolite profiles were generated via ultrahigh performance liquid chromatography-tandem mass spectroscopy (UPLC-MS/MS) by a commercial laboratory (Metabolon Inc.,

Morrisville, NC, USA) between June and September, 2017. Serum samples (100 µL) were deproteinated using methanol (500 µL) precipitation with vigorous shaking for 2 min followed by centrifugation. The methanol contained four recovery standards (DL-2-fluorophenylglycine, tridecanoic acid, d6-cholesterol, and 4-chlorophenylalanine) used to assess extraction efficiency. The resulting extracts were dried and reconstituted in solvents compatible to each of the four UPLC-MS/MS methods. Two aliquots were reconstituted in 50 µL of 6.5 mM ammonium bicarbonate in water (pH 8) for negative ion analysis methods, and two were reconstituted using 50 µL 0.1% formic acid in water (pH 3.5) for the positive ion analysis methods. Reconstitution solvents contained internal standards (d7-glucose, d3-leucine, d8-phenylalanine, d5-tryptophan, d5-hippuric acid, Br-phenylalanine, d5-indole acetic acid, amitriptyline, d9-progesterone) to assess instrument performance and to aid in chromatographic alignment. Analytic controls consisting of pooled samples, generated from small volumes of each experimental sample, were analyzed simultaneously with experimental samples to serve as a technical replicate throughout the dataset. Methanol-extracted water samples served as process blanks. Experimental samples were randomized across the platform run with quality control samples spaced evenly among the injections.

For UPLC-MS/MS, all methods utilized a Waters ACQUITY ultraperformance liquid chromatographer and a Thermo Scientific Q Exactive high resolution/accurate mass spectrometer interfaced with a heated electrospray ionization source and Orbitrap mass analyzer operated at 35,000 mass resolution. One aliquot was analyzed using acidic positive ion conditions optimized for hydrophilic compounds. In this method, the extract was gradient eluted from a C18 column (Waters UPLC BEH C18-2.1 $\times$ 100 mm, 1.7 µm) using water and methanol containing 0.05% perfluoropentanoic acid (PFPA) and 0.1% formic acid (FA). Another aliquot was analyzed using acidic positive ion conditions chromatographically optimized for hydrophobic compounds. In this method, the extract was gradient eluted from the same C18 column using methanol, acetonitrile, water, 0.05% PFPA, and 0.01% FA. Another aliquot was analyzed using basic negative ion optimized conditions using a separate dedicated C18 column (Waters UPLC BEH C18-2.1 $\times$ 100 mm, 1.7 µm). The basic extracts were gradient-eluted from the column using methanol and water with 6.5 mM ammonium bicarbonate at pH 8. The fourth aliquot was analyzed via negative ionization following elution from an HILIC column (Waters UPLC BEH Amide 2.1 $\times$ 150 mm, 1.7 µm) using a gradient consisting of water and acetonitrile with 10 mM ammonium formate, pH 10.8. The MS analysis alternated between MS and data-dependent MSn scans using dynamic exclusion. The scan covered 70–1000 *m/z*. Data extraction, compound identification, and data processing were performed using a proprietary software platform by Metabolon Inc. Compounds were identified by comparison to library entries of purified and authenticated standards. Metabolites were quantified by measuring the area-under-the-curve of the chromatographic peak.

2.3. Statistical Analysis

Categorical variables from the demographics data were compared using the Fisher's exact or chi-squared test. The normality of numerical demographic and clinical variables were assessed using Shapiro–Wilk tests and by examining histograms. For data that were normally distributed, *t*-tests were used to compare means, and for those with non-normal distributions, nonparametric Wilcoxon rank-sum tests were used. Results were considered statistically significant when $p < 0.05$. Relative abundances of serum metabolites were normalized using median-scaling, missing values were imputed with the sample set minimum, and the data were log-transformed. Unsupervised analysis consisted of principal component analysis (PCA) and hierarchical clustering of Euclidian distances. Welch's two-sample *t*-test was used to detect serum metabolites that varied between the EPI and healthy control groups. Metabolite set enrichment analysis (MSEA) was used to identify metabolic subpathways enriched in metabolites that varied between groups. Metabolites were ranked by the product of the $-\log(p\text{-value})$ from the *t*-tests and the log2-fold-change in the metabolites between groups. This resulted in a ranked vector where the top of the vector contained significantly variable metabolites with relative abundances that were increased, and the bottom of the vector contained significantly variable metabolites that were decreased in the EPI group compared with healthy controls. Next, MSEA was implemented by using the "fgsea" R package (version 1.22.0) to calculate normalized enrichment scores (NES) for each metabolic subpathway [22,23]. Subpathways with a NES > 0 were significantly upregulated, and those with a NES < 0 were significantly downregulated in the EPI group compared with healthy controls.

The Benjamini–Hochberg false discovery rate (FDR) procedure was used to calculate FDR-corrected *p*-values (q-values) for multiple comparisons in the metabolomics data [24,25]. We sought to maximize the discovery of features that vary in association with EPI while controlling for false discovery within reasonable bounds. Metabolomics results were considered statistically significant when $p < 0.05$ and q < 0.2. All statistical analyses were performed in the R language for statistical computing (version 4.2.1) using the RStudio integrated development environment running on macOS (Monterey 12.5).

3. Results

3.1. Demographic and Clinical Data

Serum samples from 30 dogs were analyzed for this study, including 20 dogs with EPI and 10 healthy controls. There were no statistically significant differences in age, breed, or reproductive status between the EPI and healthy control groups (Table 1). The persistence of clinical signs associated with EPI were common in dogs with EPI following treatment with PERT. Diarrhea was reported in 40% (n = 8) and persistent weight loss in 15% (n = 3) of dogs in the EPI group. Other clinical signs that are considered atypical for EPI were also reported in the EPI group, including decreased appetite in 10% (n = 2) and vomiting in 30% (n = 6). Dogs with EPI had received a variety of adjunct therapies in addition to PERT. Antibiotics were administered in 35% (n = 7) of dogs with EPI, including metronidazole in 5% (n = 1) and tylosin in 30% (n = 6). Cobalamin and folate supplementation was administered to 70% (n = 14) and 40% (n = 8) of dogs with EPI, respectively. Other adjunctive medications administered to dogs with EPI included gastric-acid-suppressing drugs in 30% (n = 6) and probiotics in 70% (n = 14). As the presence of the active clinical signs of gastrointestinal disease and administration of gastrointestinal medication were an exclusion criterion for healthy dogs, no statistical comparisons were performed between the EPI and healthy control groups. Except for one dog in the EPI group that was fed a home-prepared diet, all dogs included in this study were fed diets that were formulated to meet the nutritional levels established by the AAFCO Dog Food Nutrient Profiles (https://www.aafco.org; accessed 7 June 2023) for maintenance or animal feeding tests using AAFCO procedures that substantiated that the diet provides complete and balanced nutrition for the maintenance of adult dogs. Four dogs in the EPI group were fed mixtures of different commercial diets that were formulated to meet the AAFCO profiles. With the exception of the dog fed a home-prepared diet and those fed mixtures of commercial diets, all dogs were fed dry kibble diets. Mean ($\pm$ SD) dietary fat (as-fed basis) was significantly ($p = 0.012$) higher in dogs in the EPI group (15.4% $\pm$ 1.7) compared with the healthy controls (13.3% $\pm$ 2.1), but there were no other significant differences in dietary protein ($p = 0.08$) or crude fiber ($p = 0.95$) between the groups.

Serum biomarkers for exocrine pancreatic and gastrointestinal function were measured in the sera of dogs included in this study: cTLI, cobalamin, and folate (Table 1). Serum concentrations of cTLI (median, IQR) were significantly lower in dogs with EPI (1.0 µg/L, 1.0–1.1) compared with the healthy controls (18.8 µg/L, 15.7–29.1; $p < 0.001$). After excluding dogs that had received supplementation with cobalamin or folate, serum cobalamin (median, IQR) concentrations were significantly lower in dogs with EPI (231 ng/L, 207–313) compared with the healthy controls (411 ng/L, 297–593, $p = 0.013$), but there were no statistically significant differences in serum folate concentrations between the groups ($p = 0.39$).

Table 1. Patient population statistics. Categorical variables compared using Fisher's exact or chi-squared tests. Variables with normal distributions are presented as means and standard deviations (SD) and are otherwise presented as medians and interquartile ranges (IQRs). Comparisons of serum cobalamin and folate omit dogs receiving supplementation. Laboratory reference intervals (RIs) are listed for cTLI, cobalamin, and folate.

	Healthy	EPI	*p*-Value
Age (years)			0.36
Mean (SD)	5.0 ($\pm$ 1.74)	4.3 ($\pm$ 2.0)	
Breed			0.62
Akita	0 (0%)	1 (5%)	
Australian Shepherd	0 (0%)	2 (10%)	
Border Collie	0 (0%)	1 (5%)	
Cavalier King Charles Spaniels	0 (0%)	1 (5%)	
German Shepherd	4 (40%)	9 (45%)	
Labrador Retriever	1 (10%)	1 (5%)	
Mixed Breed	4 (40%)	4 (20%)	
Pit Bull Terrier	1 (10%)	0 (0%)	
West Highland White Terrier	0 (0%)	1 (5%)	
Sex			0.16
Spayed Female	3 (30%)	11 (55%)	
Intact Male	0 (0%)	2 (10%)	
Neutered Male	7 (70%)	7 (35%)	
Serum cTLI (RI: 5.7–45.2 µg/L)			<0.001
Median (IQR)	18.8 (15.7–29.1)	1.0 (1.0–1.1)	
Serum cobalamin (RI: 251–908 ng/L)			0.01
Median (IQR)	411 (297–593)	231 (207–313)	
Serum folate (RI: 7.7–24.4 ng/L)			0.39
Mean (SD)	12.1 ($\pm$5.2)	13.9 ($\pm$4.5)	

3.2. Untargeted Metabolomics Analysis

The untargeted serum metabolome dataset contained 759 named biochemicals. Among metabolic superpathways, lipids were the most numerous (n = 355), followed by amino acids (n = 187), xenobiotics (n = 79), nucleotides (n = 47), peptides (n = 31), cofactors and vitamins (n = 29), carbohydrates (n = 21), and energy (n = 10) metabolites. PCA separated samples from dogs with EPI from healthy controls along the first principal component (Figure 1A), consistent with metabolome-wide differences in the abundance of serum biochemicals among groups. Similarly, hierarchical clustering of metabolite relative abundances based on Euclidian distances revealed the clustering of most samples from the healthy controls away from those with EPI (Figure 1B).

We sought to detect individual metabolites that varied in association with EPI by comparing relative abundances of serum metabolites between EPI and the healthy controls. After controlling for false discovery (q < 0.2), the relative abundances of 114 serum metabolites varied significantly between the groups, with 76 being increased and 38 being decreased in dogs with EPI compared with the healthy controls. A selection of significantly variable serum metabolites with a large effect size is presented in Table 2. The complete output is presented in Supplementary File S2, and the plots of all significantly variable metabolites are presented in Supplementary File S3.

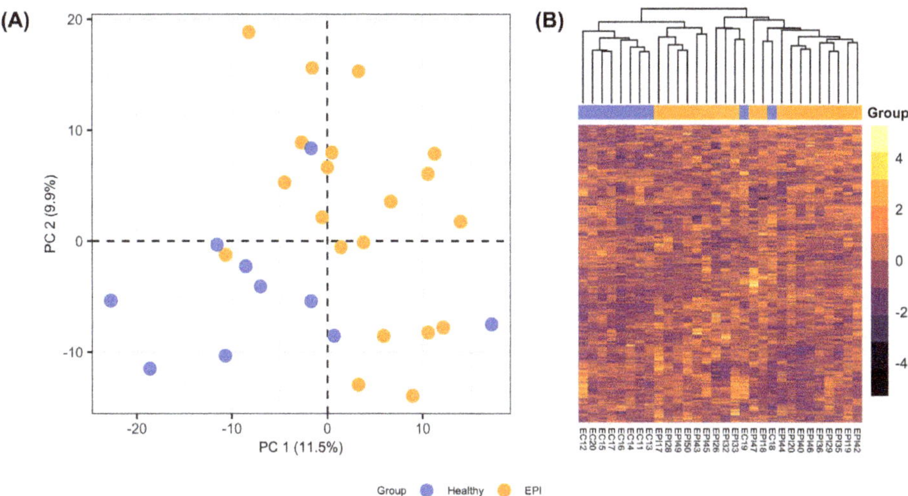

Figure 1. Unsupervised analysis of serum metabolite profiles. Samples are annotated by color according to their group: EPI (orange), healthy (blue). (**A**) Principal component analysis of relative metabolite abundances. (**B**) Statistical heatmap of relative metabolite abundances using Euclidian distances and hierarchical clustering.

Table 2. Significantly variable serum metabolites. Significantly variable serum metabolites with large effect size (| log2-fold-change | > 0.6; corresponding > 1.5 fold difference in means). *p*-values derived from Welch's *t*-test. q-value, false discovery rate-adjusted *p*-values; log2FC, log2 of the fold difference in means between dogs with EPI and healthy controls.

Biochemical	Subpathway	*p*-Value	q-Value	log2FC
Amino Acids				
Alpha-ketoglutaramate	Glutamate Metabolism	0.001	0.019	0.99
Cysteinylglycine Disulfide	Glutathione Metabolism	0.001	0.023	−1.31
Cysteine-glutathione Disulfide		0.003	0.043	−0.93
4-guanidinobutanoate	Guanidino and Acetamido Metabolism	0.002	0.034	1.65
Homocarnosine	Histidine Metabolism	0.001	0.017	1.24
5-aminovalerate	Lysine Metabolism	0.002	0.034	0.99
N-acetyl-cadaverine		0.001	0.024	1.19
Cystine	Methionine, Cysteine, s-Adenosylmethionine, and Taurine Metabolism	0.000	<0.001	−2.29
Phenyllactate	Phenylalanine Metabolism	<0.001	0.010	0.93
Phenylpyruvate		<0.001	0.011	1.31
4-hydroxyphenylacetate		<0.001	0.000	2.09
Phenol Sulfate	Tyrosine Metabolism	0.003	0.042	−1.93
4-hydroxyphenylpyruvate		0.001	0.023	0.83
4 hydroxyphenylacetatoylcarnitine		<0.001	0.001	3.61
Pro-hydroxy-pro	Urea cycle; Arginine and Proline Metabolism	0.001	0.023	−1.37
Argininate		0.001	0.014	0.66
2-oxoarginine		<0.001	0.001	2.44

Table 2. *Cont.*

Biochemical	Subpathway	*p*-Value	q-Value	log2FC
Carbohydrates				
1,5-anhydroglucitol	Glycolysis, Gluconeogenesis, and Pyruvate Metabolism	<0.001	0.006	−0.73
Glycerate		<0.001	0.001	1.03
Ribose	Pentose Metabolism	<0.001	<0.001	1.97
Cofactors and Vitamins				
Threonate	Ascorbate and Aldarate Metabolism	<0.001	0.002	−0.98
Alpha-CEHC Sulfate	Tocopherol Metabolism	<0.001	0.010	−2.03
Pyridoxal	Vitamin B6 Metabolism	0.002	0.028	−0.79
Lipids				
2-hydroxydecanoate	Fatty Acid, Monohydroxy	0.001	0.023	1.48
1-arachidonoyl-GPA (20:4)	Lysophospholipid	<0.001	0.008	1.02
1-oleoyl-GPA (18:1)		<0.001	<0.001	2.27
1-palmitoyl-GPA (16:0)		<0.001	<0.001	2.38
Heptanoate (7:0)	Medium Chain Fatty Acid	0.003	0.046	0.80
Choline	Phospholipid Metabolism	<0.001	0.001	0.62
1-(1-enyl-palmitoyl)-2-oleoyl-GPE (P-16:0/18:1)	Plasmalogen	<0.001	0.006	0.81
1-(1-enyl-stearoyl)-2-oleoyl-GPE (P-18:0/18:1)		<0.001	0.011	0.92
Sphingomyelin (d18:1/25:0, d19:0/24:1, d20:1/23:0, d19:1/24:0)	Sphingolipid Metabolism	0.001	0.020	−1.02
Sphingomyelin (d18:1/14:0, d16:1/16:0)		0.001	0.025	0.65
Nucleotides				
Xanthine	Purine Metabolism, (Hypo)Xanthine/Inosine-containing	<0.001	0.001	1.53
Adenosine 5′-monophosphate	Purine Metabolism, Adenine-containing	<0.001	0.001	−2.89
Adenosine		<0.001	0.004	−2.45
Dihydroorotate	Pyrimidine Metabolism, Orotate-containing	<0.001	0.001	−1.17
Uracil	Pyrimidine Metabolism, Uracil-containing	<0.001	0.008	1.33
Peptides				
4-hydroxyphenylacetylglutamine	Acetylated Peptides	<0.001	0.001	1.93
Phenylacetylthreonine		0.002	0.028	1.96
4-hydroxyphenylacetylglycine		0.000	0.001	2.10
Leucylglutamine	Dipeptide	0.001	0.024	1.90
Xenobiotics				
Tartronate (hydroxymalonate)	Bacterial/Fungal	<0.001	0.008	0.94
Perfluorooctanesulfonic Acid	Chemical	0.002	0.034	−2.11
S-(3-hydroxypropyl)mercapturic Acid		0.001	0.026	0.76
Hydroquinone Sulfate	Drug–Topical Agents	0.002	0.028	−1.86
Quinate	Food Component/Plant	<0.001	0.006	3.16

The amino acid and lipid superpathways contained the greatest number of metabolites that varied significantly between the groups (Figure 2A). Metabolite set enrichment analysis was performed to identify metabolic subpathways that were enriched among significantly variable metabolites (the complete results are in Supplementary File S4). After controlling for multiple comparisons, the "Phenylalanine Metabolism" subpathway (NES = 1.42, $p = 0.005$, q = 0.17) was found to be increased, and the "Purine Metabolism, Adenine containing" subpathway (NES = -1.57, $p = 0.002$, q = 0.12) was decreased in dogs with EPI (Figure 2B).

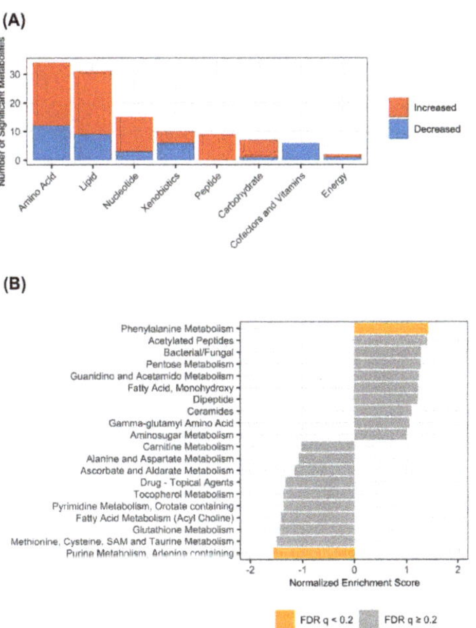

Figure 2. Metabolic pathway enrichment among serum metabolites. (**A**) Counts of significantly variable metabolites from each metabolic superpathway. Each bar represents the number of significantly variable metabolites in each superpathway. The red bars show the number of metabolites that were increased in dogs with EPI compared with healthy controls. The blue bars show the number of metabolites that were decreased in dogs with EPI compared with healthy controls. (**B**) Metabolite set enrichment analysis of metabolic subpathways (MSEA). MSEA was used to detect metabolic subpathways enriched among metabolites with high (positive NES) and low (negative NES) stability. The 20 subpathways with the highest absolute value of NES are shown. Bars are annotated using the FDR-adjusted p-value (q-value), such that grey bars represent pathways with q-value ≥ 0.2 and orange bars represent subpathways with q-value < 0.2.

4. Discussion

The purpose of this investigation was to identify differences in serum metabolomes between healthy dogs and dogs with EPI. Unsupervised analysis with PCA and hierarchical clustering of Euclidian distances revealed the separation of samples from dogs with EPI and the healthy controls, consistent with metabolome-wide differences in serum metabolite profiles between the groups. The relative abundances of 143 serum metabolites varied in association with EPI ($p < 0.05$), and 114 of these were considered statistically significant after controlling for false discovery (q < 0.2). Pathway enrichment analysis identified two metabolic subpathways that were differentially enriched among significantly variable metabolites. The "Purine Metabolism, Adenine Containing" subpathway was downregulated in dogs with EPI. Within this subpathway, adenosine and adenosine 5′-monophosphate (AMP) were significantly lower in the serum of dogs with EPI. Previous studies have identified adenosine receptors in pancreatic acinar and ductal cells, and purines are implicated in the regulation of exocrine pancreatic secretions in rodents, humans, and dogs [26–30]. Additionally, adenosine has been found to exert an anti-inflammatory effect on intestinal epithelial cells, and it is plausible that the decreased availability of adenosine could contribute to clinical signs of gastrointestinal dysfunction that persist after PERT due to the dysregulation of intestinal mucosal inflammatory responses [31].

Several gut microbial metabolites were found to vary significantly in the sera of dogs with EPI compared with the healthy controls. The "Phenylalanine Metabolism" subpathway was significantly upregulated in association with EPI. This subpathway contains several gut microbial metabolites, including phenylpyruvate, 4-hydroxyphenylacetate, and phenyllactate, which were all significantly increased in the serum of dogs with EPI compared with the healthy controls. These biochemicals are microbial catabolites of phenylalanine that are generated by bacteria in the genera *Lactobacillus* and *Bifidobacterium* [32–37]. Previous studies of fecal microbiomes identified enrichment in genes associated with phenylalanine metabolism and increased abundances of *Lactobacillus* and *Bifidobacterium* in dogs with EPI [38]. The acetylated peptides phenylacetylglutamine and 4-hydroxyphenylacetylglutamine were significantly increased in dogs with EPI. A recent study integrating the deep metagenomic sequencing of gut microbiota with untargeted serum metabolomics in humans revealed strong, inverse correlations between serum concentrations of these acetylated peptides and the abundance of *Faecalibacterium prausnitzii*, an important commensal bacterium that maintains host health [39]. The sulfated benzoate derivates, 3-methoxycatechol sulfate and 4-hydroxycatechol sulfate, were decreased, and o-cresol-sulfate was increased in the sera of dogs with EPI. In a previous metagenomic study of human feces, the genes involved in benzoate catabolism via catechol were affiliated with diverse clades of bacteria including Bacteroidetes, Proteobacteria, Firmicutes, and Chloroflexi [40]. Notably, o-cresol sulfate is a well-characterized uremic toxin highlighting the interconnected nature of gut microbial and host metabolic systems in dogs. Consistent with our finding of the variation in numerous gut microbial metabolites in association with EPI, previous studies have identified microbiota dysbiosis in dogs with EPI characterized by small intestinal bacterial overgrowth, higher fecal abundance of *E. coli*, *Lactobacillus*, and *Bifidobacterium* and lower fecal abundance of *Fusobacterium*, *C. hiranonis*, and *Faecalibacterium* [38,41–44]. Interestingly, differences in fecal microbiomes in some dogs with EPI persist following treatment, and fecal microbiomes of dogs with EPI following the initiation of PERT are significantly different from healthy dogs and similar to those with EPI [38,44]. Consequently, it seems likely that these differences in the serum relative abundances of microbial metabolites in dogs with EPI are due to changes in gut microbiome composition, metabolic function, or both. As all dogs enrolled in this study were treated with PERT, our findings suggest that pancreatic enzyme supplementation may not be sufficient to restore eubiosis in dogs with EPI. These results also suggest that phenylalanine derivatives and acetylated peptides may be promising markers for enteric microbiota dysbiosis in dogs with EPI.

The presence of enteric microbiota dysbiosis in dogs with EPI is likely secondary to the malabsorption of dietary macromolecules due to the insufficient secretion of pancreatic digestive enzymes. However, it is possible that enteric microbiota participate in the pathogenesis or progression of PAA. In experiments in the 1970s, PAA-like lesions were induced in rats fed a liquid elemental diet [45,46]. The PAA-like lesions were rare in germ-free rats but invariably present in conventional rats. Germ-free rats purposefully contaminated with cecal contents from conventional mice also developed PAA-like lesions, but those contaminated with sterilized cecal contents did not. These findings suggest that PAA was induced by nutritional/metabolic disturbances, the impacts of which were at least partially conditioned by the presence of commensal enteric microbiota. Whether enteric microbiota participate in the pathogenesis of PAA or its progression to EPI in dogs is unknown.

Some of the metabolic disturbances we identified in dogs with EPI are similar to those observed during fasting or starvation, including greater relative abundances of urea cycle metabolites, purines, pyrimidines, and fatty acids [47]. Urea cycle amino acid derivatives (arginate, homoarginine, and 2-oxoarginine) were significantly lower, and N-acetyl-cadaverine (a lysine derivative) was greater in dogs with EPI, consistent with the increased rate of deamination reactions due to proteolysis. Greater relative abundances of free fatty acids, docosahexaenoylcarnitine (an acylcarnitine), and α-ketoglutaramate (a ketogenic amino acid) are consistent with the mobilization (lipolysis) of peripheral adipose stores and increased mitochondrial beta-oxidation reactions. Interestingly, some of the changes in microbial metabolites discussed above also indicate a state of malnourishment. In particular, increased plasma phenyllactate and urinary p-cresol have been identified in undernourished mice [35]. Collectively, these findings are consistent with the presence of a catabolic state in dogs with EPI, characterized by the mobilization of bodily stores of protein and adipose tissue, to maintain energy homeostasis.

Kynurenine, a derivative of tryptophan, was significantly greater in the serum of dogs with EPI (Supplementary Files S2 and S3). This metabolite is noteworthy owing to its association with intestinal mucosal inflammation in humans and rodent models of inflammatory bowel disease (IBD). Kynurenine is a mammalian catabolite of tryptophan that is generated in hepatic and peripheral tissues, primarily intestinal epithelial cells and leukocytes. In peripheral tissues, kynurenine is synthesized by indoleamine 2,3-dioxygenase-1 (IDO-1), the rate-limiting step in the peripheral kynurenine pathway [48]. The increased expression of IDO-1 and increased circulating kynurenine concentrations are associated with endoscopic inflammation and disease activity and predictive of outcomes in humans with IBD [49–51]. Thus, greater relative abundances of kynurenine are suggestive of the presence of intestinal mucosal inflammation in dogs with EPI, a factor that could contribute to the pathophysiology of gastrointestinal clinical signs, especially those atypical of EPI (e.g., vomiting, anorexia). Additional studies of kynurenine and other tryptophan catabolites in EPI are currently underway to confirm and further elucidate the importance of this finding.

We compared pancreatic and gastrointestinal biomarkers and identified significantly lower serum cobalamin in dogs with EPI compared with healthy controls. Cobalamin deficiency has been documented in up to 67% of dogs with EPI and is a negative prognostic indicator for therapeutic response [4,5]. The persistence of altered serum cobalamin in this population of dogs with EPI treated with PERT emphasizes the necessity to monitor serum cobalamin concentrations during treatment. Persistent cobalamin deficiency following PERT suggests that oral pancreatic enzyme supplementation alone is insufficient to restore cobalamin homeostasis. A previous study identified lower concentrations of lipid-soluble vitamins in this same population of dogs with EPI, consistent with persistent deficiency of micronutrient absorption in dogs with EPI treated with PERT [3]. We identified evidence of other micronutrient disturbances involving vitamin B6 and vitamin C, as evidenced by decreased pyridoxal and threonate, respectively. The causes for the altered metabolism of these water-soluble vitamins is unknown and should be investigated in future studies.

Our findings should be interpreted in light of several important limitations. First, this was an untargeted study intended to reveal novel metabolic markers and associations that are plausibly involved in the pathophysiology of EPI. Direct causal associations cannot be determined from any of these results. Many hundreds of metabolites were measured in a relatively small number of dogs, increasing the risk of discovering falsely significant features. We attempted to control for false discovery, but it is likely that some of the results we report as significant are false positives. Controlling the FDR at q < 0.2 means that 22/114 significant metabolites were likely to be false discoveries. We considered this to be an acceptable rate of false discovery given the objectives of the study. Future, hypothesis-driven investigations should confirm our findings using targeted analytic methods. As the serum samples from the dogs with EPI were collected in primary care veterinary clinics, preanalytical conditions could have impacted our results. We provided detailed instructions for primary care veterinarians to collect, process, store, and ship the samples. However, the samples were collected and processed by different personnel and in different laboratories. We cannot exclude the possibility that differences in the personnel, methods, and equipment (e.g., centrifuges) used to process the samples or a deviation from the sample collection protocol could have influenced our results. The sample sizes of the EPI and healthy control groups were different. We cannot exclude the possibility that the difference in group sizes could have influenced our results. Future investigations should assess serum metabolite concentrations using larger numbers of animals with similar sample sizes, and our results could inform such future investigations.

Regarding the patient population, we were unable to control environmental factors, including the dogs' geographic locations or diets. This was a field study of animals with spontaneously occurring EPI, and some patients were receiving therapeutic diets as a component of therapy. Owing to this, we were not able to control the brands, nutrient compositions, quantities, or feeding frequencies of the diets fed to dogs enrolled in this study. Previous investigations have revealed the significant effects of diet on the serum metabolomes of dogs and other animals [52–54]. Though we did not standardize the diets of dogs in this study, we compared the macronutrient composition (protein, fat, and crude fiber) of their diets to understand whether there were significant differences between the EPI and healthy control groups. We identified a small, but statistically significant, difference in the dietary proportions of fat between the groups, with the diets of dogs in the EPI group containing a higher proportion of fat compared with the healthy controls. There were no significant differences in the dietary proportions of protein or crude fiber between the groups. However, because energy-basis nutrient profiles were not available for all diets, macronutrient data on an as-fed basis was analyzed which does not account for differences in moisture content or energy density between the diets. Data related to the quantities and frequencies of feedings were not collected, so their impacts cannot be assessed. Thus, we cannot exclude the possibility that differences in diet could have influenced our findings. Future investigations are needed to understand the influence of different diets on the serum metabolomes of dogs with EPI; these future studies should aim to include assessments of diets on an energy basis. There were small differences in the mean ages of the EPI and healthy controls groups, though the differences were not statistically significant. Serum metabolomes are known to vary with age in dogs and other mammals [55–57]. To understand whether differences in age were likely to have influenced our results, we attempted to detect correlations among relative abundances of serum metabolites and age. Of 114 serum metabolites that varied significantly between the EPI and healthy control groups, only 9 were significantly correlated with the age (Supplementary File S5). It is unlikely that the small differences in age between groups had a significant impact on our findings.

Treatments administered to dogs in the EPI group could have influenced our findings. The brands, strengths, and dosages of oral pancreatic enzyme extracts administered to dogs in the EPI group were not standardized. A study in piglets in which EPI was experimentally induced by pancreatic duct ligation, suggests that PERT can result in the normalization of enteric microbiota dysbiosis [58]. Thus, it is possible that some of our findings could have been influenced by inadequate PERT, resulting in some dogs with EPI having persistent digestive dysfunction and associated dysbiosis. While we attempted to control for the potential impact of PERT on serum metabolomes by administering pancreatic enzyme extracts to the healthy controls, it is likely that dogs with EPI respond differently to PERT compared with healthy dogs. It is also plausible that differences between the dogs with EPI and healthy controls could have been influenced by the duration of treatment. Dogs with EPI had been treated for months–years with PERT, whereas the healthy controls received enzyme extracts for only 14 days. We compared serum metabolite profiles in the healthy dogs before and after oral pancreatic enzyme administration, but there were no significant differences in metabolite abundances between the two groups (Supplementary File S6). Though we detected persistent clinical signs of gastrointestinal dysfunction in dogs with

EPI, the clinical assessment of dogs with EPI was not among the objectives of this investigation. For this reason, we did not attempt to compare their serum metabolomes to dogs with a resolution of clinical signs. Lacking clinical assessments (e.g., weight, body condition, fecal quality scoring) before and after the initiation of PERT, we were not able to make accurate assessments of the clinical responses to PERT in dogs with EPI. The determination as to whether dogs had persistent clinical signs was based on subjective assessments made by the dogs' owners and provided to the investigators during historical interviews. Owing to these factors, we concluded that our clinical data were insufficient to compare serum metabolomes between dogs with EPI that had resolved vs. persistent clinical signs. Future investigations should assess changes in the metabolomes of dogs following the initiation of PERT using more rigorous clinical assessments collected before and after the initiation of PERT. Based on historical interviews and reviews of the medical records, antibiotics, probiotics, and gastric-acid-suppressing drugs were administered to some dogs in the EPI group within the 30 days prior to sample collection. The exact brands, dosages, and administration frequencies of these treatments were not known for all dogs. These medications are known to impact enteric microbial communities in ways that could have affected our findings, especially in relation to the metabolites derived from enteric microbiota [59–61]. Targeted follow-up studies in dogs with EPI not exposed to antibiotics and probiotics are needed to confirm our findings.

5. Conclusions

This investigation identified relative abundances of 114 serum metabolites that varied significantly between dogs with EPI and healthy controls after pancreatic enzyme replacement therapy. The differences in amino acids and lipids were consistent with increased proteolysis and lipolysis, indicating a persistent catabolic state in dogs with EPI. The relative abundances of gut microbial metabolites were altered in dogs with EPI, consistent with disruptions in gut microbial communities. Increased kynurenine suggests the presence of intestinal inflammation in dogs with EPI. Our findings have revealed evidence of persistent gastrointestinal dysfunction, enteric microbiota dysbiosis, and altered lipid, amino acid, nucleotide, and peptide metabolism in dogs with EPI receiving PERT. The extent to which these metabolic alterations participate in the pathogenesis or pathophysiology of EPI is unknown. These findings should be interpreted with caution and should be subject to future, hypothesis-driven research to confirm and further elucidate their pathophysiologic significance. Future studies should investigate the metabolic disturbances we have identified in larger populations of dogs with EPI while attempting to standardize diets and therapeutic strategies. Additionally, efforts to understand whether the persistence of gastrointestinal clinical signs and dysbiosis in dogs with EPI following the initiation of PERT is related to inadequate treatment, a concurrent chronic enteropathy, or an unidentified facet of its pathophysiology are needed.

Supplementary Materials: The following supporting information can be downloaded at: https://www.mdpi.com/article/10.3390/ani13142313/s1. File S1: Information about diet and pancreatic enzyme histories; File S2: Complete output of *t*-test comparing serum metabolites between dogs with EPI and healthy controls; File S3: Boxplots of all significantly variable serum metabolites; File S4: Complete output of metabolite set enrichment analysis; File S5: Correlations among serum metabolite abundances and age; File S6: Comparisons of serum metabolomes in healthy dogs before and after pancreatic enzyme supplementation.

Author Contributions: Conceptualization, P.C.B., S.I.R., D.A.W.; methodology, P.C.B., S.I.R., K.S.S., M.A.M., M.D.R., D.A.W.; formal analysis, P.C.B.; data curation, P.C.B.; writing—original draft preparation, P.C.B.; writing—review and editing, P.C.B., S.I.R., K.S.S., M.A.M., M.D.R., D.A.W.; visualization, P.C.B.; supervision, D.A.W.; project administration, D.A.W.; funding acquisition, P.C.B. All authors have read and agreed to the published version of the manuscript.

Funding: This investigation was funded by a grant program for veterinary medical residents sponsored by Nestle Purina PetCare, Inc. and a generous donation from Paula and John Gatens. Funding was awarded to one of the primary authors (PCB) and the principal investigator (DAW). The funders had no role in study design, data collection and analysis, decision to publish, or preparation of the manuscript.

Institutional Review Board Statement: This investigation was approved by the University of Illinois Institutional Animal Care and Use Committee (IACUC #14284; approved 12 December 2014).

Informed Consent Statement: All samples from dogs in the EPI groups were collected by primary care veterinarians from client-owned dogs with spontaneously occurring EPI. Samples from dogs in the healthy control group were collected from client-owned dogs lacking a history or clinical signs of gastrointestinal dysfunction. All owners provided informed consent.

Data Availability Statement: All data and R code necessary to replicate this analysis are located in a GitHub repository (https://github.com/pcbarko/K9_EPI_Serum_Metabolome, accessed on 28 June 2023).

Acknowledgments: The authors thank Olesia Kennedy, Epi4Dogs Foundation Inc., and Paula and John Gatens for their invaluable support for this investigation. The authors would also like to acknowledge Diane Sloan for donating the pancreatin used in this investigation. Without their generous assistance, this investigation would not have been possible.

Conflicts of Interest: The authors declare no conflict of interest.

References

1. Westermarck, E.; Wiberg, M. Exocrine Pancreatic Insufficiency in the Dog: Historical Background, Diagnosis, and Treatment. *Top. Companion Anim. Med.* **2012**, *27*, 96–103. [CrossRef]
2. Batt, R.M. Exocrine Pancreatic Insufficiency. *Vet. Clin. N. Am. Small Anim. Pract.* **1993**, *23*, 595–608. [CrossRef]
3. Barko, P.C.; Williams, D.A. Serum concentrations of lipid-soluble vitamins in dogs with exocrine pancreatic insufficiency treated with pancreatic enzymes. *J. Vet. Intern. Med.* **2018**, *32*, 1600–1608. [CrossRef]
4. Soetart, N.; Rochel, D.; Drut, A.; Jaillardon, L. Serum cobalamin and folate as prognostic factors in canine exocrine pancreatic insufficiency: An observational cohort study of 299 dogs. *Vet. J.* **2019**, *243*, 15–20. [CrossRef]
5. Chang, C.; Lidbury, J.A.; Suchodolski, J.S.; Steiner, J.M. Effect of oral or injectable supplementation with cobalamin in dogs with hypocobalaminemia caused by chronic enteropathy or exocrine pancreatic insufficiency. *J. Vet. Intern. Med.* **2022**, *36*, 1607–1621. [CrossRef]
6. Wiberg, M.E.; Nurmi, A.-K.; Westermarck, E. Serum Trypsinlike Immunoreactivity Measurement for the Diagnosis of Subclinical Exocrine Pancreatic Insufficiency. *J. Vet. Intern. Med.* **1999**, *13*, 426–432. [CrossRef]
7. Wiberg, M.E.; Westermarck, E. Subclinical exocrine pancreatic insufficiency in dogs. *J. Am. Vet. Med. Assoc.* **2002**, *220*, 1183–1187. [CrossRef]
8. Wiberg, M. Pancreatic acinar atrophy in German shepherd dogs and rough-coated Collies. Etiopathogenesis, diagnosis and treatment. A review. *Vet. Q.* **2004**, *26*, 61–75. [CrossRef]
9. Westermarck, E.; Batt, R.M.; Vaillant, C.; Wiberg, M. Sequential study of pancreatic structure and function during development of pancreatic acinar atrophy in a German shepherd dog. *Am. J. Vet. Res.* **1993**, *54*, 1088–1094.
10. Williams, D.A.; Batt, R.M. Sensitivity and specificity of radioimmunoassay of serum trypsin-like immunoreactivity for the diagnosis of canine exocrine pancreatic insufficiency. *J. Am. Vet. Med. Assoc.* **1988**, *192*, 195–201.
11. Williams, D.A.; Batt, R.M. Diagnosis of canine exocrine pancreatic insufficiency by the assay of serum trypsin-like immunoreactivity. *J. Small Anim. Pract.* **1983**, *24*, 583–588. [CrossRef]
12. Batchelor, D.J.; Noble, P.-J.M.; Taylor, R.H.; Cripps, P.J.; German, A.J. Prognostic Factors in Canine Exocrine Pancreatic Insufficiency: Prolonged Survival is Likely if Clinical Remission is Achieved. *J. Vet. Intern. Med.* **2007**, *21*, 54–60. [CrossRef]
13. German, A.J. Exocrine Pancreatic Insufficiency in the Dog: Breed Associations, Nutritional Considerations, and Long-term Outcome. *Top. Companion Anim. Med.* **2012**, *27*, 104–108. [CrossRef]
14. Wiberg, M.; Lautala, H.; Westermarck, E. Response to long-term enzyme replacement treatment in dogs with exocrine pancreatic insufficiency—PubMed. *J. Am. Vet. Med. Assoc.* **1998**, *213*, 86–90.
15. Wiberg, M.E.; Saari, S.A.; Westermarck, E.; Meri, S. Cellular and humoral immune responses in atrophic lymphocytic pancreatitis in German shepherd dogs and rough-coated collies. *Vet. Immunol. Immunopathol.* **2000**, *76*, 103–115. [CrossRef]
16. Wiberg, M.E.; Saari, S.A.M.; Westermarck, E. Exocrine Pancreatic Atrophy in German Shepherd Dogs and Rough-coated Collies: An End Result of Lymphocytic Pancreatitis. *Vet. Pathol.* **1999**, *36*, 530–541. [CrossRef]
17. Westermarck, E.; Saari, S.; Wiberg, M. Heritability of Exocrine Pancreatic Insufficiency in German Shepherd Dogs. *J. Vet. Intern. Med.* **2010**, *24*, 450–452. [CrossRef]

18. Tsai, K.L.; Noorai, R.E.; Starr-Moss, A.N.; Quignon, P.; Rinz, C.J.; Ostrander, E.A.; Steiner, J.M.; Murphy, K.E.; Clark, L.A. Genome-wide association studies for multiple diseases of the German Shepherd Dog. *Mamm. Genome* **2011**, *23*, 203–211. [CrossRef]

19. Evans, J.M.; Tsai, K.L.; Starr-Moss, A.N.; Steiner, J.M.; Clark, L.A. Association ofDLA-DQB1alleles with exocrine pancreatic insufficiency in Pembroke Welsh Corgis. *Anim. Genet.* **2015**, *46*, 462–465. [CrossRef]

20. Tsai, K.L.; Starr-Moss, A.N.; Venkataraman, G.M.; Robinson, C.; Kennedy, L.J.; Steiner, J.M.; Clark, L.A. Alleles of the major histocompatibility complex play a role in the pathogenesis of pancreatic acinar atrophy in dogs. *Immunogenetics* **2013**, *65*, 501–509. [CrossRef]

21. Wishart, D.S. Metabolomics for Investigating Physiological and Pathophysiological Processes. *Physiol. Rev.* **2019**, *99*, 1819–1875. [CrossRef]

22. Sergushichev, A. Algorithm for cumulative calculation of gene set enrichment statistic. *Sci. Technol. J. Inf. Technol. Mech. Opt.* **2016**, *16*, 956. [CrossRef]

23. Korotkevich, G.; Sukhov, V.; Budin, N.; Shpak, B.; Artyomov, M.N.; Sergushichev, A. Fast gene set enrichment analysis. *bioRxiv* **2021**, 060012. [CrossRef]

24. Benjamini, Y.; Hochberg, Y. Controlling the False Discovery Rate: A Practical and Powerful Approach to Multiple Testing. *J. Royal Statistical Soc. Ser. B Methodol.* **1995**, *57*, 289–300. [CrossRef]

25. Storey, J.D.; Bass, A.J.; Dabney, A.; Robinson, D. Qvalue: Q-Value Estimation for False Discovery Rate Control. 2020. Available online: http://github.com/jdstorey/qvalue (accessed on 28 June 2023).

26. Hayashi, M.; Inagaki, A.; Novak, I.; Matsuda, H. The adenosine A2B receptor is involved in anion secretion in human pancreatic duct Capan-1 epithelial cells. *Pflugers Arch.* **2016**, *468*, 1171–1181. [CrossRef] [PubMed]

27. Iwatsuki, K. Subtypes of adenosine receptors on pancreatic exocrine secretion in anaesthetized dogs. *Fundam. Clin. Pharmacol.* **2000**, *14*, 203–208. [CrossRef] [PubMed]

28. Novak, I.; Hede, S.E.; Hansen, M.R. Adenosine receptors in rat and human pancreatic ducts stimulate chloride transport. *Pflugers Arch.* **2008**, *456*, 437–447. [CrossRef]

29. Burnstock, G.; Novak, I. Purinergic signalling in the pancreas in health and disease. *J. Endocrinol.* **2012**, *213*, 123–141. [CrossRef]

30. Hayashi, M. Expression of Adenosine Receptors in Rodent Pancreas. *Int. J. Mol. Sci.* **2019**, *20*, 5329. [CrossRef]

31. Ye, J.H.; Rajendran, V.M. Adenosine: An immune modulator of inflammatory bowel diseases. *World J. Gastroenterol.* **2009**, *15*, 4491–4498. [CrossRef]

32. Beloborodova, N.; Bairamov, I.; Olenin, A.; Shubina, V.; Teplova, V.; Fedotcheva, N. Effect of phenolic acids of microbial origin on production of reactive oxygen species in mitochondria and neutrophils. *J. Biomed. Sci.* **2012**, *19*, 89. [CrossRef]

33. Li, X.; Jiang, B.; Pan, B.; Mu, W.; Zhang, T. Purification and Partial Characterization of *Lactobacillus* Species SK007 Lactate Dehydrogenase (LDH) Catalyzing Phenylpyruvic Acid (PPA) Conversion into Phenyllactic Acid (PLA). *J. Agric. Food Chem.* **2008**, *56*, 2392–2399. [CrossRef] [PubMed]

34. Jia, J.; Mu, W.; Zhang, T.; Jiang, B. Bioconversion of Phenylpyruvate to Phenyllactate: Gene Cloning, Expression, and Enzymatic Characterization of d- and ll-Lactate Dehydrogenases from Lactobacillus plantarum SK002. *Appl. Biochem. Biotechnol.* **2010**, *162*, 242–251. [CrossRef] [PubMed]

35. Preidis, G.A.; Ajami, N.J.; Wong, M.C.; Bessard, B.C.; Conner, M.E.; Petrosino, J.F. Microbial-Derived Metabolites Reflect an Altered Intestinal Microbiota during Catch-Up Growth in Undernourished Neonatal Mice. *J. Nutr.* **2016**, *146*, 940–948. [CrossRef] [PubMed]

36. Nierop Groot, M.N.; de Bont, J.A.M. Conversion of Phenylalanine to Benzaldehyde Initiated by an Aminotransferase in Lactobacillus plantarum. *Appl. Environ. Microbiol.* **1998**, *64*, 3009–3013. [CrossRef]

37. Visconti, A.; Le Roy, C.I.; Rosa, F.; Rossi, N.; Martin, T.C.; Mohney, R.P.; Li, W.; de Rinaldis, E.; Bell, J.T.; Venter, J.C.; et al. Interplay between the human gut microbiome and host metabolism. *Nat. Commun.* **2019**, *10*, 4505. [CrossRef]

38. Isaiah, A.; Parambeth, J.C.; Steiner, J.M.; Lidbury, J.A.; Suchodolski, J.S. The fecal microbiome of dogs with exocrine pancreatic insufficiency. *Anaerobe* **2017**, *45*, 50–58. [CrossRef]

39. Dekkers, K.F.; Sayols-Baixeras, S.; Baldanzi, G.; Nowak, C.; Hammar, U.; Nguyen, D.; Varotsis, G.; Brunkwall, L.; Nielsen, N.; Eklund, A.C.; et al. An online atlas of human plasma metabolite signatures of gut microbiome composition. *Nat. Commun.* **2022**, *13*, 5370. [CrossRef]

40. Yadav, M.; Lomash, A.; Kapoor, S.; Pandey, R.; Chauhan, N.S. Mapping of the benzoate metabolism by human gut microbiome indicates food-derived metagenome evolution. *Sci. Rep.* **2021**, *11*, 5561. [CrossRef]

41. Williams, D.A.; Batt, R.M.; McLean, L. Bacterial overgrowth in the duodenum of dogs with exocrine pancreatic insufficiency. *J. Am. Vet. Med. Assoc.* **1987**, *191*, 201–206.

42. Simpson, K.; Morton, D.; Sørensen, S.; McLEAN, L.; Riley, J.; Batt, R. Biochemical changes in the jejunal mucosa of dogs with exocrine pancreatic insufficiency following pancreatic duct ligation. *Res. Vet. Sci.* **1989**, *47*, 338–345. [CrossRef] [PubMed]

43. Simpson, K.W.; Batt, R.M.; Jones, D.; Morton, D.B. Effects of exocrine pancreatic insufficiency and replacement therapy on the bacterial flora of the duodenum in dogs. *Am. J. Vet. Res.* **1990**, *51*, 203–206. [PubMed]

44. Blake, A.B.; Guard, B.C.; Honneffer, J.B.; Lidbury, J.A.; Steiner, J.M.; Suchodolski, J.S. Altered microbiota, fecal lactate, and fecal bile acids in dogs with gastrointestinal disease. *PLoS ONE* **2019**, *14*, e0224454. [CrossRef] [PubMed]

45. Geever, E.F.; Seifter, E.; Levenson, S.M. Pancreatic pathology, chemically defined liquid diets and bacterial flora in the rat. *Br. J. Exp. Pathol.* **1970**, *51*, 341–347. [PubMed]
46. Levenson, S.M.M.; Kan, D.M.; Gruber, C.B.; Crowley, L.B.; Jaffe, E.R.M.; Nakao, K.M.; Geever, E.F.M.; Seifter, E. Hemolytic Anemia and Pancreatic Acinar Atrophy and Fibrosis Conditioned by "Elemental" Liquid Diets and the Ordinary Intenstinal Microflora. *Ann. Surg.* **1971**, *174*, 469–510. [CrossRef]
47. Kondoh, H.; Teruya, T.; Yanagida, M. Metabolomics of human fasting: New insights about old questions. *Open Biol.* **2020**, *10*, 200176. [CrossRef]
48. Haq, S.; Grondin, J.A.; Khan, W.I. Tryptophan-derived serotonin-kynurenine balance in immune activation and intestinal inflammation. *FASEB J.* **2021**, *35*, e21888. [CrossRef]
49. Gupta, N.K.; Thaker, A.I.; Kanuri, N.; Riehl, T.E.; Rowley, C.W.; Stenson, W.F.; Ciorba, M.A. Serum Analysis of Tryptophan Catabolism Pathway: Correlation with Crohn's Disease Activity. *Inflamm. Bowel Dis.* **2012**, *18*, 1214–1220. [CrossRef]
50. Sofia, M.A.; Ciorba, M.A.; Meckel, K.; Lim, C.K.; Guillemin, G.J.; Weber, C.R.; Bissonnette, M.; Pekow, J.R. Tryptophan Metabolism through the Kynurenine Pathway is Associated with Endoscopic Inflammation in Ulcerative Colitis. *Inflamm. Bowel Dis.* **2018**, *24*, 1471–1480. [CrossRef]
51. Nikolaus, S.; Schulte, B.; Al-Massad, N.; Thieme, F.; Schulte, D.M.; Bethge, J.; Rehman, A.; Tran, F.; Aden, K.; Häsler, R.; et al. Increased Tryptophan Metabolism Is Associated with Activity of Inflammatory Bowel Diseases. *Gastroenterology* **2017**, *153*, 1504–1516. [CrossRef]
52. Deng, P.; Jones, J.C.; Swanson, K.S. Effects of dietary macronutrient composition on the fasted plasma metabolome of healthy adult cats. *Metabolomics* **2013**, *10*, 638–650. [CrossRef]
53. Lyu, Y.; Liu, D.; Nguyen, P.; Peters, I.; Heilmann, R.M.; Fievez, V.; Hemeryck, L.Y.; Hesta, M. Differences in Metabolic Profiles of Healthy Dogs Fed a High-Fat vs. a High-Starch Diet. *Front. Vet. Sci.* **2022**, *9*, 801863. [CrossRef] [PubMed]
54. Wu, G.D.; Compher, C.; Chen, E.Z.; Smith, S.A.; Shah, R.D.; Bittinger, K.; Chehoud, C.; Albenberg, L.G.; Nessel, L.; Gilroy, E.; et al. Comparative metabolomics in vegans and omnivores reveal constraints on diet-dependent gut microbiota metabolite production. *Gut* **2016**, *65*, 63–72. [CrossRef] [PubMed]
55. Puurunen, J.; Ottka, C.; Salonen, M.; Niskanen, J.E.; Lohi, H. Age, breed, sex and diet influence serum metabolite profiles of 2000 pet dogs. *R. Soc. Open Sci.* **2022**, *9*, 211642. [CrossRef]
56. Yue, T.; Tan, H.; Shi, Y.; Xu, M.; Luo, S.; Weng, J.; Xu, S. Serum Metabolomic Profiling in Aging Mice Using Liquid Chromatography—Mass Spectrometry. *Biomolecules* **2022**, *12*, 1594. [CrossRef]
57. Yu, Z.; Zhai, G.; Singmann, P.; He, Y.; Xu, T.; Prehn, C.; Römisch-Margl, W.; Lattka, E.; Gieger, C.; Soranzo, N.; et al. Human serum metabolic profiles are age dependent. *Aging Cell* **2012**, *11*, 960–967. [CrossRef]
58. Ritz, S.; Hahn, D.; Wami, H.T.; Tegelkamp, K.; Dobrindt, U.; Schnekenburger, J. Gut microbiome as a response marker for pancreatic enzyme replacement therapy in a porcine model of exocrine pancreas insufficiency. *Microb. Cell Factories* **2020**, *19*, 221. [CrossRef]
59. Whittemore, J.C.; Price, J.M.; Moyers, T.; Suchodolski, J.S. Effects of Synbiotics on the Fecal Microbiome and Metabolomic Profiles of Healthy Research Dogs Administered Antibiotics: A Randomized, Controlled Trial. *Front. Vet. Sci.* **2021**, *8*, 665713. [CrossRef]
60. Stavroulaki, E.M.; Suchodolski, J.S.; Xenoulis, P.G. Effects of antimicrobials on the gastrointestinal microbiota of dogs and cats. *Vet. J.* **2023**, *291*, 105929. [CrossRef]
61. Garcia-Mazcorro, J.F.; Suchodolski, J.S.; Jones, K.R.; Clark-Price, S.C.; Dowd, S.E.; Minamoto, Y.; Markel, M.; Steiner, J.M.; Dossin, O. Effect of the proton pump inhibitor omeprazole on the gastrointestinal bacterial microbiota of healthy dogs. *FEMS Microbiol Ecol.* **2012**, *80*, 624–636. [CrossRef]

Review

Pathophysiology, Diagnosis, and Management of Canine Intestinal Lymphangiectasia: A Comparative Review

Sara A. Jablonski

Department of Small Animal Clinical Sciences, College of Veterinary Medicine, Michigan State University, East Lansing, MI 48824, USA; sjw@msu.edu

Simple Summary: Canine intestinal lymphangiectasia (IL) is a disorder characterized by dilation, obstruction, and/or dysfunction of the lymphatic vessels within the small intestine. Dogs with IL often suffer from diarrhea, weight loss, vomiting, and fluid accumulations secondary to protein loss from the intestine. This review compiles the current knowledge of the pathophysiology, diagnosis and management of disorders of the intestinal lymphatic vasculature in dogs and humans and aims to (1) improve understanding of these disorders and (2) identify areas where research is needed to improve outcomes for dogs with IL.

Abstract: Intestinal lymphangiectasia was first described in the dog over 50 years ago. Despite this, canine IL remains poorly understood and challenging to manage. Intestinal lymphangiectasia is characterized by variable intestinal lymphatic dilation, lymphatic obstruction, and/or lymphangitis, and is a common cause of protein-losing enteropathy in the dog. Breed predispositions are suggestive of a genetic cause, but IL can also occur as a secondary process. Similarly, both primary and secondary IL have been described in humans. Intestinal lymphangiectasia is definitively diagnosed via intestinal histopathology, but other diagnostic results can be suggestive of IL. Advanced imaging techniques are frequently utilized to aid in the diagnosis of IL in humans but have not been thoroughly investigated in the dog. Management strategies differ between humans and dogs. Dietary modification is the mainstay of therapy in humans with additional pharmacological therapies occasionally employed, and immunosuppressives are rarely used due to the lack of a recognized immune pathogenesis. In contrast, corticosteroid and immunosuppressive therapies are more commonly utilized in canine IL. This review aims toward a better understanding of canine IL with an emphasis on recent discoveries, comparative aspects, and necessary future investigations.

Keywords: canine; intestinal lymphangiectasia; protein-losing enteropathy

Citation: Jablonski, S.A. Pathophysiology, Diagnosis, and Management of Canine Intestinal Lymphangiectasia: A Comparative Review. *Animals* **2022**, *12*, 2791. https://doi.org/10.3390/ani12202791

Academic Editors: Aarti Kathrani and Romy M. Heilmann

Received: 1 September 2022
Accepted: 13 October 2022
Published: 15 October 2022

Publisher's Note: MDPI stays neutral with regard to jurisdictional claims in published maps and institutional affiliations.

1. Overview of the Intestinal Lymphatic Vasculature

The lymphatic system of the human and canine gastrointestinal tract plays crucial roles in the removal of interstitial fluid, transport of dietary fats, and immunoregulation. Despite these critical functions and contributions to intestinal homeostasis, historically the lymphatic vasculature has been poorly studied and a general understanding of lymphatic biology is lacking. In the last several decades, however, improved recognition of these critical functions has led to a surge in interest in the study of the lymphatic vasculature [1].Identification of specific markers for lymphatic endothelial cells (LECs) and novel imaging techniques have led to enhanced understanding of the functioning of the intestinal lymphatics in both health and disease, as well as the consideration of novel therapeutic targets for diseases involving the intestinal lymphatic vasculature [1,2].

Transport through the lymphatic system is unidirectional, designed to return fluid to the blood circulation. Interstitial fluid formed in tissues is first collected into blind-ended, non-contractile vessels known as lymphatic capillaries, or lacteals. In the intestine, lacteals are located exclusively in the villi, and typically reach 60–70% of the villus length, and in health, take up no more than 25% of the villus width. Lacteals are surrounded by a cage-like structure of venous and arterial blood capillaries. Collection of fluid into these capillaries is facilitated by LECs with discontinuous button-like junctions and overlapping flaps that function as unidirectional valves. Lymph fluid flows from the lacteals to a connected network of submucosal lymphatics at the base of the villi [2–4]. It is generally believed that lacteals and the submucosal lymphatic network do not contain smooth muscle, however, some studies suggest the presence of longitudinally oriented muscle cells associated with the lacteals [5], which may act to constrict the lacteal and propel fluid. The lymphatic network in the muscularis layer of the small intestine is seemingly not connected to the lacteals and submucosal lymphatic network, however, both systems drain into the lymphatic collecting vessels near the mesenteric border of the small intestine. The LECs of the lymphatic collecting vessels contain zipper-like junctions at the cell borders with no openings and are comprised of functional units known as lymphangions. Lymphangions contain smooth muscle and valves, allowing for spontaneous contraction and transport of lymph downstream to lymph nodes, the thoracic duct, and back into blood circulation [2,3] (Figure 1).

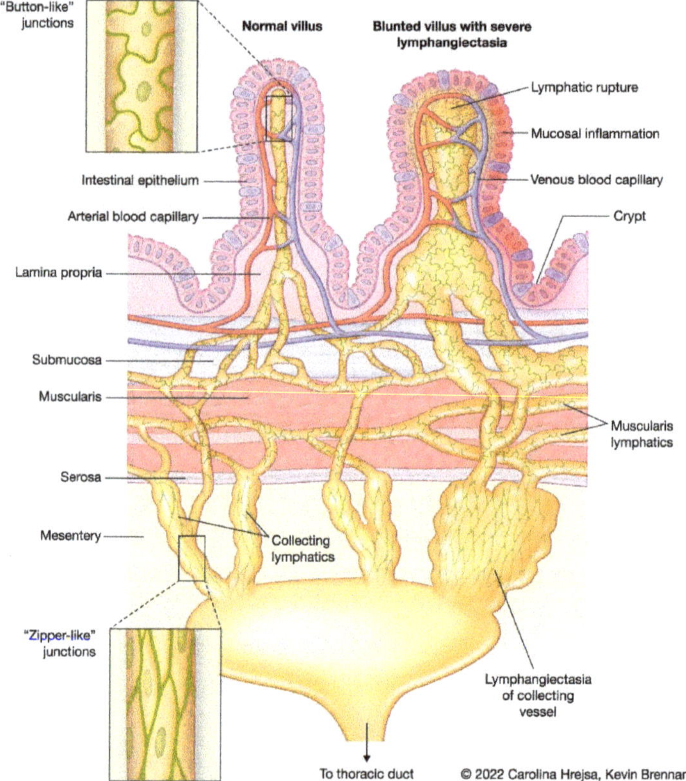

Figure 1. Schematic showing a normal villus with a healthy lacteal, cage-like structure of venous and arterial capillaries, and draining submucosal, muscularis, and collecting lymphatic vasculature. The differences between the cell junctions of the lacteal system versus the collecting lymphatics are depicted (insets). A blunted villus with severe lacteal dilation and lymphatic rupture, and lymphangiectasia of the draining lymphatic vasculature is shown for comparison.

In addition to returning fluid to the blood vasculature, lymphatics are the primary transporter of lipid and lipid-soluble substances (including lipid-soluble vitamins) from the intestine to the blood. The specific molecular methods that regulate uptake of chylomicrons into lacteals are not well understood, however it is known that this task cannot be completed through portal blood absorption [1,6–8].

The lymphatic vasculature is critical to the transport of intestinal immune and inflammatory cells, and interactions between LECs and leukocytes have been demonstrated to influence immune cell migration, thus modulating the immune response. Thus, dysfunction in the lymphatics can lead to alterations of immune cell trafficking [3,7,9]. Lymphatic endothelial cells also appear to have roles in prenatal lymphatic patterning and post-natal control of formation of new lymphatics, or lymphangiogenesis [1,10–12]. Finally, the intestinal lymphatic vasculature is influenced by and an influencer of gut microbiota health, epithelial integrity and barrier function [9].

Lymphatic pumping is influenced by contraction of skeletal muscle, respiratory movements, pulsation of the blood vasculature, variations in central venous pressure, and a variety of intrinsic factors. Intrinsic factors that affect lymphatic pumping include cholecystokinin, glucagon, serotonin, dopamine, substance-P and bradykinin, all known to increase lymphatic pumping, and nitric oxide, vasoactive intestinal peptide, prostacyclin, acetylcholine, and anti-diuretic hormone, which decrease lymphatic pumping. Oxygen levels in lymphatics are relatively low, similar to venous blood, which may make lymphatics especially vulnerable to injury during ischemic stress [1,13].

2. Pathophysiology

2.1. Lymphatic Disturbances and Consequences

Disorders of human and canine intestinal lymphatics can include lymphangiectasia (pathologic dilation of lymph vessels), lymphatic vasculature obstruction, lymphangiogenesis (formation of lymphatic vessels from pre-existing ones), and lymphatic dysfunction [9]. Importantly, these changes can be present in all layers of the small intestine and the mesentery (Figure 1). Many human disorders of the intestinal lymphatics present early in childhood, and in dogs, many breed predispositions exist, both of which suggest disorders of the intestinal lymphatics can be primary or congenital. Interestingly, some intestinal lymphatic conditions that are considered congenital can present later in life than expected, and IL can sometimes be discovered incidentally, which may suggest that conditions can exist asymptomatically or until a stimulus exacerbates it. In one mouse model of lymphatic insufficiency, mice were clinically normal for 1.5 years until acute enteritis was chemically induced. Following recovery from acute enteritis, mice with lymphatic insufficiency had persistent inflammation as well as morphologic changes to their small intestine when compared with control mice [14]. This has led researchers and clinicians to suspect that lymphatic lesions may exist incidentally in some individuals until an inflammatory process exposes them.

Lymphangiogenesis can be induced by intestinal inflammation and is mediated by vascular endothelial growth factor (VEGF) [15]. It has been described in both experimental and clinical IBD, and largely is considered an adaptive rather than maladaptive change [9] though strategies to reduce lymphatic sprouting are sometimes successfully utilized in cases of IL. Effective lymphatic pumping and transport is altered in the presence of intestinal inflammation [16], but the mode of alteration and exact reasons why the lymphatics are affected is not well understood.

Additional mechanisms that are suggested to result in IL include increased hydrostatic pressure secondary to intestinal mucosal infiltrates or increased venous pressure at the level of the thoracic duct secondary to a variety of disorders [17–19]. Thus, it has long been understood that lymphatic abnormalities can occur as a consequence of intestinal inflammation, which, due to the functions of lymphatics, can worsen underlying disease processes.

Major consequences of lymphatic disturbances include hypoproteinemia, immunological deficiencies, and vitamin D deficiency, which can result in hypocalcemia and tetany. Hypoproteinemia often results in cavity effusions (peritoneal, pleural, and rarely, pericardial), and peripheral edema [9,12,20]. Immune abnormalities that have been associated with IL in humans include lymphopenia, hypogammaglobulinemia, and selective loss of

CD4 cells [21,22]. The full implications of these quantitative immune abnormalities are not well understood, however, opportunistic infections have been reported in humans with primary intestinal lymphangiectasia (PIL) [23]. Immune suppression is sometimes pursued in cases of canine IL with PLE and could be considered counterintuitive or even harmful if a similar phenomenon occurs in dogs.

2.2. Intestinal Lymphatic Disorders in Humans

Primary intestinal lymphangiectasia, eponymously referred to as Waldmann's disease, results from malformed lymphatic vessels. It is commonly diagnosed in children prior to 3 years of age, however it can also present in adolescence or adulthood. In one review, the mean age of symptom onset was 13.3 years and 8.5 years until diagnosis [24]. However, one literature review found 49 cases from 46 case reports of PIL in adults in which onset of symptoms occurred after the patient's 18th birthday and median age at diagnosis was 43 [23]. Patients commonly develop hypoalbuminemia, hypogammaglobulinemia, lymphopenia, hypocalcemia, ascites, peripheral edema, gastrointestinal signs including diarrhea, and impaired development/growth. In 84 patients with PIL edema of the lower limbs (78%), diarrhea (62%), ascites (41%) were the most common clinical signs [24]. The most frequently observed biochemical abnormalities include hypoalbuminemia and hypogammaglobulinemia (73%), lymphopenia (63%), anemia (33%), and ionized hypocalcemia (25%) [23]. Histologically PIL is recognized as distended and obstructed lymphatics. Primary intestinal lymphangiectasia is chronic and debilitating requiring long-term therapy and can lead to life-threatening complications [20,25,26]. Intestinal lymphatic hypoplasia is a major differential diagnosis for PIL, though it is often apparent in infancy. This disorder is characterized by significantly fewer than normal lymphatics, with no obvious lymphangiectasia or lymphatic obstruction. It can be differentiated from PIL as patients with lymphatic hypoplasia typically lack lymphopenia, and the lack of lymphatics can be confirmed with the use of immunohistochemical labeling of lymphatic endothelial cells [20,27].

A variety of other syndromes of widespread dysfunction of lymphatics have been identified in humans, with varying effects on the intestinal lymphatic vasculature. Yellow nail syndrome (YNS) is a rare condition characterized by thickened yellow nails, lymphedema, and respiratory tract disease. The most accepted etiology is dysfunction of the lymphatic system, specifically abnormal lymphatic drainage. Reports of IL as a component of YNS have been described [28,29]. Noonan syndrome is a congenital condition of dysplasia of the lymphatics [30], and Nonne-Milroy syndrome is a hereditary form of lymphedema [31]. Klippel-Trenaunay syndrome is a rare congenital malformation of blood and lymph vessels [23]. Intestinal lymphangiectasia has been described in at least one case of Turner syndrome [32]. Additionally, IL has been reported in cases of Hennekam's syndrome, which is a congenital autosomal recessive lymphedema, that also commonly features pleural lymphangiectasia and facial abnormalities [33].

An array of conditions can result in secondary IL in humans including connective tissue disorders, sarcoidosis, neoplasia, chromosomal abnormalities, intestinal volvulus, and inflammatory bowel diseases (IBD). It can also occur as a consequence of cardiac disease or portal hypertension [20,34]. Additionally, elevated venous pressure and subsequent dilation of lymphatics can occur in children following the Fontan procedure, a cardiac surgery performed to treat children with only a single effective ventricle [35]. Notably, lymphatic abnormalities have been increasingly recognized as contributing to the pathogenesis of IBD. Importantly, impaired lymphatic drainage due to dilation, obstruction, and dysfunction results in the accumulation of fluid and infiltrating immune cells, thus setting up chronic edema and inflammation [8,9,15]. In humans with IBD, increased lymphangiectasia, lymphadenopathy and lymphatic obstruction correlate with a poor prognosis. Whether the lymphatic abnormalities are a consequence or cause of the inflammation is debated, with some gastroenterologists even suggesting that lymphatic dysfunction is the root of Crohn's disease (CD) [36]. Regardless, the importance of the intestinal lymphatic vasculature in fluid balance, the immune response, and even the health of the gut microbiota makes them worthy of further study in the pathogenesis and treatment of IBD.

2.3. Intestinal Lymphatic Disorders in Dogs

Intestinal lymphatic disorders in dogs are less well characterized when compared to human disorders. Intestinal lymphangiectasia in dogs was first reported in 1968 [37] and is defined by varying degrees of lymphatic dilation with or without lymphatic obstruction and lymphangitis, and can be found diffusely, segmentally, or focally in the intestine. Hypoproteinemia develops due to loss of protein and lipid-laden lymph into the intestinal lumen, resulting in the syndrome of protein-losing enteropathy (PLE) [18,19,38–43]. A recent review of PLE syndrome in dogs reported that 214/469 (46%) dogs with PLE were diagnosed with IL [18]. Several breed predispositions to IL have been identified, including the soft-coated wheaten terrier, Chinese shar-pei, Norwegian lundehund, Maltese, Rottweiler, and Yorkshire terrier [17,18,44–46]. This suggests that IL may occur as a primary condition and/or a genetic susceptibility exists. However, IL in dogs can also occur as a consequence of another condition that results in altered lymphatic flow, such as chronic inflammatory enteropathies (CIE). Chronic inflammatory enteropathies are intestinal tract conditions characterized by clinical signs of GI tract disease of at least three weeks duration, the reasonable exclusion of infectious, neoplastic, mechanical, and extra-GI causes, and histopathologic documentation of intestinal inflammation. Importantly, CIE and IL occur together commonly in dogs with hypoproteinemia, with one study noting that 76% of dogs with CIE and low serum albumin had concurrent IL [47]. Additionally, dogs with lymphoplasmacytic enteritis (LPE), a form of CIE, and endoscopically visualized "white-spots" (thought to represent intestinal villi distended with lymph), were more likely to be hypoproteinemic than dogs without white spots, and the presence of white spots was correlated with lymphatic dilation histologically [48]. Other disorders that can result in IL through the alteration of lymph flow in dogs include neoplasia, such as intestinal lymphoma or adenocarcinoma, and although they have not been reported in the literature, theoretically as a consequence of pericarditis, pericardial effusion, right-sided heart failure, and portal hypertension.

Clinical signs of IL in dogs include diarrhea, weight loss, decreased appetite, and vomiting, though GI signs are absent in a small percentage of cases. In one study of 17 dogs, diarrhea, anorexia, lethargy, vomiting, and weight loss were the most common signs in 100%, 82%, 76%, 65%, and, 47%, respectively, [19]. Signs associated with hypoalbuminemia are common including ascites, respiratory difficulty or cough due to pleural effusion, and peripheral edema. Less common clinical signs include tetany or seizures secondary to hypocalcemia, and signs secondary to thromboembolic disease [17,18,38,49]. Two separate studies of a total of 39 dogs with IL report the most common biochemical abnormalities as hypoalbuminemia (87%), hypocalcemia (68%), hypoglobulinemia (54%), hypocholesterolemia (51%), and lymphopenia (46%). Hypocobalaminemia is also common [19,50]. Decreased serum concentrations of 25(OH)D are also common [51]. Histologic evaluation reveals lymphatic dilation, however other lesions are common including inflammatory infiltrates, dilated or cystic intestinal crypts, villous atrophy, and edema [50].

Focal intestinal lipogranulomatous lymphangitis (LGL) is a discrete form of lymphangiectasia reported in a small number of dogs. In one study, 8/10 reported cases occurred in French bulldogs [52]. These cases are distinguished from typical IL cases by the presence of a small intestinal mass often involving the adjoining mesentery. Masses are typically found in the distal jejunum or ileum, and histology reveals transmural granulomatous inflammation typically with extensive lipogranulomas that involve the mesentery and the muscularis and serosal layers of the small intestine [52–54].

Intestinal lymphatic hypoplasia has been suspected and reported in 3 dogs with PLE. Immunohistochemical labeling of lymphatic endothelial cells revealed a lack of villous lacteals in these dogs when compared to healthy controls [55]. Due to the small number of cases reported, the significance of this disorder is unclear, though it may be underrecognized. Finally, IL has been reported as a component of generalized lymphangiectasis in a Great Dane and associated with chylothorax in at least two dogs [56,57].

3. Diagnosis

In both humans and dogs, definitive diagnosis of IL is made by histologic assessment of intestinal biopsies in a patient with compatible clinical signs and biochemical findings [17,18,20,34]. Direct visualization of the mucosa and various imaging modalities can also support the diagnosis. Advanced imaging of the intestinal lymphatics is much better described in humans vs. dogs.

The hallmark histologic finding in patients with IL is villous lacteal dilation, which can be described as mild, moderate or marked. In dogs, the World Small Animal Veterinary Association (WSAVA) issued guidelines that defines mild lacteal dilation as the central lacteal representing more than 25% but less than 50% of the width of the villous when sectioned longitudinally. Moderate lacteal dilation is when the central lacteal occupies 50–75% the width of the villous, and in marked lacteal dilation the lacteal dilates up to 100% of the villous. Marked villous lacteal dilation commonly causes a "club-shaped" appearance to the villous [58] (Figure 2). Patients with IL often have additional histologic lesions including various degrees and types of inflammatory infiltrates, crypt lesions, and other morphologic changes to the small intestine. Crypt lesions (dilated crypts filled with mucus and sloughed epithelium) were present in 5/17 (29%) of dogs with IL [19] and in 34/469 (7.2%) of dogs with PLE [18]. Concurrent inflammatory infiltrates are common as a consequence of lymph leakage and nonfunctional lymphatic vessels [47]. Because IL can also occur secondary to inflammation in the small intestine, determination of the primary or more significant process in patients with both IL and inflammation can be challenging. Intestinal lymphangiectasia may be limited to or worse in the ileum when compared to the duodenum in dogs, thus when IL is a differential it is important to obtain ileal biopsies where possible [59,60]. Importantly, IL can be diffuse, segmental, or focal, and though it usually affects the villous lacteal, in some cases it affects the deeper parts of the intestinal wall, such as the submucosa, muscularis, and serosal segments [50,60,61]. A study utilizing immunohistochemical (IHC) labeling of LECs in endoscopically obtained intestinal biopsies of dogs with PLE found that some dogs with lymphangiectasia in the proprial mucosa did not have accompanying villous lymphangiectasia (Figure 3). Thus, IL could be missed if only the villi are examined [60]. Similarly, in some cases of CD and PIL, dilated lymphatics are identified in the deeper mucosa, submucosa, and muscularis layers of the intestine, but not in the superficial mucosa/villous lacteals [62–64]. A more recent study evaluated for IL by labeling LECs with IHC in full-thickness small intestinal biopsies of dogs with lymphoplasmacytic enteritis (LPE) and lipogranulomatous lymphangitis (LGL). This study revealed IL in all layers of the small intestine, including the submucosa, muscularis, and mesentery, both in dogs with LPE and LGL. With the exception of one LGL case, dilated lymphatic vessels were observed in both the villus lacteal and deeper layers of the intestine, suggesting endoscopic biopsies should be sufficient to make the diagnosis in most cases [65]. However, the possibility still remains that IL can go unrecognized in superficial biopsies of the SI, and that it can be limited to areas of the intestine not routinely sampled endoscopically (e.g., jejunum). Despite this, obtaining biopsies endoscopically is still the preferred method in dogs with suspected IL, as it is considered safest, and is typically diagnostic. In humans with suspected PIL, use of IHC for diagnosis is considered routine, typically with D2-40+ antibody. Thus, consideration should be given to more widespread use of LEC labeling for accurate identification of the lymphatic vasculature in cases of suspected canine IL.

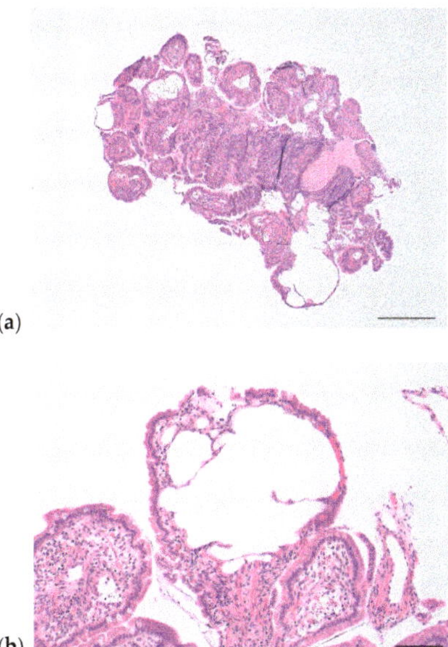

Figure 2. Photomicrographs of intestinal lymphangiectasia in a 5-year-old female, spayed soft-coated wheaten terrier. (**a**) Low-power image showing numerous moderate to markedly dilated lymphatics. Scale bar = 500 μm. (**b**) High-power image of markedly dilated villus lacteal. Scale bar = 100 μm.

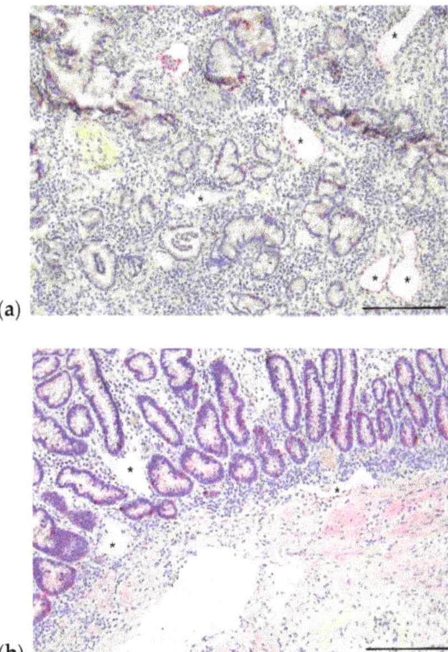

Figure 3. Photomicrographs of apparent lymphangiectasia in deeper layers of small intestine in two dogs with PLE without a diagnosis of lymphangiectasia on routine histologic examination. Intestinal lymphatics indicated by asterisks and immunohistochemically labeled with LYVE 1. (**a**) Apparent mucosal lymphangiectasia in a 9-year-old, male, castrated Australian shepherd dog with PLE; ectatic lacteals denoted by asterisks. Scale bar = 200 μm. (**b**) Apparent mucosal and submucosal lymphangiectasia in a 6-year-old, male, castrated Bernese mountain dog with PLE; ectatic lacteals denoted by asterisks. Scale bar = 200 μm.

Diagnosing IL in endoscopically obtained intestinal biopsies has limitations. Thus, in some cases gross endoscopic exam, additional endoscopic techniques, and a variety of imaging techniques can be used to support the diagnosis. In dogs, endoscopic examination of the SI can reveal pinpoint to coalescing "white spots," thought to represent dilated lacteals with or without lymphatic leakage (white streaks) (Figure 4). However, one study found that these lesions are only moderately sensitive and not specific for the diagnosis of IL [48] perhaps due to subjective interpretation. In humans, the intestinal mucosa in cases of PIL is commonly referred to as having a snow-flake appearance [64]. Interestingly, one study categorized PIL cases as "white-villi type" versus "non-white villi type" and compared their clinical characteristics and therapeutic responses. Investigators found that prior to treatment patients with "non-white type" PIL had significantly worse serum albumin and fecal α-1-antitrypsin clearance compared to patients with endoscopically visible white-tipped villi. Patients with the "white-villi type" classification had lower serum immunoglobulin A and M concentrations compared to patients without endoscopically visible white-tipped villi. The corticosteroid response was better in the patients with the "non-white villi type" classification [66]. More recently, a group of investigators retrospectively classified the endoscopic features of 123 humans with PIL into four types: nodular-type, granular-type, vesicular-type, and edematous-type. Importantly, not all patients with a histologic diagnosis of IL had characteristic white-villus changes noted endoscopically. Histologically, patients with edematous type had lymphatic dilation in the submucosa, but no obvious lymphatic dilation in the lamina propria [67]. Similar classifications in the dog would explain why some dogs in the above mentioned IHC LEC study had IL in the submucosa but not in the lamina propria [60], and why some dogs without apparent histologic evidence of IL in endoscopic obtained biopsies appear to respond to therapy for IL [68]. However, further study is needed to evaluate whether similar discrete types may occur in canine IL.

Histologic diagnosis of IL in dogs with endoscopically obtained biopsies is limited to the section of SI that can be accessed and thus sampled. Double balloon enteroscopy (DBE) is a technique that allows extensive inspection of the small bowel. It is performed with the use of two balloons that are inflated and deflated in alternating sequence to allow the endoscope to progress through the SI. This technique is considered by some to be the gold-standard for diagnosis of PIL in humans with one study reporting diagnosis being obtained by gastroduodenoscopy in 60% of cases, ileocolonoscopy in 24% of cases and enteroscopy in 100% of cases [23]. In another study, endoscopy successfully diagnosed IL in 86% of cases, with the additional 14% of cases requiring video capsule endoscopy or enteroscopy for diagnosis [24]. Notably, enteroscopy can be accompanied by a higher rate of complications, including the possibility of post-procedural pancreatitis [69]. Double balloon enteroscopy has been reported in 14 laboratory dogs in two separate studies and was successful in the majority of cases with no complications encountered [70,71]. Further study is necessary to determine what patient population can be safely examined and biopsied with DBE and whether it has notable benefit compared to traditional endoscopy for the diagnosis of IL in the dog.

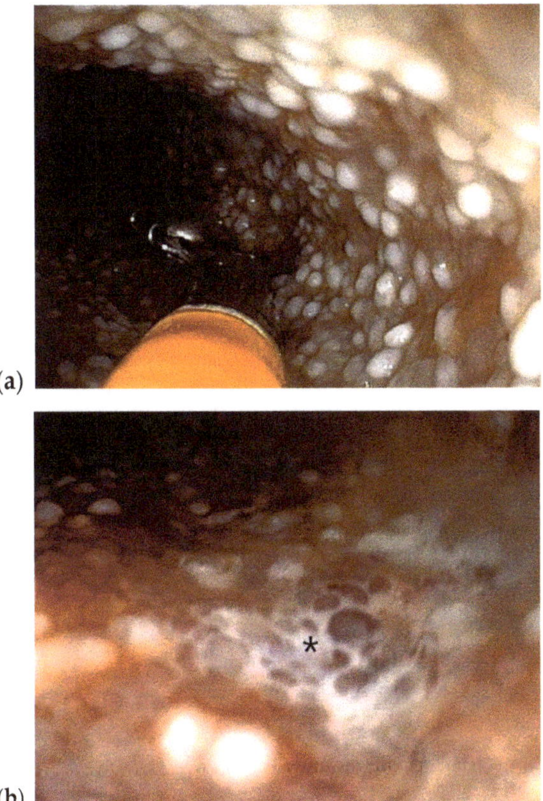

Figure 4. Duodenoscopy of a 5-year-old female, spayed soft-coated wheaten terrier with histologically diagnosed intestinal lymphangiectasia and the clinical syndrome of PLE (**a**) Multiple pinpoint to coalescing "white spots" consistent with dilated intestinal lymphatics are grossly visible during endoscopic exam and biopsy (**b**) leakage of lymph (shown with asterisk) is apparent on the mucosal surface of the small intestine.

Video capsule endoscopy (VCE) has be used for complete evaluation of the small bowel in humans with suspected PIL [72–74]. In one case of a 14-month-old child with PIL, VCE discovered regional IL of the proximal jejunum to distal ileum that traditional gastro-duodenoscopy and ileocolonoscopy did not reveal [73]. Other imaging modalities used to evaluate for IL in humans include scintigraphy, computed tomography lymphangiography (CTL), magnetic resonance lymphangiography (MRL), and dynamic contrast (DC) MRL [75–78]. Dynamic contrast MRL may hold the most promise. Importantly, injection of contrast into the distal limb lymphatic vasculature, inguinal lymph nodes or metatarsal paw pads in dogs is unlikely to adequately image the intestinal lymphatics as contrast will follow the path of least resistance and proceed anterograde through the progressively larger lymphatic vessels to the thoracic duct [76]. A study utilizing DCMRL in humans with PLE found that intra-nodal DCMRL was unable to demonstrate intestinal lymphatic abnormalities, however intrahepatic and intramesenteric DCMRL demonstrated enteric lymphatic abnormalities and leakage in the majority of patients. The authors concluded that DCMRL with multiple injection sites allows mapping of the abdominal lymphatic system [76]. In dogs with IL, transabdominal ultrasound may reveal hyperechoic mucosal striations, which are reported to be associated with lacteal dilation [79]. Oral administration of corn oil may help improve visualization of lacteals sonographically [80]. Technetium-labeled serum albumin scintigraphy has been used to localize protein loss in dogs with PLE, allowing for partial enterectomies to be performed in two dogs with focal disease [81]. The author attempted CTL via intra-metatarsal pad injection for visualization of the intestinal lymphatics in two dogs with histologically confirmed IL. The intestinal lymphatics, IL and or lymph leakage were not able to be visualized with this technique.

Elevated fecal α1-protease inhibitor levels or α1-protease inhibitor clearance can support the diagnosis in both humans and dogs. Serum and fecal canine α1- protease inhibitor concentrations have been demonstrated to reflect the severity of lacteal dilation in dogs. Serum-to-fecal α1-protease inhibitor ratio were lower in dogs with more severe lacteal and crypt lesions and more accurate in dogs with hypoalbuminemia [82]. Additionally, concurrent IL can be suspected in dogs with PLE and chronic inflammatory enteropathy (CIE) and lymphoplasmacytic enteritis (LPE) based on decreasing serum albumin, cholesterol and 25(OH)D concentrations [47,51,82,83]. Recently, serum C-reactive protein, bacterial lipopolysaccharide, and cleaved cytokeratin 18, and both serum and fecal zonulin were elevated in dogs with IL when compared to dogs with other gastrointestinal conditions [84]. Although nitric oxide is known to relax lymphatic vessels, there was no significant difference in inducible nitric oxide synthase between IL-positive and IL-negative tissues [85].

Finally, and crucially, the diagnosis of IL must include an investigation to determine whether the disease may be secondary to another cause. If another cause cannot be found, the assumption is the IL is the primary abnormality. Disorders that may result in secondary IL in the dog include local GI disorders such as chronic inflammatory enteropathies and neoplasia (e.g., intestinal lymphoma), and theoretically extra-GI disorders including portal hypertension, constrictive pericarditis, and right-sided heart disease.

4. Management

If IL is suspected or confirmed to be secondary to another cause, initial management should directly address the underlying cause. Subsequently, monitoring should be performed to determine if the clinical syndrome associated with the IL has resolved with treatment of the purported cause. In some cases, however, it may be necessary to address the IL directly, even if it is considered secondary. In one prospective study evaluating response to a dietary adjustment in dogs with steroid-refractory inflammatory PLE, 8/10 dogs experienced a complete remission following dietary adjustment. Although the reason for the response is unknown, 7/8 dogs that responded were switched to a more fat-restricted diet than they had previously been consuming [68]. Thus, these dogs may have required therapy directed at IL in order to see clinical improvement from their primarily inflammatory PLE. In humans with Crohn's disease, many studies describe concurrent lymphangiectasia, lymphangitis, lymphangiogenesis, and other lymphatic lesions [9,12,36,63]. Laboratory models suggest that normalization of lymphatic structure and function improves intestinal inflammation [8,14,16,86,87]. In one study, 61% of patients with CD not experiencing symptom relief despite various combinations of pharmaceuticals and dose escalations achieved clinical remission following institution of an exclusionary diet that was also low in fat and contained a moderate amount of soluble fiber [88]. Investigations are underway to determine if therapies directly targeting lymphatics may be useful in patients with Crohn's disease.

4.1. Management of PIL in Humans

Lifelong adherence to a low-fat, high-protein, medium-chain-triglyceride (MCT) substituted diet is the cornerstone of therapy in humans with PIL. Dietary fat induces dilation of lymphatics in health. Therefore, decreasing dietary fat lessens the strain on diseased lymphatics and decreases excessive dilation reducing the risk of lymphatic rupture. Because they are absorbed directly into the portal circulation in humans, MCTs do not require the lymphatic system for absorption, making them ideal for the provision of adequate dietary fat and calories in a person with IL [23,89]. A literature search reviewing the evidence for MCT diet in patients with PIL found that 17/27 (63%) of MCT diet treated cases had complete resolution of symptoms compared to 10/28 (36%) of cases not treated with MCT. Feeding an MCT supplemented diet also improved mortality in this group of patients [90]. However, not all cases of PIL respond to dietary therapy alone. Information is limited on exactly what percentage of human patients with PIL are responsive to dietary changes alone, as most reports are of individual cases or small case series [91–96] (Table 1). Of thirty-eight cases of PIL treated with diet, 24 (63%) had clinical and biochemical improvement in response to diet, and children were more likely to show improvement when treated with diet when compared to adults [24]. In another study of 28 children with PIL, dietary therapy was successful in 79% of patients, with 6 children requiring additional therapies to achieve clinical control [97]. Patients responding to diet appear to largely require this therapy permanently, with one study reporting that clinical and biochemical abnormalities returned in patients following the withdrawal of low fat diet [98]. Partial or total parenteral nutrition (TPN) can be considered in patients failing to respond to traditional dietary changes, however one study showed enteral therapy to be an equally effective substitute for TPN in patients with IL [99]. Many gastroenterologists advocate for continuing this dietary approach in patients with PIL regardless of apparent response, as response may be inadequate but not entirely absent, and this disorder remains difficult to treat and associated with significant mortality.

Second-line therapies in humans with PIL include radiologic interventions, surgery, and pharmaceuticals (Table 1). For focal cases of IL identified by scintigraphy or lymphangiography, surgical resection of affected bowel [100–105] and lymphatic embolizations have been utilized successfully. In one recent report, a 15-year-old male suffering from PIL for 7 years was treated with a series of two glue embolization's of a leaking lymphatic channel in the duodenum. Following treatment, his serum albumin concentration improved from 1.9 g/dL to 5.0 g/dL over a period of 8 months [106]. Embolization is also commonly performed after the Fontan procedure or in patients with right heart failure and IL [64,107,108]. For more extensive disease, small numbers of reports have described positive responses to corticosteroids, octreotide, sirolimus or everolimus, propranolol [109,110], and tranexamic acid [97,111]. These therapies are often given in combination with dietary treatment. Corticosteroids have be used successfully in patients with IL as a complication of inflammatory disease, the Fontan procedure, but are not typically utilized in patients with PIL [112–114]. In patients with PIL and an incomplete response to diet, therapy with octreotide may be attempted. Octreotide is a long-acting somatostatin analogue that is thought to decrease intestinal fat absorption as well as inhibit gastrointestinal vasoactive peptides and stimulate the autonomic nervous system [23]. Multiple published case reports describe treatment of PIL with octreotide, sometimes as a sole therapy and other times in combination with other therapies [97,115–123]. While many of these case reports describe treatment as successful, octreotide may not be helpful in cases with extensive lymphangiectasia [64]. In those cases, sirolimus or everolimus may be attempted [124,125]. Sirolimus and everolimus inhibit mTOR, a key enzyme for cell growth and angiogenesis. Thus, sirolimus directly suppresses lymphatic sprouting and proliferation. Everolimus works similarly but has improved pharmacokinetic properties. Ozeki et al. [125] discovered that mTOR expression was increased in tissues affected by PIL and used everolimus to successfully treat a 12-year-old boy with PIL.

Table 1. Summary of selected therapies for treatment of PIL in humans.

Therapy	Mechanism of Action	Citations
Low-fat, high-protein, MCT diet	Decreases excessive lymphatic dilation and rupture	[23,91–96]
Octreotide	Decreased fat absorption Inhibits GI vasoactive peptides Induces splanchnic vasoconstriction	[97,115–123]
Sirolimus or everolimus	Suppress lymphatic sprouting and proliferation	[124,125]
Propanolol	Reduces expression of vascular endothelial growth factor	[109,110]
Tranexamic acid	Normalization of fibrinolytic activity (increased fibrinolytic activity leads to protein loss)	[97,111]
Surgical resection	Direct removal of affected tissue	[100–105]
Lymphatic embolization	Address focal leakage of lymphatic vasculature	[64,106–108]

Albumin infusions, supplementation of fat soluble vitamins, managing nutritional deficiencies are also commonly used in the management of IL in humans. Vitamin D is typically supplemented in the form of 25(OH)D, 1,5(OH)$_2$D, alfacalcidiol, or vitamin D3.

4.2. Management of IL in Dogs

An important impediment to the proper management of IL in dogs is identification of the disease process and the recognition of its contribution to the clinical syndrome in the patient. In veterinary medicine, endoscopic exam and biopsy of the small intestine is commonly declined by the PLE dog's guardian, and even when biopsy is performed IL can be missed or its contribution to the disease process underappreciated. Thus, it is critical for veterinary clinicians to know breed predispositions for IL and understand that significant negative correlations between lymphatic disease and serum albumin have been demonstrated in multiple studies even in cases where inflammation predominates histologically [47,60,83]. Therefore, IL should be considered a possibility in any dog suffering from PLE. Similar to Craven and Washabau [18], this author believes that because PLE is a life-threatening disorder with a high rate of mortality, the safest approach may be to assume all processes (ie, lymph fluid loss, increased intestinal permeability, mucosal injury) are occurring in a PLE patient and treat accordingly, in particular in severe cases or those not responding to therapy. It is also crucial that veterinary clinicians understand PLE as a heterogenous disease; a syndrome caused by a variety of disorders. Thus, individualized therapy is recommended rather than one standard approach.

When canine IL is confirmed or suspected, the most important part of the patient's management is the implementation of a low-fat diet. There is no consensus on the definition of "low-fat" in veterinary medicine, however the fat content of current commercially available, highly digestible, low fat diets ranges from 17 to 26 g fat/Mcal ME (1.7 to 2.6 g/100 kcal). In dogs with moderate to marked IL, or dogs with refractory PLE, an ultra-low fat diet with < 15 g fat/Mcal ME (1.5 g/100 kcal) may be needed. In those cases, it will be necessary to provide a veterinary-nutritionist formulated home-prepared diet, which is formulated with consideration of the patient's entire disease process and dietary fat content of the previous diet. Other important factors that a veterinary nutritionist would consider in cases of IL and PLE include fiber content, food volume, frequency of feeding and the food form [126]. Dietary choices are further complicated in cases of PLE where the patient is suffering from both inflammatory enteritis and IL and it is unclear which process is the driving force. These patients may benefit from a fat-restricted diet with a novel or hydrolyzed protein source, for which few commercial options currently exist. A board-certified veterinary nutritionist could help in the selection of a commercially available therapeutic diet, if one is thought to be suitable. In cases where a commercial diet is considered unsuitable, a home-prepared diet formulated by a board certified veterinary nutritionist may be the safest approach as the nutritionist can consider all the patient's disease processes and their individual contributions to enteric protein loss. Several studies have reported a positive response to diet in canine patients with IL or heterogenous PLE that may include IL [49,68,127,128] (Table 2). Nagata et al. reported that dogs responding to diet as their only therapy for PLE generally had canine chronic enteropathy clinical activity index (CCECAI) of less than 8. Importantly, optimization of diet may be important even in dogs not initially thought to be diet-responsive, as one study reported improvement in clinical signs and serum albumin following a diet change in dogs with steroid-refractory inflammatory PLE [68]. Additionally, dogs should not be classified as unresponsive to food based on a single dietary trial, but rather multiple diet trials may be necessary. Similar to humans, dogs with IL might benefit from MCT since they are classically underconditioned but need to be fed a low-fat diet. However, the benefits of MCT in dogs with IL have not been critically assessed and they can affect palatability of the diet [126]. One study suggested that MCTs are still absorbed into the lymphatic system in dogs, therefore would not help to reduce lymph flow [129]. Assisted enteral feeding was significantly associated with a positive outcome in dogs with PLE [130]. The placement of an enteral feeding tube should be strongly considered in patients unwilling to accept recommended diets.

Table 2. Studies reporting on dietary treatment of suspected or confirmed IL in dogs.

Authors (Citation) Study Type	Dogs	Results
Okanishi et al. JVIM 2014;28:809–817 [49] Retrospective	24 dogs with unresponsive or relapsed histologically confirmed IL	19/24 (79%) dogs responded satisfactorily to dietary fat restriction
Rudinsky et al. JSAP 2017;58:103–108 [127] Retrospective	11 Yorkshire terriers with PLE, 4 with histologically confirmed IL	Clinical signs resolved completely in 8 dogs with dietary therapy alone
Nagata et al. JVIM 2020;34:659–668 [128] Retrospective	33 dogs with PLE, 25 with histologically confirmed IL	17/21 IL dogs treated with ultra-low fat diet had partial or complete response
Jablonski Wennogle et al. JSAP 2021;62:756–764 [68] Prospective	12 dogs with steroid-refractory PLE, 4 with histologically confirmed IL	8/10 dogs had complete remission with dietary change; 7/8 had dietary fat lowered
Olson and Zimmer. JAVMA 1978;173:271–274 [39] Case report	1.5 year old, female, Doberman pincher with IL and PLE	Remission of clinical signs and improvement of serum albumin with dietary therapy
Jones et al. NZ Vet J 1984;32:213–216 [40] Case report	8 month old, female, mixed breed dog with IL and PLE	Full recovery with dietary management alone

Glucocorticoids are commonly prescribed to canine patients with both IL and PLE. It is important to note that glucocorticoids are very infrequently used in humans with IL, and there is no evidence for an autoimmune or immune-mediated etiopathogenesis in cases of IL. However, dogs with IL can develop lymphangitis and secondary enteritis that may benefit from anti-inflammatory doses of steroids, and some dogs with PLE may be suffering from both IL and immune-driven inflammatory enteritis [17]. In dogs with IL it is suggested to avoid immunosuppressive doses of steroids (>1 mg/kg, PO, q24), as some of the side effects associated with higher doses of steroids including muscle protein catabolism, hyperlipidemia, and thrombosis are particularly detrimental to dogs with PLE [18]. Though some clinicians may be tempted to increase doses of glucocorticoids due to concern for failure of drug absorption, a study performed in humans with Crohn's disease found that adequate serum prednisolone levels were achieved despite concern for malabsorption. This study population included some patients who had a failure to improve despite treatment with prednisolone [131]. Since there is no suspected immune basis for IL, immunosuppressive therapies are not recommended or warranted, unless the dog is similarly afflicted by inflammatory enteritis suspected to be secondary to an overzealous immune system. It is pivotal to remember, however, that lymph is a local tissue irritant, so inflammation can occur as a consequence of IL and may not be the result of immune system dysfunction [19]. In one study of 43 dogs diagnosed with PLE, 14 of which had histologically confirmed IL, dogs treated with dietary therapy alone had significantly better outcomes than those treated with immunosuppressive therapy [132]. In another study of 31 dogs with PLE, 10 of which had lacteal dilation noted histologically, the addition of a secondary immunosuppressive agent to glucocorticoid therapy did not result in shorter time to improvement of albumin or improved outcome when compared to glucocorticoids alone [133].

Dogs with IL and secondary PLE should also be treated with supplementation of deficiencies (e.g., cobalamin), supportive therapies to improve appetite and combat nausea, if suspected, and thromboprophylaxis. Dogs with PLE are classified as at "high risk" for thrombosis based on the 2022 CURATIVE guidelines and thromboprophylaxis is recommended [134]. Until more information regarding the best thromboprophylactic approach in dogs with PLE is available, the use of clopidogrel or a factor Xa inhibitor (e.g., rivaroxaban) is reasonable. Similarly, the benefit of or best approach to treatment of hypovitaminosis D with or without ionized hypocalcemia, secondary hyperparathyroidism, and/or ionized hypomagnesemia in dogs with IL is unclear. Ongoing studies may help to clarify this. In the meantime, the use of calcitriol for dogs with IL and significant ionized hypocalcemia is recommended provided close monitoring of serum calcium, phosphorus, and 25(OH)D is performed. Canine albumin infusions, removal of problematic effusions, and therapies to address dysbiosis may also be considered on case-by-case basis. Colloids can be useful for oncotic support in patients that develop refractory effusions.

The use of other pharmaceutical therapies reported to be of benefit in human IL have not yet been published in dogs. However, octreotide has been used in dogs with IL by the authors and others at a 5–10 µg/kg, subcutaneously, q8–12 hours. If a positive benefit is noted, octreotide can be tapered to the lowest effective dose and frequency, and continued long-term. Anecdotal reports describe varying responses, with some dogs experiencing no improvement and others achieving apparent complete clinical and biochemical remission. More information is needed before this therapy can be routinely recommended in cases of IL. However, the use of octreotide can be considered in cases refractory or achieving incomplete responses to other therapies, including the previous administration of a novel protein, ultra-low fat diet.

5. Conclusions and Future Directions

Despite being first described over 50 years ago, canine IL remains poorly understood with unacceptably high mortality rates. There is a need for research addressing various gaps in knowledge including ways to predict IL and improve diagnosis with the use of biochemical parameters and novel imaging and procedural techniques. Intestinal lymphangiectasia should be considered as a contributing factor in any case of canine PLE. Historically, immunosuppressive agents have been used in the management of IL despite limited evidence of their benefit in humans. Many recent studies support dietary management as the foundation of therapy. Investigations are needed to determine the best dietary approaches for individual cases of IL, and whether dietary treatment alone or dietary therapy in combination with anti-inflammatory glucocorticoids is superior in specific cases of IL. Investigations are also needed to explore the role of treatment of vitamin deficiencies (e.g., vitamin D) and to evaluate novel therapies (e.g., octreotide) in dogs with IL.

Funding: This research received no external funding.

Institutional Review Board Statement: Not applicable.

Informed Consent Statement: Not applicable.

Data Availability Statement: Not applicable.

Acknowledgments: Thank you to Victoria Watson, DVM, PhD, DACVP for providing photomicrographs depicting IL. The author would also like to acknowledge Carolina Hrejsa and Kevin Brennan for their assistance in figure development.

Conflicts of Interest: The author declares no conflict of interest.

References

1. Alexander, J.S.; Ganta, V.C.; Jordan, P.A.; Witte, M.H. Gastrointestinal Lymphatics in Health and Disease. *Pathophysiol. Off. J. Int. Soc. Pathophysiol.* **2010**, *17*, 315–335. [CrossRef] [PubMed]
2. Goswami, A.K.; Khaja, M.S.; Downing, T.; Kokabi, N.; Saad, W.E.; Majdalany, B.S. Lymphatic Anatomy and Physiology. *Semin. Interv. Radiol.* **2020**, *37*, 227–236. [CrossRef] [PubMed]
3. Miller, M.J.; McDole, J.R.; Newberry, R.D. Microanatomy of the Intestinal Lymphatic System. *Ann. N. Y. Acad. Sci.* **2010**, *1207* (Suppl. S1), E21–E28. [CrossRef] [PubMed]
4. Cifarelli, V.; Eichmann, A. The Intestinal Lymphatic System: Functions and Metabolic Implications. *Cell Mol. Gastroenterol. Hepatol.* **2019**, *7*, 503–513. [CrossRef] [PubMed]
5. Ohtani, O. Three-Dimensional Organization of Lymphatics and Its Relationship to Blood Vessels in Rat Small Intestine. *Cell Tissue Res.* **1987**, *248*, 365–374. [CrossRef] [PubMed]
6. Dixon, J.B. Lymphatic Lipid Transport: Sewer or Subway? *Trends Endocrinol. Metab. Tem* **2010**, *21*, 480–487. [CrossRef] [PubMed]
7. Ge, Y.; Li, Y.; Gong, J.; Zhu, W. Mesenteric Organ Lymphatics and Inflammatory Bowel Disease. *Ann. Anat. Anat. Anz. Off. Organ Anat. Ges.* **2018**, *218*, 199–204. [CrossRef] [PubMed]
8. Hokari, R.; Tomioka, A. The Role of Lymphatics in Intestinal Inflammation. *Inflamm. Regen.* **2021**, *41*, 25. [CrossRef] [PubMed]
9. Zhang, L.; Ocansey, D.K.W.; Liu, L.; Olovo, C.V.; Zhang, X.; Qian, H.; Xu, W.; Mao, F. Implications of Lymphatic Alterations in the Pathogenesis and Treatment of Inflammatory Bowel Disease. *Biomed. Pharmacother. Biomed. Pharmacother.* **2021**, *140*, 111752. [CrossRef]
10. Oliver, G.; Sosa-Pineda, B.; Geisendorf, S.; Spana, E.P.; Doe, C.Q.; Gruss, P. Prox 1, a Prospero-Related Homeobox Gene Expressed during Mouse Development. *Mech. Dev.* **1993**, *44*, 3–16. [CrossRef]
11. Podgrabinska, S.; Braun, P.; Velasco, P.; Kloos, B.; Pepper, M.S.; Skobe, M. Molecular Characterization of Lymphatic Endothelial Cells. *Proc. Natl. Acad. Sci. USA.* **2002**, *99*, 16069–16074. [CrossRef] [PubMed]
12. Alexander, J.S.; Chaitanya, G.V.; Grisham, M.B.; Boktor, M. Emerging Roles of Lymphatics in Inflammatory Bowel Disease. *Ann. N. Y. Acad. Sci.* **2010**, *1207* (Suppl. S1), E75–E85. [CrossRef] [PubMed]
13. Chakraborty, S.; Davis, M.J.; Muthuchamy, M. Emerging Trends in the Pathophysiology of Lymphatic Contractile Function. *Semin. Cell Dev. Biol.* **2015**, *38*, 55–66. [CrossRef] [PubMed]
14. Davis, R.B.; Kechele, D.O.; Blakeney, E.S.; Pawlak, J.B.; Caron, K.M. Lymphatic Deletion of Calcitonin Receptor-like Receptor Exacerbates Intestinal Inflammation. *JCI Insight* **2017**, *2*, e92465. [CrossRef] [PubMed]
15. Liao, S.; von der Weid, P.-Y. Inflammation-Induced Lymphangiogenesis and Lymphatic Dysfunction. *Angiogenesis* **2014**, *17*, 325–334. [CrossRef] [PubMed]
16. Wu, T.F.; MacNaughton, W.K.; von der Weid, P.-Y. Lymphatic Vessel Contractile Activity and Intestinal Inflammation. *Mem. Inst. Oswaldo Cruz* **2005**, *100* (Suppl. S1), 107–110. [CrossRef] [PubMed]
17. Dossin, O.; Lavoué, R. Protein-Losing Enteropathies in Dogs. *Vet. Clin. North Am. Small Anim. Pract.* **2011**, *41*, 399–418. [CrossRef] [PubMed]
18. Craven, M.D.; Washabau, R.J. Comparative Pathophysiology and Management of Protein-Losing Enteropathy. *J. Vet. Intern. Med.* **2019**, *33*, 383–402. [CrossRef]
19. Kull, P.A.; Hess, R.S.; Craig, L.E.; Saunders, H.M.; Washabau, R.J. Clinical, Clinicopathologic, Radiographic, and Ultrasonographic Characteristics of Intestinal Lymphangiectasia in Dogs: 17 Cases (1996–1998). *J. Am. Vet. Med. Assoc.* **2001**, *219*, 197–202. [CrossRef] [PubMed]
20. Lopez, R.N.; Day, A.S. Primary Intestinal Lymphangiectasia in Children: A Review. *J. Paediatr. Child Health* **2020**, *56*, 1719–1723. [CrossRef] [PubMed]
21. Magdo, H.S.; Stillwell, T.L.; Greenhawt, M.J.; Stringer, K.A.; Yu, S.; Fifer, C.G.; Russell, M.W.; Schumacher, K.R. Immune Abnormalities in Fontan Protein-Losing Enteropathy: A Case-Control Study. *J. Pediatr.* **2015**, *167*, 331–337. [CrossRef] [PubMed]
22. Fuss, I.J.; Strober, W.; Cucchiara, B.A.; Pearlstein, G.R.; Bossuyt, X.; Brown, M.; Fleisher, T.A.; Horgan, K. Intestinal Lymphangiectasia, a Disease Characterized by Selective Loss of Naive CD45RA+ Lymphocytes into the Gastrointestinal Tract. *Eur. J. Immunol.* **1998**, *28*, 4275–4285. [CrossRef]

23. Huber, R.; Semmler, G.; Mayr, A.; Offner, F.; Datz, C. Primary Intestinal Lymphangiectasia in an Adult Patient: A Case Report and Review of Literature. *World J. Gastroenterol.* **2020**, *26*, 7707–7718. [CrossRef] [PubMed]
24. Wen, J.; Tang, Q.; Wu, J.; Wang, Y.; Cai, W. Primary Intestinal Lymphangiectasia: Four Case Reports and a Review of the Literature. *Dig. Dis. Sci.* **2010**, *55*, 3466–3472. [CrossRef]
25. Waldmann, T.A.; Steinfeld, J.L.; Dutcher, T.F.; Davidson, J.D.; Gordon, R.S. The Role of the Gastrointestinal System in "Idiopathic Hypoproteinemia". *Gastroenterology* **1961**, *41*, 197–207. [CrossRef]
26. Braamskamp, M.J.A.M.; Dolman, K.M.; Tabbers, M.M. Clinical Practice. Protein-Losing Enteropathy in Children. *Eur. J. Pediatr.* **2010**, *169*, 1179–1185. [CrossRef]
27. Hardikar, W.; Smith, A.L.; Chow, C.W. Neonatal Protein-Losing Enteropathy Caused by Intestinal Lymphatic Hypoplasia in Siblings. *J. Pediatr. Gastroenterol. Nutr.* **1997**, *25*, 217–221. [CrossRef]
28. Cheslock, M.; Harrington, D.W. Yellow Nail Syndrome. In *StatPearls*; StatPearls Publishing: Treasure Island, FL, USA, 2022.
29. Malek, N.P.; Ocran, K.; Tietge, U.J.; Maschek, H.; Gratz, K.F.; Trautwein, C.; Wagner, S.; Manns, M.P. A Case of the Yellow Nail Syndrome Associated with Massive Chylous Ascites, Pleural and Pericardial Effusions. *Z. Gastroenterol.* **1996**, *34*, 763–766.
30. Ferrell, R.E.; Finegold, D.N. Research Perspectives in Inherited Lymphatic Disease: An Update. *Ann. N. Y. Acad. Sci.* **2008**, *1131*, 134–139. [CrossRef]
31. Irrthum, A.; Karkkainen, M.J.; Devriendt, K.; Alitalo, K.; Vikkula, M. Congenital Hereditary Lymphedema Caused by a Mutation That Inactivates VEGFR3 Tyrosine Kinase. *Am. J. Hum. Genet.* **2000**, *67*, 295–301. [CrossRef]
32. Atton, G.; Gordon, K.; Brice, G.; Keeley, V.; Riches, K.; Ostergaard, P.; Mortimer, P.; Mansour, S. The Lymphatic Phenotype in Turner Syndrome: An Evaluation of Nineteen Patients and Literature Review. *Eur. J. Hum. Genet.* **2015**, *23*, 1634–1639. [CrossRef] [PubMed]
33. Hennekam, R.C.; Geerdink, R.A.; Hamel, B.C.; Hennekam, F.A.; Kraus, P.; Rammeloo, J.A.; Tillemans, A.A. Autosomal Recessive Intestinal Lymphangiectasia and Lymphedema, with Facial Anomalies and Mental Retardation. *Am. J. Med. Genet.* **1989**, *34*, 593–600. [CrossRef] [PubMed]
34. Umar, S.B.; DiBaise, J.K. Protein-Losing Enteropathy: Case Illustrations and Clinical Review. *Am. J. Gastroenterol.* **2010**, *105*, 43–49; quiz 50. [CrossRef] [PubMed]
35. Feldt, R.H.; Driscoll, D.J.; Offord, K.P.; Cha, R.H.; Perrault, J.; Schaff, H.V.; Puga, F.J.; Danielson, G.K. Protein-Losing Enteropathy after the Fontan Operation. *J. Thorac. Cardiovasc. Surg.* **1996**, *112*, 672–680. [CrossRef]
36. Van Kruiningen, H.J.; Colombel, J.-F. The Forgotten Role of Lymphangitis in Crohn's Disease. *Gut* **2008**, *57*, 1–4. [CrossRef] [PubMed]
37. Campbell, R.S.; Brobst, D.; Bisgard, G. Intestinal Lymphangiectasia in a Dog. *J. Am. Vet. Med. Assoc.* **1968**, *153*, 1051–1054.
38. Allenspach, K.; Iennarella-Servantez, C. Canine Protein Losing Enteropathies and Systemic Complications. *Vet. Clin. N. Am. Small Anim. Pract.* **2021**, *51*, 111–122. [CrossRef]
39. Olson, N.C.; Zimmer, J.F. Protein-Losing Enteropathy Secondary to Intestinal Lymphangiectasia in a Dog. *J. Am. Vet. Med. Assoc.* **1978**, *173*, 271–274.
40. Jones, B.R.; Labuc, R.H.; Jones, J.M.; Pauli, J.V.; Arthur, D.E. Intestinal Lymphangiectasia in a Dog. *N. Z. Vet. J.* **1984**, *32*, 213–216. [CrossRef]
41. Kleint, M. [Intestinal lymphangiectasis in the dog. A literature review with a case history]. *Tierarztl. Prax.* **1994**, *22*, 165–171.
42. Mattheeuws, D.; De Rick, A.; Thoonen, H.; Van Der Stock, J. Intestinal Lymphangiectasia in a Dog. *J. Small Anim. Pract.* **1974**, *15*, 757–761. [CrossRef] [PubMed]
43. Milstein, M.; Sanford, S.E. Intestinal Lymphangiectasia in a Dog. *Can. Vet. J. Rev. Veterinaire Can.* **1977**, *18*, 127–130.
44. Simmerson, S.M.; Armstrong, P.J.; Wünschmann, A.; Jessen, C.R.; Crews, L.J.; Washabau, R.J. Clinical Features, Intestinal Histopathology, and Outcome in Protein-Losing Enteropathy in Yorkshire Terrier Dogs. *J. Vet. Intern. Med.* **2014**, *28*, 331–337. [CrossRef] [PubMed]
45. Allenspach, K.; Lomas, B.; Wieland, B.; Harris, T.; Pressler, B.; Mancho, C.; Lees, G.E.; Vaden, S.L. Evaluation of Perinuclear Anti-Neutrophilic Cytoplasmic Autoantibodies as an Early Marker of Protein-Losing Enteropathy and Protein-Losing Nephropathy in Soft Coated Wheaten Terriers. *Am. J. Vet. Res.* **2008**, *69*, 1301–1304. [CrossRef] [PubMed]
46. Littman, M.P.; Dambach, D.M.; Vaden, S.L.; Giger, U. Familial Protein-Losing Enteropathy and Protein-Losing Nephropathy in Soft Coated Wheaten Terriers: 222 Cases (1983–1997). *J. Vet. Intern. Med.* **2000**, *14*, 68–80. [CrossRef]
47. Jablonski Wennogle, S.A.; Priestnall, S.L.; Webb, C.B. Histopathologic Characteristics of Intestinal Biopsy Samples from Dogs with Chronic Inflammatory Enteropathy with and without Hypoalbuminemia. *J. Vet. Intern. Med.* **2017**, *31*, 371–376. [CrossRef]
48. García-Sancho, M.; Sainz, A.; Villaescusa, A.; Rodríguez, A.; Rodríguez-Franco, F. White Spots on the Mucosal Surface of the Duodenum in Dogs with Lymphocytic Plasmacytic Enteritis. *J. Vet. Sci.* **2011**, *12*, 165–169. [CrossRef]
49. Okanishi, H.; Yoshioka, R.; Kagawa, Y.; Watari, T. The Clinical Efficacy of Dietary Fat Restriction in Treatment of Dogs with Intestinal Lymphangiectasia. *J. Vet. Intern. Med.* **2014**, *28*, 809–817. [CrossRef]
50. Larson, R.N.; Ginn, J.A.; Bell, C.M.; Davis, M.J.; Foy, D.S. Duodenal Endoscopic Findings and Histopathologic Confirmation of Intestinal Lymphangiectasia in Dogs. *J. Vet. Intern. Med.* **2012**, *26*, 1087–1092. [CrossRef]
51. Jablonski Wennogle, S.A.; Priestnall, S.L.; Suárez-Bonnet, A.; Webb, C.B. Comparison of Clinical, Clinicopathologic, and Histologic Variables in Dogs with Chronic Inflammatory Enteropathy and Low or Normal Serum 25-Hydroxycholecalciferol Concentrations. *J. Vet. Intern. Med.* **2019**, *33*, 1995–2004. [CrossRef]

52. Lecoindre, A.; Lecoindre, P.; Cadoré, J.L.; Chevallier, M.; Guerret, S.; Derré, G.; Mcdonough, S.P.; Simpson, K.W. Focal Intestinal Lipogranulomatous Lymphangitis in 10 Dogs. *J. Small Anim. Pract.* **2016**, *57*, 465–471. [CrossRef] [PubMed]

53. Watson, V.E.; Hobday, M.M.; Durham, A.C. Focal Intestinal Lipogranulomatous Lymphangitis in 6 Dogs (2008–2011). *J. Vet. Intern. Med.* **2014**, *28*, 48–51. [CrossRef] [PubMed]

54. Van Kruiningen, H.J.; Lees, G.E.; Hayden, D.W.; Meuten, D.J.; Rogers, W.A. Lipogranulomatous Lymphangitis in Canine Intestinal Lymphangiectasia. *Vet. Pathol.* **1984**, *21*, 377–383. [CrossRef] [PubMed]

55. Malatos, J.M.; Kurpios, N.A.; Duhamel, G.E. Small Intestinal Lymphatic Hypoplasia in Three Dogs with Clinical Signs of Protein-Losing Enteropathy. *J. Comp. Pathol.* **2018**, *160*, 39–49. [CrossRef] [PubMed]

56. Fossum, T.W.; Sherding, R.G.; Zack, P.M.; Birchard, S.J.; Smeak, D.D. Intestinal Lymphangiectasia Associated with Chylothorax in Two Dogs. *J. Am. Vet. Med. Assoc.* **1987**, *190*, 61–64.

57. Fossum, T.W.; Hodges, C.C.; Scruggs, D.W.; Fiske, R.A. Generalized Lymphangiectasis in a Dog with Subcutaneous Chyle and Lymphangioma. *J. Am. Vet. Med. Assoc.* **1990**, *197*, 231–236.

58. Day, M.J.; Bilzer, T.; Mansell, J.; Wilcock, B.; Hall, E.J.; Jergens, A.; Minami, T.; Willard, M.; Washabau, R.; World Small Animal Veterinary Association Gastrointestinal Standardization Group. Histopathological Standards for the Diagnosis of Gastrointestinal Inflammation in Endoscopic Biopsy Samples from the Dog and Cat: A Report from the World Small Animal Veterinary Association Gastrointestinal Standardization Group. *J. Comp. Pathol.* **2008**, *138* (Suppl. S1), S1–S43. [CrossRef]

59. Procoli, F.; Mõtsküla, P.F.; Keyte, S.V.; Priestnall, S.; Allenspach, K. Comparison of Histopathologic Findings in Duodenal and Ileal Endoscopic Biopsies in Dogs with Chronic Small Intestinal Enteropathies. *J. Vet. Intern. Med.* **2013**, *27*, 268–274. [CrossRef]

60. Jablonski Wennogle, S.A.; Priestnall, S.L.; Suárez-Bonnet, A.; Soontararak, S.; Webb, C.B. Lymphatic Endothelial Cell Immunohistochemical Markers for Evaluation of the Intestinal Lymphatic Vasculature in Dogs with Chronic Inflammatory Enteropathy. *J. Vet. Intern. Med.* **2019**, *33*, 1669–1676. [CrossRef]

61. Kleinschmidt, S.; Meneses, F.; Nolte, I.; Hewicker-Trautwein, M. Retrospective Study on the Diagnostic Value of Full-Thickness Biopsies from the Stomach and Intestines of Dogs with Chronic Gastrointestinal Disease Symptoms. *Vet. Pathol.* **2006**, *43*, 1000–1003. [CrossRef]

62. Sura, R.; Colombel, J.-F.; Van Kruiningen, H.J. Lymphatics, Tertiary Lymphoid Organs and the Granulomas of Crohn's Disease: An Immunohistochemical Study. *Aliment. Pharmacol. Ther.* **2011**, *33*, 930–939. [CrossRef] [PubMed]

63. Van Kruiningen, H.J.; Hayes, A.W.; Colombel, J.-F. Granulomas Obstruct Lymphatics in All Layers of the Intestine in Crohn's Disease. *APMIS* **2014**, *122*, 1125–1129. [CrossRef] [PubMed]

64. Kwon, Y.; Kim, M.J. The Update of Treatment for Primary Intestinal Lymphangiectasia. *Pediatr. Gastroenterol. Hepatol. Nutr.* **2021**, *24*, 413–422. [CrossRef] [PubMed]

65. Nagahara, T.; Ohno, K.; Nagao, I.; Nakagawa, T.; Goto-Koshino, Y.; Tsuboi, M.; Chambers, J.K.; Uchida, K.; Tomiyasu, H.; Tsujimoto, H. Evaluation of the Degree and Distribution of Lymphangiectasia in Full-Thickness Canine Small Intestinal Specimens Diagnosed with Lymphoplasmacytic Enteritis and Granulomatous Lymphangitis. *J. Vet. Med. Sci.* **2022**, *84*, 566–573. [CrossRef]

66. Ohmiya, N.; Nakamura, M.; Yamamura, T.; Yamada, K.; Nagura, A.; Yoshimura, T.; Hirooka, Y.; Hirata, I.; Goto, H. Classification of Intestinal Lymphangiectasia with Protein-Losing Enteropathy: White Villi Type and Non-White Villi Type. *Digestion* **2014**, *90*, 155–166. [CrossRef] [PubMed]

67. Meng, M.-M.; Liu, K.-L.; Xue, X.-Y.; Hao, K.; Dong, J.; Yu, C.-K.; Liu, H.; Wang, C.-H.; Su, H.; Lin, W.; et al. Endoscopic Classification and Pathological Features of Primary Intestinal Lymphangiectasia. *World J. Gastroenterol.* **2022**, *28*, 2482–2493. [CrossRef] [PubMed]

68. Jablonski Wennogle, S.A.; Stockman, J.; Webb, C.B. Prospective Evaluation of a Change in Dietary Therapy in Dogs with Steroid-Resistant Protein-Losing Enteropathy. *J. Small Anim. Pract.* **2021**, *62*, 756–764. [CrossRef] [PubMed]

69. Saygili, F.; Saygili, S.M.; Oztas, E. Examining the Whole Bowel, Double Balloon Enteroscopy: Indications, Diagnostic Yield and Complications. *World J. Gastrointest. Endosc.* **2015**, *7*, 247–252. [CrossRef] [PubMed]

70. Sarria, R.; López Albors, O.; Soria, F.; Ayala, I.; Pérez Cuadrado, E.; Esteban, P.; Latorre, R. Characterization of Oral Double Balloon Endoscopy in the Dog. *Vet. J. Lond. Engl.* **2013**, *195*, 331–336. [CrossRef] [PubMed]

71. Latorre, R.; Ayala, I.; Soria, F.; Carballo, F.; Ayala, M.D.; Pérez-Cuadrado, E. Double-Balloon Enteroscopy in Two Dogs. *Vet. Rec.* **2007**, *161*, 587–590. [CrossRef] [PubMed]

72. Rivet, C.; Lapalus, M.-G.; Dumortier, J.; Le Gall, C.; Budin, C.; Bouvier, R.; Ponchon, T.; Lachaux, A. Use of Capsule Endoscopy in Children with Primary Intestinal Lymphangiectasia. *Gastrointest. Endosc.* **2006**, *64*, 649–650. [CrossRef] [PubMed]

73. van der Reijden, S.M.; van Wijk, M.P.; Jacobs, M.A.J.M.; de Meij, T.G.J. Video Capsule Endoscopy to Diagnose Primary Intestinal Lymphangiectasia in a 14-Month-Old Child. *J. Pediatr. Gastroenterol. Nutr.* **2017**, *64*, e161. [CrossRef] [PubMed]

74. Ersoy, O.; Akin, E.; Demirezer, A.; Yilmaz, E.; Solakoglu, T.; Irkkan, C.; Yurekli, O.T.; Buyukasik, S. Evaluation of Primary Intestinal Lymphangiectasia by Capsule Endoscopy. *Endoscopy* **2013**, *45* (Suppl. S2), E61–E62. [CrossRef] [PubMed]

75. Chavhan, G.B.; Amaral, J.G.; Temple, M.; Itkin, M. MR Lymphangiography in Children: Technique and Potential Applications. *Radiogr. Rev. Publ. Radiol. Soc. N. Am. Inc* **2017**, *37*, 1775–1790. [CrossRef]

76. Brownell, J.N.; Biko, D.M.; Mamula, P.; Krishnamurthy, G.; Escobar, F.; Srinivasan, A.; Laje, P.; Piccoli, D.A.; Pinto, E.; Smith, C.L.; et al. Dynamic Contrast Magnetic Resonance Lymphangiography Localizes Lymphatic Leak to the Duodenum in Protein-Losing Enteropathy. *J. Pediatr. Gastroenterol. Nutr.* **2022**, *74*, 38–45. [CrossRef] [PubMed]

77. Matsumoto, T.; Kudo, T.; Endo, J.; Hashida, K.; Tachibana, N.; Murakoshi, T.; Hasebe, T. Transnodal Lymphangiography and Post-CT for Protein-Losing Enteropathy in Noonan Syndrome. *Minim. Invasive Ther. Allied Technol. MITAT Off. J. Soc. Minim. Invasive Ther.* **2015**, *24*, 246–249. [CrossRef] [PubMed]

78. Pascal, P.; Malloizel, J.; Lairez, O.; de Volontat, M.D.; Bournet, B. Primary Intestinal Lymphangiectasia: Diagnostic Accuracy of 99mTc-Labeled Human Serum Albumin Nanocolloid SPECT/CT Before Biopsy. *Clin. Nucl. Med.* **2021**, *46*, e34–e35. [CrossRef] [PubMed]

79. Sutherland-Smith, J.; Penninck, D.G.; Keating, J.H.; Webster, C.R.L. Ultrasonographic Intestinal Hyperechoic Mucosal Striations in Dogs Are Associated with Lacteal Dilation. *Vet. Radiol. Ultrasound Off. J. Am. Coll. Vet. Radiol. Int. Vet. Radiol. Assoc.* **2007**, *48*, 51–57. [CrossRef] [PubMed]

80. Pollard, R.E.; Johnson, E.G.; Pesavento, P.A.; Baker, T.W.; Cannon, A.B.; Kass, P.H.; Marks, S.L. Effects of Corn Oil Administered Orally on Conspicuity of Ultrasonographic Small Intestinal Lesions in Dogs with Lymphangiectasia. *Vet. Radiol. Ultrasound Off. J. Am. Coll. Vet. Radiol. Int. Vet. Radiol. Assoc.* **2013**, *54*, 390–397. [CrossRef] [PubMed]

81. Engelmann, N.; Ondreka, N.; von Pückler, K.; Mohrs, S.; Sicken, J.; Neiger, R. Applicability of 99m Tc-Labeled Human Serum Albumin Scintigraphy in Dogs with Protein-Losing Enteropathy. *J. Vet. Intern. Med.* **2017**, *31*, 365–370. [CrossRef]

82. Heilmann, R.M.; Parnell, N.K.; Grützner, N.; Mansell, J.; Berghoff, N.; Schellenberg, S.; Reusch, C.E.; Suchodolski, J.S.; Steiner, J.M. Serum and Fecal Canine A1-Proteinase Inhibitor Concentrations Reflect the Severity of Intestinal Crypt Abscesses and/or Lacteal Dilation in Dogs. *Vet. J. Lond. Engl.* **2016**, *207*, 131–139. [CrossRef]

83. Rossi, G.; Cerquetella, M.; Antonelli, E.; Pengo, G.; Magi, G.E.; Villanacci, V.; Rostami-Nejad, M.; Spaterna, A.; Bassotti, G. The Importance of Histologic Parameters of Lacteal Involvement in Cases of Canine Lymphoplasmacytic Enteritis. *Gastroenterol. Hepatol. Bed Bench* **2015**, *8*, 33–41. [PubMed]

84. Rossi, G.; Gavazza, A.; Vincenzetti, S.; Mangiaterra, S.; Galosi, L.; Marchegiani, A.; Pengo, G.; Sagratini, G.; Ricciutelli, M.; Cerquetella, M. Clinicopathological and Fecal Proteome Evaluations in 16 Dogs Presenting Chronic Diarrhea Associated with Lymphangiectasia. *Vet. Sci.* **2021**, *8*, 242. [CrossRef] [PubMed]

85. Nagahara, T.; Ohno, K.; Nagao, I.; Nakagawa, T.; Goto-Koshino, Y.; Tsuboi, M.; Chambers, J.K.; Uchida, K.; Tomiyasu, H.; Tsujimoto, H. Association between Intestinal Lymphangiectasia and Expression of Inducible Nitric Oxide Synthase in Dogs with Lymphoplasmacytic Enteritis. *J. Vet. Med. Sci.* **2022**, *84*, 20–24. [CrossRef] [PubMed]

86. D'Alessio, S.; Correale, C.; Tacconi, C.; Gandelli, A.; Pietrogrande, G.; Vetrano, S.; Genua, M.; Arena, V.; Spinelli, A.; Peyrin-Biroulet, L.; et al. VEGF-C-Dependent Stimulation of Lymphatic Function Ameliorates Experimental Inflammatory Bowel Disease. *J. Clin. Investig.* **2014**, *124*, 3863–3878. [CrossRef]

87. D'Alessio, S.; Tacconi, C.; Danese, S. Targeting Lymphatics in Inflammatory Bowel Disease. *Oncotarget* **2015**, *6*, 34047–34048. [CrossRef]

88. Sigall Boneh, R.; Sarbagili Shabat, C.; Yanai, H.; Chermesh, I.; Ben Avraham, S.; Boaz, M.; Levine, A. Dietary Therapy with the Crohn's Disease Exclusion Diet Is a Successful Strategy for Induction of Remission in Children and Adults Failing Biological Therapy. *J. Crohns Colitis* **2017**, *11*, 1205–1212. [CrossRef] [PubMed]

89. Khayat, A.A. Primary Intestinal Lymphangiectasia Presenting as Limb Hemihyperplasia: A Case Report and Literature Review. *BMC Gastroenterol.* **2021**, *21*, 225. [CrossRef] [PubMed]

90. Desai, A.P.; Guvenc, B.H.; Carachi, R. Evidence for Medium Chain Triglycerides in the Treatment of Primary Intestinal Lymphangiectasia. *Eur. J. Pediatr. Surg. Off. J. Austrian Assoc. Pediatr. Surg. Al Z. Kinderchir.* **2009**, *19*, 241–245. [CrossRef] [PubMed]

91. Greco, F.; Piccolo, G.; Sorge, A.; Pavone, P.; Triglia, T.; Spina, M.; Sorge, G. [Early-onset of primary intestinal lymphangiectasia. A case report and diet treatment]. *Minerva Pediatr.* **2003**, *55*, 615–619.

92. Isa, H.M.; Al-Arayedh, G.G.; Mohamed, A.M. Intestinal Lymphangiectasia in Children. A Favorable Response to Dietary Modifications. *Saudi Med. J.* **2016**, *37*, 199–204. [CrossRef] [PubMed]

93. Lai, Y.; Yu, T.; Qiao, X.-Y.; Zhao, L.-N.; Chen, Q.-K. Primary Intestinal Lymphangiectasia Diagnosed by Double-Balloon Enteroscopy and Treated by Medium-Chain Triglycerides: A Case Report. *J. Med. Case Reports* **2013**, *7*, 19. [CrossRef] [PubMed]

94. Li, S.; Liu, X.; He, Y.; Li, Q.; Ji, L.; Shen, W.; Tong, G. Nutritional Therapy and Effect Assessment of Infants with Primary Intestinal Lymphangiectasia: Case Reports. *Medicine* **2017**, *96*, e9240. [CrossRef] [PubMed]

95. Martín, C.C.; García, A.F.-A.; Restrepo, J.M.R.; Pérez, A.S. [Sucessful dietetic-therapy in primary intestinal lymphangiectasia and recurrent chylous ascites: A case report]. *Nutr. Hosp.* **2007**, *22*, 723–725. [PubMed]

96. Ohno, S.; Nakahara, S.; Kasahara, K.; Murakami, R.; Mitsuuchi, M.; Makiguchi, Y.; Takahashi, H.; Adachi, M.; Endo, T.; Imai, K.; et al. [A case report of primary intestinal lymphangiectesia successfully treated with low fat diet]. *Nihon Shokakibyo Gakkai Zasshi Jpn. J. Gastro-Enterol.* **1997**, *94*, 767–771.

97. Prasad, D.; Srivastava, A.; Tambe, A.; Yachha, S.K.; Sarma, M.S.; Poddar, U. Clinical Profile, Response to Therapy, and Outcome of Children with Primary Intestinal Lymphangiectasia. *Dig. Dis. Basel Switz.* **2019**, *37*, 458–466. [CrossRef]

98. Vignes, S.; Bellanger, J. Primary Intestinal Lymphangiectasia (Waldmann's Disease). *Orphanet J. Rare Dis.* **2008**, *3*, 5. [CrossRef]

99. Aoyagi, K.; Iida, M.; Matsumoto, T.; Sakisaka, S. Enteral Nutrition as a Primary Therapy for Intestinal Lymphangiectasia: Value of Elemental Diet and Polymeric Diet Compared with Total Parenteral Nutrition. *Dig. Dis. Sci.* **2005**, *50*, 1467–1470. [CrossRef]

100. Chen, C.-P.; Chao, Y.; Li, C.-P.; Lo, W.-C.; Wu, C.-W.; Tsay, S.-H.; Lee, R.-C.; Chang, F.-Y. Surgical Resection of Duodenal Lymphangiectasia: A Case Report. *World J. Gastroenterol.* **2003**, *9*, 2880–2882. [CrossRef]

101. Connor, F.L.; Angelides, S.; Gibson, M.; Larden, D.W.; Roman, M.R.; Jones, O.; Currie, B.G.; Day, A.S.; Bohane, T.D. Successful Resection of Localized Intestinal Lymphangiectasia Post-Fontan: Role of (99m)Technetium-Dextran Scintigraphy. *Pediatrics* **2003**, *112*, e242–e247. [CrossRef]

102. Zhu, L.; Cai, X.; Mou, Y.; Zhu, Y.; Wang, S.; Wu, J. Partial Enterectomy: Treatment for Primary Intestinal Lymphangiectasia in Four Cases. *Chin. Med. J.* **2010**, *123*, 760–764. [PubMed]

103. Huber, T.; Paschold, M.; Eckardt, A.J.; Lang, H.; Kneist, W. Surgical Therapy of Primary Intestinal Lymphangiectasia in Adults. *J. Surg. Case Rep.* **2015**, *2015*, rjv081. [CrossRef] [PubMed]

104. Kneist, W.; Drescher, D.G.; Hansen, T.; Kreitner, K.F.; Lang, H. [Surgical therapy of segmental jejunal, primary intestinal lymphangiectasia]. *Z. Gastroenterol.* **2013**, *51*, 576–579. [CrossRef] [PubMed]

105. Mari, J.; Kovacs, T.; Pasztor, G.; Tiszlavicz, L.; Bereczki, C.; Szucs, D. Pediatric Localized Intestinal Lymphangiectasia Treated with Resection. *Int. Med. Case Rep. J.* **2019**, *12*, 23–27. [CrossRef] [PubMed]

106. Kwon, Y.; Kim, E.S.; Choe, Y.H.; Hyun, D.; Kim, M.J. Therapeutic Lymphatic Embolization in Pediatric Primary Intestinal Lymphangiectasia. *Yonsei Med. J.* **2021**, *62*, 470–473. [CrossRef] [PubMed]

107. Kylat, R.I.; Witte, M.H.; Barber, B.J.; Dori, Y.; Ghishan, F.K. Resolution of Protein-Losing Enteropathy after Congenital Heart Disease Repair by Selective Lymphatic Embolization. *Pediatr. Gastroenterol. Hepatol. Nutr.* **2019**, *22*, 594–600. [CrossRef] [PubMed]

108. Li, A.A.; Raghu, P.; Chen, A.; Triadafilopoulos, G.; Park, W. Sticky Situation: Bleeding Duodenal Lymphangiectasias Treated with Lymphatic Glue Embolization. *Dig. Dis. Sci.* **2022**, *67*, 71–74. [CrossRef]

109. Poralla, C.; Specht, S.; Born, M.; Müller, A.; Bartmann, P.; Müller, A. Treatment of Congenital Generalized Lymphangiectasia with Propranolol in a Preterm Infant. *Pediatrics* **2014**, *133*, e439–e442. [CrossRef]

110. Liviskie, C.J.; Brennan, C.C.; McPherson, C.C.; Vesoulis, Z.A. Propranolol for the Treatment of Lymphatic Malformations in a Neonate-A Case Report and Review of Literature. *J. Pediatr. Pharmacol. Ther.* **2020**, *25*, 155–162. [CrossRef]

111. MacLean, J.E.; Cohen, E.; Weinstein, M. Primary Intestinal and Thoracic Lymphangiectasia: A Response to Antiplasmin Therapy. *Pediatrics* **2002**, *109*, 1177–1180. [CrossRef]

112. Thacker, D.; Patel, A.; Dodds, K.; Goldberg, D.J.; Semeao, E.; Rychik, J. Use of Oral Budesonide in the Management of Protein-Losing Enteropathy after the Fontan Operation. *Ann. Thorac. Surg.* **2010**, *89*, 837–842. [CrossRef] [PubMed]

113. Hoashi, T.; Ichikawa, H.; Ueno, T.; Kogaki, S.; Sawa, Y. Steroid Pulse Therapy for Protein-Losing Enteropathy after the Fontan Operation. *Congenit. Heart Dis.* **2009**, *4*, 284–287. [CrossRef] [PubMed]

114. Fleisher, T.A.; Strober, W.; Muchmore, A.V.; Broder, S.; Krawitt, E.L.; Waldmann, T.A. Corticosteroid-Responsive Intestinal Lymphangiectasia Secondary to an Inflammatory Process. *N. Engl. J. Med.* **1979**, *300*, 605–606. [CrossRef] [PubMed]

115. Al Sinani, S.; Rawahi, Y.A.; Abdoon, H. Octreotide in Hennekam Syndrome-Associated Intestinal Lymphangiectasia. *World J. Gastroenterol.* **2012**, *18*, 6333–6337. [CrossRef]

116. Alshikho, M.J.; Talas, J.M.; Noureldine, S.I.; Zazou, S.; Addas, A.; Kurabi, H.; Nasser, M. Intestinal Lymphangiectasia: Insights on Management and Literature Review. *Am. J. Case Rep.* **2016**, *17*, 512–522. [CrossRef]

117. Bac, D.J.; Van Hagen, P.M.; Postema, P.T.; ten Bokum, A.M.; Zondervan, P.E.; van Blankenstein, M. Octreotide for Protein-Losing Enteropathy with Intestinal Lymphangiectasia. *Lancet Lond. Engl.* **1995**, *345*, 1639. [CrossRef]

118. Ballinger, A.B.; Farthing, M.J. Octreotide in the Treatment of Intestinal Lymphangiectasia. *Eur. J. Gastroenterol. Hepatol.* **1998**, *10*, 699–702.

119. Filik, L.; Oguz, P.; Koksal, A.; Koklu, S.; Sahin, B. A Case with Intestinal Lymphangiectasia Successfully Treated with Slow-Release Octreotide. *Dig. Liver Dis. Off. J. Ital. Soc. Gastroenterol. Ital. Assoc. Study Liver* **2004**, *36*, 687–690. [CrossRef]

120. Kuroiwa, G.; Takayama, T.; Sato, Y.; Takahashi, Y.; Fujita, T.; Nobuoka, A.; Kukitsu, T.; Kato, J.; Sakamaki, S.; Niitsu, Y. Primary Intestinal Lymphangiectasia Successfully Treated with Octreotide. *J. Gastroenterol.* **2001**, *36*, 129–132. [CrossRef]

121. Sari, S.; Baris, Z.; Dalgic, B. Primary Intestinal Lymphangiectasia in Children: Is Octreotide an Effective and Safe Option in the Treatment? *J. Pediatr. Gastroenterol. Nutr.* **2010**, *51*, 454–457. [CrossRef]

122. Suehiro, K.; Morikage, N.; Murakami, M.; Yamashita, O.; Hamano, K. Late-Onset Primary Intestinal Lymphangiectasia Successfully Managed with Octreotide: A Case Report. *Ann. Vasc. Dis.* **2012**, *5*, 96–99. [CrossRef] [PubMed]

123. Troskot, R.; Jurčić, D.; Bilić, A.; Gomerčić Palčić, M.; Težak, S.; Brajković, I. How to Treat an Extensive Form of Primary Intestinal Lymphangiectasia? *World J. Gastroenterol.* **2015**, *21*, 7320–7325. [CrossRef] [PubMed]

124. Pollack, S.F.; Geffrey, A.L.; Thiele, E.A.; Shah, U. Primary Intestinal Lymphangiectasia Treated with Rapamycin in a Child with Tuberous Sclerosis Complex (TSC). *Am. J. Med. Genet. A* **2015**, *167*, 2209–2212. [CrossRef] [PubMed]

125. Ozeki, M.; Hori, T.; Kanda, K.; Kawamoto, N.; Ibuka, T.; Miyazaki, T.; Fukao, T. Everolimus for Primary Intestinal Lymphangiectasia with Protein-Losing Enteropathy. *Pediatrics* **2016**, *137*, e20152562. [CrossRef] [PubMed]

126. Tolbert, M.K.; Murphy, M.; Gaylord, L.; Witzel-Rollins, A. Dietary Management of Chronic Enteropathy in Dogs. *J. Small Anim. Pract.* **2022**, *63*, 425–434. [CrossRef]

127. Rudinsky, A.J.; Howard, J.P.; Bishop, M.A.; Sherding, R.G.; Parker, V.J.; Gilor, C. Dietary Management of Presumptive Protein-Losing Enteropathy in Yorkshire Terriers. *J. Small Anim. Pract.* **2017**, *58*, 103–108. [CrossRef]

128. Nagata, N.; Ohta, H.; Yokoyama, N.; Teoh, Y.B.; Nisa, K.; Sasaki, N.; Osuga, T.; Morishita, K.; Takiguchi, M. Clinical Characteristics of Dogs with Food-Responsive Protein-Losing Enteropathy. *J. Vet. Intern. Med.* **2020**, *34*, 659–668. [CrossRef]

129. Jensen, G.L.; McGarvey, N.; Taraszewski, R.; Wixson, S.K.; Seidner, D.L.; Pai, T.; Yeh, Y.Y.; Lee, T.W.; DeMichele, S.J. Lymphatic Absorption of Enterally Fed Structured Triacylglycerol vs Physical Mix in a Canine Model. *Am. J. Clin. Nutr.* **1994**, *60*, 518–524. [CrossRef]
130. Economu, L.; Chang, Y.-M.; Priestnall, S.L.; Kathrani, A. The Effect of Assisted Enteral Feeding on Treatment Outcome in Dogs with Inflammatory Protein-Losing Enteropathy. *J. Vet. Intern. Med.* **2021**, *35*, 1297–1305. [CrossRef]
131. Tanner, A.R.; Halliday, J.W.; Powell, L.W. Serum Prednisolone Levels in Crohn's Disease and Coeliac Disease Following Oral Prednisolone Administration. *Digestion* **1981**, *21*, 310–315. [CrossRef]
132. Allenspach, K.; Rizzo, J.; Jergens, A.E.; Chang, Y.M. Hypovitaminosis D Is Associated with Negative Outcome in Dogs with Protein Losing Enteropathy: A Retrospective Study of 43 Cases. *BMC Vet. Res.* **2017**, *13*, 96. [CrossRef] [PubMed]
133. Salavati Schmitz, S.; Gow, A.; Bommer, N.; Morrison, L.; Mellanby, R. Diagnostic Features, Treatment, and Outcome of Dogs with Inflammatory Protein-Losing Enteropathy. *J. Vet. Intern. Med.* **2019**, *33*, 2005–2013. [CrossRef] [PubMed]
134. deLaforcade, A.; Bacek, L.; Blais, M.-C.; Boyd, C.; Brainard, B.M.; Chan, D.L.; Cortellini, S.; Goggs, R.; Hoareau, G.L.; Koenigshof, A.; et al. 2022 Update of the Consensus on the Rational Use of Antithrombotics and Thrombolytics in Veterinary Critical Care (CURATIVE) Domain 1-Defining Populations at Risk. *J. Vet. Emerg. Crit. Care 2001* **2022**, *32*, 289–314. [CrossRef] [PubMed]

Article

Effects of Canine-Obtained Lactic-Acid Bacteria on the Fecal Microbiota and Inflammatory Markers in Dogs Receiving Non-Steroidal Anti-Inflammatory Treatment

Kristin M. V. Herstad [1,*], Hilde Vinje [2], Ellen Skancke [1], Terese Næverdal [3], Francisca Corral [1], Ann-Katrin Llarena [4], Romy M. Heilmann [5], Jan S. Suchodolski [6], Joerg M. Steiner [6] and Nicole Frost Nyquist [4]

[1] Department of Companion Animal Clinical Sciences, Faculty of Veterinary Medicine, The Norwegian University of Life Sciences, 1433 Aas, Norway
[2] Faculty of Chemistry, Biotechnology and Food Sciences, The Norwegian University of Life Sciences, 1433 Aas, Norway
[3] EMPET Skedsmo Dyresykehus, 2007 Kjeller, Norway
[4] Department of Paraclinical Sciences, Faculty of Veterinary Medicine, The Norwegian University of Life Sciences, 1433 Aas, Norway
[5] Department for Small Animals, Veterinary Teaching Hospital, College of Veterinary Medicine, University of Leipzig, DE-04103 Leipzig, SN, Germany
[6] Gastrointestinal Laboratory, Department of Small Animal Clinical Sciences, College of Veterinary Medicine and Biomedical Sciences, Texas A&M University, 4474 TAMU, College Station, TX 77843, USA
* Correspondence: kristin.herstad@nmbu.no

Citation: Herstad, K.M.V.; Vinje, H.; Skancke, E.; Næverdal, T.; Corral, F.; Llarena, A.-K.; Heilmann, R.M.; Suchodolski, J.S.; Steiner, J.M.; Nyquist, N.F. Effects of Canine-Obtained Lactic-Acid Bacteria on the Fecal Microbiota and Inflammatory Markers in Dogs Receiving Non-Steroidal Anti-Inflammatory Treatment. *Animals* **2022**, *12*, 2519. https://doi.org/10.3390/ani12192519

Academic Editor: Andrea Boari

Received: 22 August 2022
Accepted: 18 September 2022
Published: 21 September 2022

Publisher's Note: MDPI stays neutral with regard to jurisdictional claims in published maps and institutional affiliations.

Simple Summary: The use of non-steroidal anti-inflammatory drugs (NSAIDs) has prolonged the longevity and well-being of dogs with osteoarthritis and other painful conditions. However, this treatment is also associated with diarrhea in dogs, but the pathogenetic mechanisms and possible prevention strategies remain unknown. This study aimed to determine whether canine-obtained lactic acid bacteria affect the frequency of diarrhea, fecal microbiota (dysbiosis index), and gastrointestinal inflammation (assessed by calprotectin and S100A12/Calgranulin C) in dogs receiving NSAIDs. Diarrhea occurred in 4/12 dogs (33%) receiving placebo and 1/10 dogs (10%) receiving canine-obtained lactic acid bacteria (LAB), but this difference was not significant. The fecal dysbiosis index, calprotectin, and S100A12 were not significantly different between dogs receiving NSAIDs and LAB and dogs receiving NSAIDs and placebo. This study suggests that LAB is safe to use in NSAID-treated dogs, but further studies are needed to determine its potential to ameliorate diarrhea and gastrointestinal inflammation in dogs receiving NSAIDs.

Abstract: Non-steroidal anti-inflammatory drugs (NSAIDs) may cause enteropathy in dogs and probiotics may be one option to prevent this. The objective of this study was to determine whether the administration of canine-obtained lactic acid bacteria (LAB) has an effect on the frequency of diarrhea, the composition of the fecal microbiota, and/or markers of gastrointestinal inflammation in dogs receiving NSAIDs when compared to dogs given NSAIDs and a placebo. A total of 22 dogs treated with NSAIDs for various clinical indications were enrolled in a seven-day randomized, double-blinded placebo-controlled interventional study. Dogs were randomized to receive either placebo or LAB, a product containing *Limosilactobacillus fermentum*, *Lacticaseibacillus rhamnosus*, and *Lactiplantibacillus plantarum*. Fecal samples were collected on days one and seven. The fecal microbiota was evaluated using the fecal dysbiosis index (DI) and individual bacterial taxa. Fecal calprotectin (CP) and S100A12/Calgranulin C concentrations were used as markers of gastrointestinal inflammation. There was a difference in frequency of diarrhea between groups, with it affecting 4/12 dogs (33%) in the placebo group and 1/10 dogs (10%) in the LAB group, but this difference did not reach statistical significance ($p = 0.32$). There was a correlation between S100A12 and CP ($p < 0.001$), and *Clostridium perfringens* correlated with S100A12 ($p < 0.015$). Neither treatment significantly affected S100A12 ($p = 0.37$), CP ($p = 0.12$), or fecal DI ($p = 0.65$). This study suggests that LAB is a safe supplement to use for short-term treatment in NSAID-treated dogs, but further studies are needed to determine its potential to prevent NSAID-induced enteropathy in dogs.

Keywords: probiotics; NSAID-induced enteropathy; intestinal dysbiosis; inflammatory biomarkers

1. Introduction

Non-steroidal anti-inflammatory drugs (NSAIDs) are the most commonly prescribed analgesics in veterinary medicine [1]. The introduction of NSAIDs has prolonged the longevity and well-being of dogs with osteoarthritis and other painful conditions [2]. However, treatment with NSAIDs may have side effects, most commonly involving the gastrointestinal (GI) tract [3,4]. NSAID-induced gastric ulcerations are related to the decreased perfusion of the gastric mucosa due to a lack of prostaglandins, followed by damage through the actions of gastric acid [4]. However, NSAID-induced lesions may also occur in the lower part of the GI tract where other mechanisms are involved. Indeed, a study using video capsule endoscopy documented that the majority (10/12, 83%) of dogs receiving long-term NSAID treatment had GI lesions involving the distal part of the small intestine [5]. The mechanisms behind these GI lesions include alterations in the adherence and mucosal invasion of intestinal microbes and intestinal dysbiosis [6]. If dysbiosis plays a role in NSAID-induced GI lesions, the modulation of the intestinal microbiota may reduce such side effects [6,7].

Several studies suggest that the intestinal microbiota may have an impact on NSAID-induced side effects in the GI tract [8–10]. For example, germ-free animals are resistant to NSAID-induced enteropathy. Facultative anaerobic bacteria, such as *Escherichia coli*, *Klebsiella* spp., and *Proteus* spp., identified in rats were reported to contribute to ulcer formation, whereas *Lactobacillus* spp. and *Bifidobacterium* spp. were found to prevent their development, possibly by repressing the establishment of ulcer-associated bacteria [11]. The degree of intestinal dysbiosis can be evaluated using the fecal dysbiosis index (DI), which is a mathematical algorithm based on the results of quantitative PCR (qPCR) assays including seven key bacterial taxa (*Faecalibacterium* spp., *Turicibacter* spp., *Streptococcus* spp., *E. coli*, *Blautia* spp., *Fusobacterium* spp., and *Clostridium hiranonis*) [12].

In dogs with chronic enteropathy, biomarkers suggestive of GI inflammation include fecal calgranulin C (S100A12) and calprotectin (CP) [13]. In dogs with chronic inflammatory enteropathy, fecal S100A12 and CP have been shown to correlate with clinical disease [14,15]. Probiotics are live microorganisms consumed orally to promote a healthy gut state [16]. Studies in humans and laboratory rodents have shown that administration of probiotics together with NSAIDs is seen as one option to reduce the risk of NSAID-induced gastroenteropathy [17–19], although one study in rats did not find favorable effects of combining NSAIDS with probiotics [20]. To the best of the authors' knowledge, no previous studies have evaluated the effects of lactic acid bacteria or probiotics in dogs treated with NSAIDs.

In this study, we aimed to evaluate the frequency of diarrhea, markers of GI inflammation (fecal CP and S100A12) and DI in NSAID-treated dogs administered LAB compared to those administered a placebo. We hypothesized that NSAID-treated dogs receiving LAB would have less diarrhea, lower fecal CP and S100A12 concentrations, and a lower fecal DI.

2. Materials and Methods

2.1. Animals

The study protocol was reviewed and approved according to the ethics committee guidelines at the Faculty of Veterinary Medicine, Norwegian University of Life Sciences (NMBU) (approval number: 14/04723-63). Written informed consent was obtained from all dog owners before participation, and they were informed that their participation in the study was voluntary.

Client-owned dogs initiating NSAID treatment regardless of the indication for treatment were included in the study. The type of NSAID used was at the veterinarian's discretion. However, dogs were not included if NSAIDs had been administered within the last three months prior to inclusion in the study. The dogs were fed their usual diets consisting of various commercial dry foods and were not restricted to any specific diet throughout the study trial. Any episodes of hyporexia/anorexia were reported. One dog in the LAB group and one dog in the placebo group received antibiotics (amoxicillin) as part of the treatment plan to manage their conditions. None of the dogs received proton pump inhibitors (PPIs).

2.2. Lactic Acid Bacillus (LAB) and Placebo Products

The LAB was designed to contain lactic acid bacteria that had been cultured from healthy dogs [21] and consisted of *Limosilactobacillus fermentum*, *Lacticaseibacillus rhamnosus*, and *Lactiplantibacillus plantarum* fermented in milk. The placebo product was powdered micro-crystallized cellulose. Dog owners were instructed to administer 1 teaspoon ($\sim$5 g) LAB/placebo (for dogs weighing > 3 kg) or $\frac{1}{2}$ teaspoon ($\sim$2.5 g) LAB/placebo (for dogs weighing < 3 kg) once daily, which for LAB corresponded to 6.2×10^8 and 3.1×10^8 colony-forming units (CFU), respectively, of each of the bacteria. The powder was either sprinkled over the food or diluted in water and given orally by syringe.

2.3. Study Design

The study was a seven-day randomized double-blinded placebo-controlled interventional trial (Figure 1). Dog owners were instructed to record their dog's appetite, fecal consistency, and any episodes of emesis daily during the seven-day trial period. Fecal consistency was recorded as either watery, loose, normal, or hard. The term "diarrhea" was used if the fecal quality was watery. All dogs were randomized to receive either LAB or placebo for seven days. The randomization process was performed by block randomization using a block size of six (random.org). In dogs that developed GI side effects, NSAID treatment was discontinued if deemed necessary by the attending veterinarian.

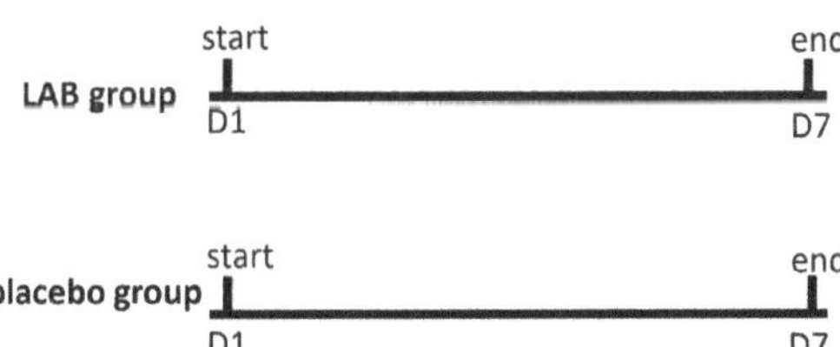

Figure 1. Overview of the study design. Dogs were randomized to receive either LAB or placebo during a seven-day interventional trial. Fecal samples were obtained on days one (D1) and seven (D7) of NSAID plus LAB or NSAID plus placebo treatment. The second fecal sample was taken on the last day of NSAID treatment in dogs that discontinued the study due to developing diarrhea. Fecal consistency was recorded daily throughout the study.

Fecal samples were collected on the first day of the study (day, D1) and at the end of the study (day, D7). When diarrhea required the discontinuation of NSAID treatment, the second fecal sample was collected on the last day of treatment. Fecal samples were collected immediately following natural defecation. The sample was further divided into two aliquots deposited into sterile plastic containers and frozen immediately, either in the owner's home freezer ($-20\,^\circ$C) and then transported on dry ice for storage at $-80\,^\circ$C at the central storage unit, or frozen immediately at $-80\,^\circ$C. Fecal samples were sent to the Gastrointestinal Laboratory at Texas A&M University (TAMU) on dry ice to measure fecal CP and S100A12 concentrations and determine the fecal DI.

2.4. Microbiota Analyses

The fecal microbiota was evaluated based on the fecal DI [12] using quantitative PCR, as described previously [22]. DI > 2 indicates intestinal dysbiosis and values between 0 and 2 were considered equivocal. We also quantified fecal abundances of *Clostridium perfringens* [23] and *Lactobacillus* spp. [24] using qPCR. Briefly, fecal DNA was extracted using the QIAmp PowerFecal Pro DNA KIT (Qiagen) and an automatic extraction system (Thermo KingFisher Flex Magnetic Particle Purification 96 PCR Isolation system), following the manufacturers' instructions. The qPCR assays were performed using a Bio-Rad C1000 Touch Thermal Cycler (Bio-Rad Laboratories, California, USA) with the following protocol: initial denaturation at 98 °C for 2 min; 35 cycles with denaturation at 98 °C for 3 s; and annealing for 3 s. All samples were run in duplicates and the average of the two samples was used for further analyses. The qPCR results were analyzed using the Bio-Rad CFX Maestro 1.1 software (Bio-Rad Laboratories). The qPCR data for the individual bacterial taxa (*Faecalibacterium* spp.; *Turicibacter* spp.; *Streptococcus* spp.; *E.coli, Blautia* spp.; *Fusobacterium* spp., *Clostridium hiranonis, Clostridium perfringens*, and *Lactobacillus* spp.) were normalized to the qPCR data for total bacteria [22].

2.5. Markers of Gastrointestinal Inflammation

Fecal CP concentrations were measured by a fully analytically validated species-specific sandwich ELISA, as described previously [15,25], and reported as ng/g [26] with the current reference interval (RI) used at the Gastrointestinal Laboratory at Texas A&M University, TX, USA (0–961 ng/g). Fecal S100A12 concentrations were measured using a fully analytically validated species-specific in-house sandwich ELISA with an RI of 2–484 ng/g [27,28].

2.6. Statistical Analyses

Statistical analyses were performed using Prism v8, GraphPad Software Inc, San Diego, CA, USA, and R software v. 2021.9.1.372 (RStudio Team (2021). Rstudio: Integrated Development for R. Rstudio, PBC, Boston, MA, USA URL http://www.rstudio.com/, accessed on 26 March 2021). Due to unequal variances between the groups, a Welch's t-test was used to test for significant differences from D1 to D7 for S100A12, CP, individual bacterial taxa, and DI between the LAB group and the placebo group. Pearson's product-moment correlation tested the correlation between inflammatory markers and bacterial taxa. Fisher's exact test was used to test for differences in the frequency of diarrhea between the LAB and placebo groups during the study period. A principal component analysis (PCA) was conducted on the differences between D1 and D7 for fecal DI, bacterial taxa, fecal CP, and fecal S100A12, using the functions "prcomp" and "autoplot" in the ggfortify package in R. Statistical significance for all tests was set at $p < 0.05$.

3. Results

3.1. Demographic and Clinical Factors

A total of 22 dogs were enrolled in the study, of which 10 dogs received LAB and 12 received placebo. The study population consisted of dogs of various breeds, both sexes, and different ages. Dogs in the LAB group were between 4 months and 14 years of age with a median of 6 years, while dogs in the placebo group were between 1 and 10 years of age with a median of 5.9 years (Table 1). The dogs received various commercial diets and one dog in each group received an antibiotic (amoxicillin). However, as detected by Cook's distance, this did not influence the final result.

Robenacoxib was used in 7 out of 10 dogs (70%) in the LAB group and 8 out of 12 dogs (66%) in the placebo group, whereas meloxicam was given to 2 out of 10 (20%) in the LAB group and 3 out of 12 (25%) in the placebo group. The type of NSAID being administered was not recorded for one dog in the LAB group and one dog in the placebo group. The reason for NSAID prescription was an orthopedic condition in 3/10 dogs (30%) in the LAB group and 3/12 dogs (25%) in the placebo group. A surgical procedure under general anesthesia had been performed in 8/10 dogs (80%) in the LAB group and 10/12 dogs (83%) in the placebo group. Of these dogs, diarrhea occurred in 1/8 (12.5%) in the LAB group and 3/10 (30%) in the placebo group.

Table 1. Overview of demographic factors, treatments, and occurrence of diarrhea in dogs included in the study.

Test Product	Breed	Age (Years)	Sex	Anesthesia	Reason for NSAID Treatment	Name of NSAID Treatment *	Occurrence of Diarrhea	Discontinued Treatment
Placebo	Finnish Lapphund	3	F	yes	Removal of benign skin tumor	Robenacoxib	yes	no
Placebo	German Short-haired Pointer	3	F	yes	Removal of benign skin tumor	Robenacoxib	no	no
Placebo	Pug	6	M	yes	Dental procedure	Robenacoxib	no	no
Placebo	Cavalier King Charles Spaniel	4	M	no	Benign prostate hypertrophy	Robenacoxib	yes	no
Placebo	Mixed breed	UN	F	yes	Mastectomy	Meloxicam	yes	yes (day 4)
Placebo	Pointer dog	2	F	yes	Pyometra surgery	Meloxicam	no	no
Placebo	Miniature Dachshund	10	F	no	Osteoarthritis	Meloxicam	no	no
Placebo	Dachshund	8	F	yes	Hemilaminectomy	UN	no	no
Placebo	Mixed breed	10	F	yes	Mastectomy	Robenacoxib	no	no
Placebo	Cocker Spaniel	1	F	yes	Patella luxation surgery	Robenacoxib	yes	yes (day 3)
Placebo	Jack Russel Terrier	8	F	yes	Hemilaminectomy	Robenacoxib	no	no
Placebo	Danish–Swedish Farmdog	10	F	yes	TPLO surgery	Robenacoxib	no	no
LAB	Alaskan Malamute	2	M	yes	Dental procedure	Robenacoxib	no	no
LAB	Cocker Spaniel	2	M	yes	Castration	Robenacoxib	no	no
LAB	Shih Tzu	4	F	yes	Pyometra surgery	Meloxicam	no	no
LAB	Finish Lapphund	10	M	yes	Removal of benign skin tumor	UN	no	no
LAB	Pomeranian	0.3	M	yes	Bone fracture surgery	Meloxicam	no	no
LAB	English Setter	10	F	yes	Removal of benign skin tumor	Robenacoxib	no	no
LAB	English Bulldog	7	F	yes	Pyometra surgery	Robenacoxib	yes	yes (day 3)
LAB	Alaskan Husky	14	F	no	Osteoarthritis	Robenacoxib	no	no
LAB	Medium Poodle	6	M	yes	Cystotomy due to urolithiasis	Robenacoxib	no	no
LAB	Drentsche Patrijshond	5	F	no	Diffuse pain related to the skeleton	Robenacoxib	no	no

Abbreviations: UN, not recorded; LAB, lactic acid bacteria treatment; TPLO, tibial plateau leveling osteotomy. * The doses were calculated based on the individual dog's body weight and given enterally once daily. Dogs undergoing surgery were treated with parenteral NSAIDs for the first 24–48 h.

Of the dogs receiving placebo, 4/12 dogs (33%) developed diarrhea, while this only occurred in 1/10 dogs (10%) in the LAB group. However, this difference did not reach statistical significance (Fisher's exact test, odds ratio: 0.24, $p = 0.32$). The NSAID treatment was discontinued because of the severity of diarrhea in one of the dogs receiving LAB and two of the dogs receiving placebo. Two dogs receiving LAB vomited on D2 and D7, respectively, whereas one dog receiving placebo vomited on D7 after initiating NSAID treatment. The dogs had normal appetite throughout the study.

3.2. PCA Analyses

The PCA analysis on the change from D1 to D7 revealed no clear distinction between dogs receiving LAB vs. placebo, but indicated that changes in fecal CP and S100A12 concentrations between D1 and D7 in the individual dogs are strongly associated with the abundances of *C. perfringens*. These span the principal component 2 (PC2), which explains 22.3% of the variation in our dataset. The bacterial taxa, except for *E. coli*, can be interpreted as in contrast with the fecal DI, and span the PC1, explaining 29.3% of the variation in the data (Figure 2). The correlation between fecal S100A12 and CP was significant (Pearson's correlation coefficient = 0.63, $p < 0.001$). The abundance of *C. perfringens* correlated significantly with S100A12 (Pearson's correlation coefficient = 0.510, $p = 0.015$), but not with CP (Pearson's correlation coefficient = 0.379, $p = 0.084$).

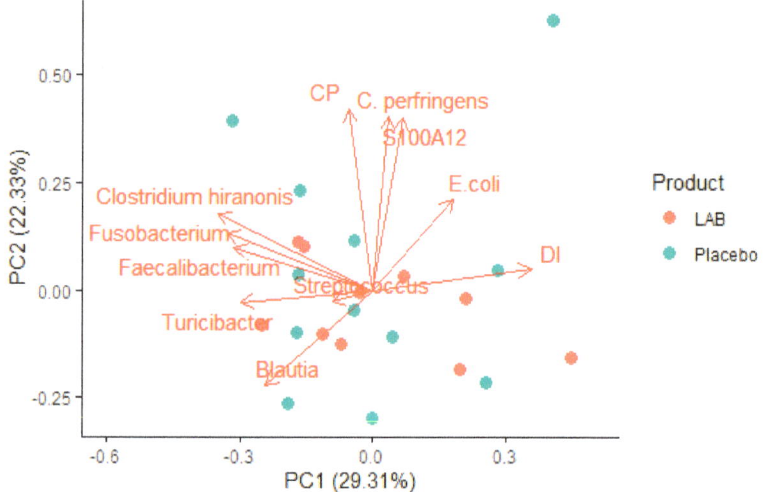

Figure 2. PCA plot showing the changes in the different variables from D1 to D7 in individual dogs given LAB vs. placebo. Fecal CP and S100A12 are strongly correlated and span out on the principal component (PC) 2, which explains 22.3% of the variation in the data, whereas all the bacterial taxa, except *E. coli*, span out on the PC1, explaining 29.3% of the variation in the data.

3.3. CP and S100A12 Concentrations, DI and Bacterial Taxa in Dogs Receiving LAB vs. Placebo

The CP concentration in the three dogs with the highest levels at D1 in the LAB group reduced their levels to below 50 ng/g at D7. The same three dogs also had the highest S100A12 concentration at D1, which dropped to a level below 20 ng/g at D7. In the placebo group, five dogs had a CP concentration above the RI (0–961 ng/g) at D7 and the same five dogs had the highest S100A12 concentrations at D7 (Figure 3). There were no significant differences in the levels of S100A12 ($p = 0.37$) or CP ($p = 0.12$) between dogs receiving LAB vs. placebo.

Neither LAB nor placebo had any significant effect on DI ($p = 0.65$) or any of the bacterial taxa during the study period (all $p > 0.05$) (Figure 4).

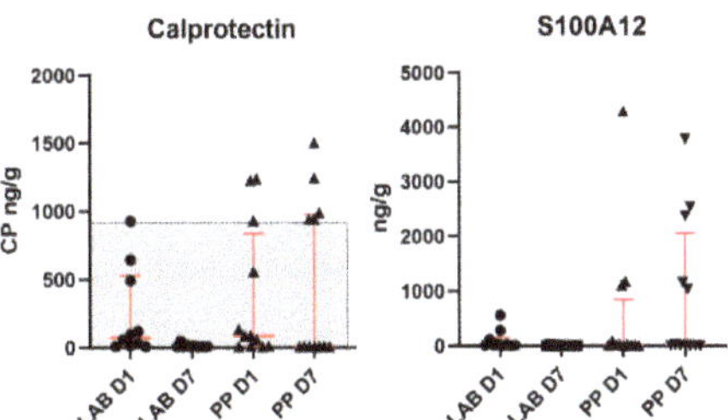

Figure 3. The plot shows the interindividual variation of fecal calprotectin (CP) concentrations and S100A12 in dogs given LAB or placebo on D1 and D7. Red lines show the medians and interquartile ranges, and the grey shaded area corresponds to the reference interval for CP (0–961 ng/g, Gastrointestinal Laboratory, Texas A&M University, TX, USA). Abbreviations: LAB, lactic acid bacteria; PP, placebo product.

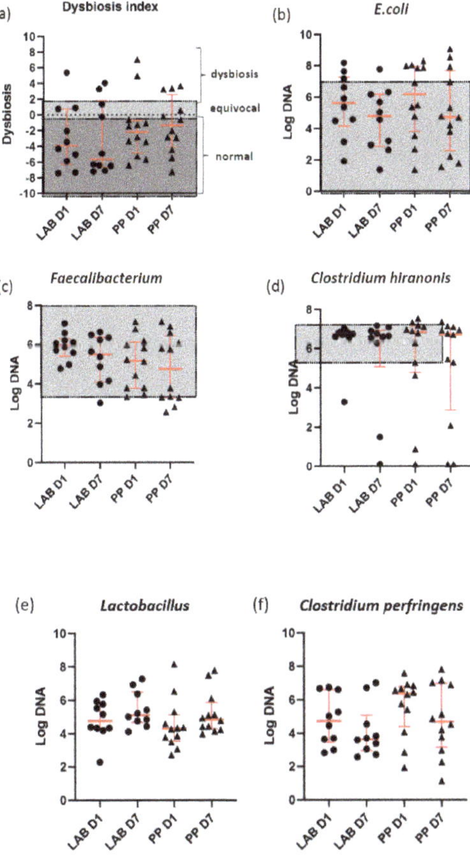

Figure 4. Dot plots showing (**a**) the canine fecal dysbiosis index (DI) and fecal abundances for (**b**) *E. coli*, (**c**) *Faecalibacterium*, (**d**) *Clostridium hiranonis*, (**e**) *Lactobacillus*, and (**f**) *Clostridium perfringens* for both groups of dogs on D1 and D7. Medians and interquartile ranges are indicated by red lines and the grey shaded areas correspond to the respective reference intervals. Abbreviations: LAB, lactic acid bacteria; PP, placebo product.

4. Discussion

As far as the authors are aware, no previous studies have investigated the potential protective effects of orally administered canine-obtained lactic acid bacteria or probiotics in dogs given NSAIDs. Among the dogs given the placebo, four dogs (33%) developed diarrhea compared to only one dog (10%) in the LAB group. This difference was not significant and indicates that LAB is safe to use in dogs. However, larger studies are needed to determine whether LAB can prevent diarrhea in dogs given NSAIDs. We found that S100A12 and CP were strongly correlated, as was also demonstrated in a previous study of dogs with chronic inflammatory enteropathy [29]. Five dogs given placebo had increased CP concentrations above the upper limit of the RI at D7, whereas all dogs given LAB had negligible CP concentrations at D7. Fecal CP has been shown to be a sensitive screening marker for NSAID-induced enteropathy in human patients [30], even with short-term treatment (seven days) [31]. Moreover, a study showed that humans treated with NSAIDs and probiotics had decreased fecal CP concentrations compared to those given NSAIDs and placebo [32]. However, another study did not find a beneficial effect of probiotics in humans given NSAIDs [33].

As for CP, S100A12 has also been shown to be increased in dogs with chronic inflammatory enteropathy [29], and has been used in humans to separate patients with inflammatory bowel disease from those with irritable bowel disease [34]. S100A12 may therefore potentially be useful as screening marker for NSAID-induced enteropathy. We found that the changes in S100A12 concentrations and *C. perfringens* abundance between D1 and D7 were significantly correlated, suggesting that this bacterial taxon may play a role in NSAID-induced enteropathy. Its role in inflammation is not a new phenomenon, as *C. perfringens* is associated with acute hemorrhagic diarrhea in dogs [35–37], where its pathogenetic potential is linked to the production of *netE* and *netF* toxins [38].

Previous studies have found that the DI can change in response to diet [39,40], and it can be useful for differentiating dogs with chronic enteropathy from healthy dogs [12]. A previous study of dogs with diarrhea found that the LAB product used could resolve diarrhea and reduce fecal abundances of *C. perfringens* and *Enterococcus faecium* [41], indicating that LAB may potentially cause changes in the bacterial populations and improve gut health. Although we did not find any significant change in DI between the groups, there might be changes in the microbiota composition that would be detected using high-throughput sequencing methods and, ideally, DNA shotgun sequencing.

There was no significant change in the fecal abundance of *C. hiranonis*, a key bacterium in bile acid metabolism, in dogs receiving LAB vs. placebo. Interestingly, studies in rodents and cell culture systems have found that the enterohepatic circulation of NSAIDs is associated with higher levels of secondary bile acids, which can damage intestinal cells [42,43]. However, whether the dogs in this study also had changes in absolute fecal primary and secondary bile acid concentrations was not determined.

Dogs receiving LAB in our study did not have a higher fecal abundances of *Lactobacillus* than the dogs in the placebo group. Previous studies have demonstrated higher numbers of these bacteria in fecal samples from dogs given LAB, and the modulated intestinal microbiota was characterized by an increased number of other variants of lactic-acid bacteria [44]. It is possible that NSAID-induced changes in the composition of the microbiota take longer to develop, albeit short-term NSAID use was found to change the intestinal microbiota in humans and rats [45]. Furthermore, in our study, dogs were initiated on LAB and NSAID simultaneously. It is possible that the LAB product would have had a more pronounced effect if given prior to initiating NSAID treatment. A recent study demonstrated that mice given oral probiotics containing lactic acid bacteria five days before NSAID treatment showed enrichment of colonic anaerobes and Lactobacilli, whereas the total abundance of Enterobacter decreased and ameliorated GI inflammation was detected compared to the controls [19].

We cannot exclude a confounding effect of general anesthesia, surgery, and/or other medications contributing to the development of GI signs, as we could not include a control group of dogs not receiving NSAIDs due to ethical constraints. However, no differences were detected between the groups of dogs with diarrhea that underwent a surgical procedure (1/3 dogs in the LAB group and 3/9 dogs in the placebo group). Thus, NSAID treatment may exacerbate GI signs and potential GI lesions in these dogs, regardless of the initiating factors. The dogs in our study were fed different diets; thus, we cannot rule out the influence of diet on the dogs' microbiota composition. However, all dogs were fed commercial dry food diets, no dogs had a change of diet during the trial period, and no dogs ate raw meat or home-cooked diets which may influence microbiota composition [46,47].

NSAID-induced enteropathy may not be associated with clinical signs, and diarrhea may not be a precise indicator for this condition [5]. In humans, clinical signs of NSAID-induced enteropathy are nonspecific and, in addition to diarrhea, may include signs of iron-deficiency anemia, GI protein loss, indigestion, constipation, and abdominal pain [6], but affected patients can also be asymptomatic [48]. Therefore, further studies are needed to determine how NSAID-induced enteropathy manifests in dogs. For this purpose, capsule endoscopy can be used to identify gastrointestinal lesions in NSAID-treated dogs. These findings should be correlated with clinical signs such as diarrhea, vomiting, anorexia/hyporexia and weight loss, and inflammatory markers (calprotectin and S100A12). It is also unknown which type of probiotic bacteria and at what dose would be most beneficial to use [7].

Notably, as this work was based on a small number of dogs, a larger study is required to document the effect of lactic acid bacteria on GI health in dogs.

5. Conclusions

This study did not find a significant difference in the frequency of diarrhea or change in the DI or individual bacteria taxa in NSAID-treated dogs given LAB vs. placebo. Further studies are needed to evaluate the potential of lactic-acid bacteria to ameliorate adverse GI effects induced by NSAIDs.

Author Contributions: Conceptualization, N.F.N., E.S. and K.M.V.H.; methodology, R.M.H., J.S.S. and J.M.S.; software, H.V. and K.M.V.H.; validation, K.M.V.H., H.V., A.-K.L., E.S. and N.F.N.; formal analysis, H.V. and K.M.V.H.; investigation, K.M.V.H., T.N., F.C. and A.-K.L.; data curation, K.M.V.H. and H.V.; writing—original draft preparation, K.M.V.H.; writing—review and editing, K.M.V.H., H.V, E.S., T.N., F.C., A.-K.L., R.M.H., J.S.S., J.M.S. and N.F.N.; visualization, H.V. and K.M.V.H.; supervision, N.F.N.; project administration, K.M.V.H.; funding acquisition, J.S.S. and J.M.S. All authors have read and agreed to the published version of the manuscript.

Funding: The canine-obtained lactic acid bacteria (LAB) product used in this study was provided by Vetcare, Finland. This product is commercially available as Canius®.

Institutional Review Board Statement: The study protocol was reviewed and approved according to the ethics committee guidelines at the Faculty of Veterinary Medicine, Norwegian University of Life Sciences (NMBU) (approval number: 14/04723-63).

Informed Consent Statement: Written informed consent was obtained from all dog owners before participation, and they were informed that their participation in the study was voluntary.

Data Availability Statement: All data are included in the manuscript.

Acknowledgments: The authors would like to thank Martine H. Rusås, Kari Elisabeth Ringvall, and Trude Fjugstad, who contributed to the collection of fecal samples.

Conflicts of Interest: Dr. Suchodolski and Dr. Steiner work for the GI Lab at Texas A&M University, which provides the dysbiosis index as well as the measurement of fecal calprotectin and S100A12 on a fee for service basis. R.M.H. served as co-editor for this Special Issue but was not involved in the editing of this manuscript.

References

1. Lascelles, B.D.; McFarland, J.M.; Swann, H. Guidelines for safe and effective use of NSAIDs in dogs. *Vet. Ther.* **2005**, *6*, 237–251. [PubMed]
2. Aragon, C.L.; Hofmeister, E.H.; Budsberg, S.C. Systematic review of clinical trials of treatments for osteoarthritis in dogs. *J. Am. Vet. Med. Assoc.* **2007**, *230*, 514–521. [CrossRef] [PubMed]
3. Hunt, J.R.; Dean, R.S.; Davis, G.N.; Murrell, J.C. An analysis of the relative frequencies of reported adverse events associated with NSAID administration in dogs and cats in the United Kingdom. *Vet. J.* **2015**, *206*, 183–190. [CrossRef] [PubMed]
4. Monteiro-Steagall, B.P.; Steagall, P.V.; Lascelles, B.D. Systematic review of nonsteroidal anti-inflammatory drug-induced adverse effects in dogs. *J. Vet. Intern. Med.* **2013**, *27*, 1011–1019. [CrossRef] [PubMed]
5. Mabry, K.; Hill, T.; Tolbert, M.K. Prevalence of gastrointestinal lesions in dogs chronically treated with nonsteroidal anti-inflammatory drugs. *J. Vet. Intern. Med.* **2021**, *35*, 853–859. [CrossRef]
6. Shin, S.J.; Noh, C.K.; Lim, S.G.; Lee, K.M.; Lee, K.J. Non-steroidal anti-inflammatory drug-induced enteropathy. *Intest Res* **2017**, *15*, 446–455. [CrossRef]
7. Montalto, M.; Gallo, A.; Gasbarrini, A.; Landolfi, R. NSAID enteropathy: Could probiotics prevent it? *J. Gastroenterol.* **2013**, *48*, 689–697. [CrossRef]
8. Otani, K.; Tanigawa, T.; Watanabe, T.; Shimada, S.; Nadatani, Y.; Nagami, Y.; Tanaka, F.; Kamata, N.; Yamagami, H.; Shiba, M.; et al. Microbiota plays a key role in non-steroidal anti-inflammatory drug-induced small intestinal damage. *Digestion* **2017**, *95*, 22–28. [CrossRef]
9. Makivuokko, H.; Tiihonen, K.; Tynkkynen, S.; Paulin, L.; Rautonen, N. The effect of age and non-steroidal anti-inflammatory drugs on human intestinal microbiota composition. *Br. J. Nutr.* **2010**, *103*, 227–234. [CrossRef]
10. Rogers, M.A.M.; Aronoff, D.M. The influence of non-steroidal anti-inflammatory drugs on the gut microbiome. *Clin. Microbiol. Infect.* **2016**, *22*, 178.e171–178.e179. [CrossRef]
11. Uejima, M.; Kinouchi, T.; Kataoka, K.; Hiraoka, I.; Ohnishi, Y. Role of intestinal bacteria in ileal ulcer formation in rats treated with a nonsteroidal antiinflammatory drug. *Microbiol. Immunol.* **1996**, *40*, 553–560. [CrossRef] [PubMed]
12. AlShawaqfeh, M.K.; Wajid, B.; Minamoto, Y.; Markel, M.; Lidbury, J.A.; Steiner, J.M.; Serpedin, E.; Suchodolski, J.S. A dysbiosis index to assess microbial changes in fecal samples of dogs with chronic inflammatory enteropathy. *FEMS Microbiol. Ecol.* **2017**, *93*. [CrossRef] [PubMed]
13. Heilmann, R.M.; Nestler, J.; Schwarz, J.; Grützner, N.; Ambrus, A.; Seeger, J.; Suchodolski, J.S.; Steiner, J.M.; Gurtner, C. Mucosal expression of S100A12 (calgranulin C) and S100A8/A9 (calprotectin) and correlation with serum and fecal concentrations in dogs with chronic inflammatory enteropathy. *Vet. Immunol. Immunopathol.* **2019**, *211*, 64–74. [CrossRef] [PubMed]
14. Heilmann, R.M.; Volkmann, M.; Otoni, C.C.; Grützner, N.; Kohn, B.; Jergens, A.E.; Steiner, J.M. Fecal S100A12 concentration predicts a lack of response to treatment in dogs affected with chronic enteropathy. *Vet. J.* **2016**, *215*, 96–100. [CrossRef]
15. Heilmann, R.M.; Berghoff, N.; Mansell, J.; Grützner, N.; Parnell, N.K.; Gurtner, C.; Suchodolski, J.S.; Steiner, J.M. Association of fecal calprotectin concentrations with disease severity, response to treatment, and other biomarkers in dogs with chronic inflammatory enteropathies. *J. Vet. Intern. Med.* **2018**, *32*, 679–692. [CrossRef]
16. Hill, C.; Guarner, F.; Reid, G.; Gibson, G.R.; Merenstein, D.J.; Pot, B.; Morelli, L.; Canani, R.B.; Flint, H.J.; Salminen, S.; et al. Expert consensus document. The International Scientific Association for Probiotics and Prebiotics consensus statement on the scope and appropriate use of the term probiotic. *Nat. Rev. Gastroenterol. Hepatol.* **2014**, *11*, 506–514. [CrossRef]
17. Gotteland, M.; Cruchet, S.; Verbeke, S. Effect of *Lactobacillus* ingestion on the gastrointestinal mucosal barrier alterations induced by indometacin in humans. *Aliment. Pharmacol. Ther.* **2001**, *15*, 11–17. [CrossRef]
18. Santiago-López, L.; Hernández-Mendoza, A.; Vallejo-Cordoba, B.; Mata-Haro, V.; Wall-Medrano, A.; González-Córdova, A.F. Milk Fermented with *Lactobacillus fermentum* Ameliorates Indomethacin-Induced Intestinal Inflammation: An Exploratory Study. *Nutrients* **2019**, *11*, 1610. [CrossRef]
19. Monteros, M.J.M.; Galdeano, C.M.; Balcells, M.F.; Weill, R.; De Paula, J.A.; Perdigón, G.; Cazorla, S.I. Probiotic lactobacilli as a promising strategy to ameliorate disorders associated with intestinal inflammation induced by a non-steroidal anti-inflammatory drug. *Sci. Rep.* **2021**, *11*, 571. [CrossRef]
20. Kamil, R.; Geier, M.S.; Butler, R.N.; Howarth, G.S. *Lactobacillus rhamnosus* GG exacerbates intestinal ulceration in a model of indomethacin-induced enteropathy. *Dig. Dis. Sci.* **2007**, *52*, 1247–1252. [CrossRef]
21. Beasley, S.S.; Manninen, T.J.; Saris, P.E. Lactic acid bacteria isolated from canine faeces. *J. Appl. Microbiol.* **2006**, *101*, 131–138. [CrossRef] [PubMed]
22. Sung, C.H.; Marsilio, S.; Chow, B.; Zornow, K.A.; Slovak, J.E.; Pilla, R.; Lidbury, J.A.; Steiner, J.M.; Park, S.Y.; Hong, M.P.; et al. Dysbiosis index to evaluate the fecal microbiota in healthy cats and cats with chronic enteropathies. *J. Feline Med. Surg.* **2022**, *24*, e1–e12. [CrossRef]
23. Minamoto, Y.; Dhanani, N.; Markel, M.E.; Steiner, J.M.; Suchodolski, J.S. Prevalence of *Clostridium perfringens*, *Clostridium perfringens* enterotoxin and dysbiosis in fecal samples of dogs with diarrhea. *Vet. Microbiol.* **2014**, *174*, 463–473. [CrossRef] [PubMed]
24. Blake, A.B.; Guard, B.C.; Honneffer, J.B.; Lidbury, J.A.; Steiner, J.M.; Suchodolski, J.S. Altered microbiota, fecal lactate, and fecal bile acids in dogs with gastrointestinal disease. *PLoS ONE* **2019**, *14*, e0224454. [CrossRef] [PubMed]

25. Grützner, N.; Heilmann, R.M.; Suchodolski, J.S.; Steiner, J.M.; Holzenburg, A. Cold-microwave enhanced enzyme-linked immunosorbent assays–a path to high-throughput clinical diagnostics. *Anal. Biochem.* **2014**, *457*, 65–73. [CrossRef]
26. Jones, S.M.; Gaier, A.; Enomoto, H.; Ishii, P.; Pilla, R.; Price, J.; Suchodolski, J.; Steiner, J.M.; Papich, M.G.; Messenger, K.; et al. The effect of combined carprofen and omeprazole administration on gastrointestinal permeability and inflammation in dogs. *J. Vet. Intern. Med.* **2020**, *34*, 1886–1893. [CrossRef]
27. Heilmann, R.M.; Lanerie, D.J.; Ruaux, C.G.; Grutzner, N.; Suchodolski, J.S.; Steiner, J.M. Development and analytic validation of an immunoassay for the quantification of canine S100A12 in serum and fecal samples and its biological variability in serum from healthy dogs. *Vet. Immunol. Immunopathol.* **2011**, *144*, 200–209. [CrossRef]
28. Heilmann, R.M.; Cranford, S.M.; Ambrus, A.; Grützner, N.; Schellenberg, S.; Ruaux, C.G.; Suchodolski, J.S.; Steiner, J.M. Validation of an enzyme-linked immunosorbent assay (ELISA) for the measurement of canine S100A12. *Vet. Clin. Pathol.* **2016**, *45*, 135–147. [CrossRef]
29. Heilmann, R.M.; Grellet, A.; Allenspach, K.; Lecoindre, P.; Day, M.J.; Priestnall, S.L.; Toresson, L.; Procoli, F.; Grützner, N.; Suchodolski, J.S.; et al. Association between fecal S100A12 concentration and histologic, endoscopic, and clinical disease severity in dogs with idiopathic inflammatory bowel disease. *Vet. Immunol. Immunopathol.* **2014**, *158*, 156–166. [CrossRef]
30. Tibble, J.A.; Sigthorsson, G.; Foster, R.; Scott, D.; Fagerhol, M.K.; Roseth, A.; Bjarnason, I. High prevalence of NSAID enteropathy as shown by a simple faecal test. *Gut* **1999**, *45*, 362–366. [CrossRef]
31. Meling, T.R.; Aabakken, L.; Røseth, A.; Osnes, M. Faecal calprotectin shedding after short-term treatment with non-steroidal anti-inflammatory drugs. *Scand. J. Gastroenterol.* **1996**, *31*, 339–344. [CrossRef] [PubMed]
32. Montalto, M.; Gallo, A.; Curigliano, V.; D'Onofrio, F.; Santoro, L.; Covino, M.; Dalvai, S.; Gasbarrini, A.; Gasbarrini, G. Clinical trial: The effects of a probiotic mixture on non-steroidal anti-inflammatory drug enteropathy—A randomized, double-blind, cross-over, placebo-controlled study. *Aliment. Pharmacol. Ther.* **2010**, *32*, 209–214. [CrossRef] [PubMed]
33. Mäkelä, S.M.; Forssten, S.D.; Kailajärvi, M.; Langén, V.L.; Scheinin, M.; Tiihonen, K.; Ouwehand, A.C. Effects of *Bifidobacterium animalis* ssp. lactis 420 on gastrointestinal inflammation induced by a nonsteroidal anti-inflammatory drug: A randomized, placebo-controlled, double-blind clinical trial. *Br. J. Clin. Pharmacol.* **2021**, *87*, 4625–4635. [CrossRef] [PubMed]
34. Kaiser, T.; Langhorst, J.; Wittkowski, H.; Becker, K.; Friedrich, A.W.; Rueffer, A.; Dobos, G.J.; Roth, J.; Foell, D. Faecal S100A12 as a non-invasive marker distinguishing inflammatory bowel disease from irritable bowel syndrome. *Gut* **2007**, *56*, 1706–1713. [CrossRef]
35. Herstad, K.M.V.; Trosvik, P.; Haaland, A.H.; Haverkamp, T.H.A.; de Muinck, E.J.; Skancke, E. Changes in the fecal microbiota in dogs with acute hemorrhagic diarrhea during an outbreak in Norway. *J. Vet. Intern. Med.* **2021**, *35*, 2177–2186. [CrossRef]
36. Guard, B.C.; Barr, J.W.; Reddivari, L.; Klemashevich, C.; Jayaraman, A.; Steiner, J.M.; Vanamala, J.; Suchodolski, J.S. Characterization of microbial dysbiosis and metabolomic changes in dogs with acute diarrhea. *PLoS ONE* **2015**, *10*, e0127259. [CrossRef]
37. Suchodolski, J.S.; Markel, M.E.; García-Mazcorro, J.F.; Unterer, S.; Heilmann, R.M.; Dowd, S.E.; Kachroo, P.; Ivanov, I.; Minamoto, Y.; Dillman, E.M.; et al. The fecal microbiome in dogs with acute diarrhea and idiopathic inflammatory bowel disease. *PLoS ONE* **2012**, *7*, e51907. [CrossRef]
38. Sindern, N.; Suchodolski, J.S.; Leutenegger, C.M.; Mehdizadeh Gohari, I.; Prescott, J.F.; Proksch, A.L.; Mueller, R.S.; Busch, K.; Unterer, S. Prevalence of *Clostridium perfringens netE* and *netF* toxin genes in the feces of dogs with acute hemorrhagic diarrhea syndrome. *J. Vet. Intern. Med.* **2019**, *33*, 100–105. [CrossRef]
39. Pilla, R.; Suchodolski, J.S. The Gut Microbiome of Dogs and Cats, and the Influence of Diet. *Vet. Clin. North Am. Small Anim. Pract.* **2021**, *51*, 605–621. [CrossRef]
40. Pilla, R.; Gaschen, F.P.; Barr, J.W.; Olson, E.; Honneffer, J.; Guard, B.C.; Blake, A.B.; Villanueva, D.; Khattab, M.R.; AlShawaqfeh, M.K.; et al. Effects of metronidazole on the fecal microbiome and metabolome in healthy dogs. *J. Vet. Intern. Med.* **2020**, *34*, 1853–1866. [CrossRef]
41. Gomez-Gallego, C.; Junnila, J.; Mannikko, S.; Hameenoja, P.; Valtonen, E.; Salminen, S.; Beasley, S. A canine-specific probiotic product in treating acute or intermittent diarrhea in dogs: A double-blind placebo controlled efficacy study. *Vet. Microbiol.* **2016**, *197*, 122–128. [CrossRef] [PubMed]
42. Dial, E.J.; Darling, R.L.; Lichtenberger, L.M. Importance of biliary excretion of indomethacin in gastrointestinal and hepatic injury. *J. Gastroenterol. Hepatol.* **2008**, *23*, e384–e389. [CrossRef] [PubMed]
43. Zhou, Y.; Dial, E.J.; Doyen, R.; Lichtenberger, L.M. Effect of indomethacin on bile acid-phospholipid interactions: Implication for small intestinal injury induced by nonsteroidal anti-inflammatory drugs. *Am. J. Physiol. Gastrointest. Liver Physiol.* **2010**, *298*, G722–G731. [CrossRef] [PubMed]
44. Manninen, T.J.; Rinkinen, M.L.; Beasley, S.S.; Saris, P.E. Alteration of the canine small-intestinal lactic acid bacterium microbiota by feeding of potential probiotics. *Appl. Environ. Microbiol.* **2006**, *72*, 6539–6543. [CrossRef]
45. Wang, X.; Tang, Q.; Hou, H.; Zhang, W.; Li, M.; Chen, D.; Gu, Y.; Wang, B.; Hou, J.; Liu, Y.; et al. Gut Microbiota in NSAID Enteropathy: New Insights From Inside. *Front. Cell. Infect. Microbiol.* **2021**, *11*, 679396. [CrossRef]
46. Herstad, K.M.V.; Gajardo, K.; Bakke, A.M.; Moe, L.; Ludvigsen, J.; Rudi, K.; Rud, I.; Sekelja, M.; Skancke, E. A diet change from dry food to beef induces reversible changes on the faecal microbiota in healthy, adult client-owned dogs. *BMC Vet. Res.* **2017**, *13*, 147. [CrossRef]

47. Sandri, M.; Dal Monego, S.; Conte, G.; Sgorlon, S.; Stefanon, B. Raw meat based diet influences faecal microbiome and end products of fermentation in healthy dogs. *BMC Vet. Res.* **2017**, *13*, 65. [CrossRef]
48. Armstrong, C.P.; Blower, A.L. Non-steroidal anti-inflammatory drugs and life threatening complications of peptic ulceration. *Gut* **1987**, *28*, 527–532. [CrossRef]

Review

The World of Organoids: Gastrointestinal Disease Modelling in the Age of 3R and One Health with Specific Relevance to Dogs and Cats

Georg Csukovich *, Barbara Pratscher and Iwan Anton Burgener

Small Animal Internal Medicine, Vetmeduni, 1210 Vienna, Austria
* Correspondence: georg.csukovich@vetmeduni.ac.at

Simple Summary: One Health is a concept that describes the interplay between humans, animals, and the environment. This interaction is becoming increasingly important as researchers try to address it in a laboratory setting. This has led to the development of new and highly sophisticated research methods paving the way for animal-free research methods. Within this context, the development of mini-organs, so-called 'organoids', is of great significance. These organoids represent entire organs on a laboratory scale and can be established from stem cells. Subsequently, organoids are used to model certain disease states and the interaction of the host with specific harmful organisms. With this review, we give an overview of what disease modelling approaches have already been carried out in the past and where the field might be heading in the future. In the context of One Health, we consider animal models whenever possible, putting a focus on gastrointestinal diseases.

Abstract: One Health describes the importance of considering humans, animals, and the environment in health research. One Health and the 3R concept, i.e., the replacement, reduction, and refinement of animal experimentation, shape today's research more and more. The development of organoids from many different organs and animals led to the development of highly sophisticated model systems trying to replace animal experiments. Organoids may be used for disease modelling in various ways elucidating the manifold host–pathogen interactions. This review provides an overview of disease modelling approaches using organoids of different kinds with a special focus on animal organoids and gastrointestinal diseases. We also provide an outlook on how the research field of organoids might develop in the coming years and what opportunities organoids hold for in-depth disease modelling and therapeutic interventions.

Keywords: one health; 3R; organoids

Citation: Csukovich, G.; Pratscher, B.; Burgener, I.A. The World of Organoids: Gastrointestinal Disease Modelling in the Age of 3R and One Health with Specific Relevance to Dogs and Cats. *Animals* **2022**, *12*, 2461. https://doi.org/10.3390/ani12182461

Academic Editors: Aarti Kathrani and Romy M. Heilmann

Received: 28 July 2022
Accepted: 14 September 2022
Published: 18 September 2022

Publisher's Note: MDPI stays neutral with regard to jurisdictional claims in published maps and institutional affiliations.

1. Introduction

The concept of 'One Health' has become increasingly important over the last few years. In contrast to specific scientific disciplines such as human medicine, veterinary medicine, or environmental sciences, One Health is an approach taking more than one of these factors into account [1]. This also includes the political implications of the surveillance of diseases and the prevention thereof and not only scientific research on pathogens and their interaction with host organisms. COVID-19, for instance, is a very prominent and current example. SARS-CoV-2 infections are diagnosed in humans as well as many different species of animals [2], and viral particles can be found in wastewater [3]. Referring to the global problem of SARS-CoV-2 infections for humans, animals, and environmental contamination, one can appreciate the importance of One Health in a global context. In vitro research methods can neither fully model these complex interactions nor entirely replace animal experimentation but are of great importance to reduce the need for animals in today's research.

More than 60 years ago, researchers were looking for ways to reduce pain and distress for laboratory animals. In 1959, Russel and Burch first explained the principle of the three

Rs (3R), i.e., Replacement, Reduction, and Refinement of animal experimentation [4]. Since then, the 3R principle has been implicitly included in animal welfare laws in the United States of America [5] as well as in Europe [6], and researchers are obliged to consider these laws when planning and carrying out experiments involving live animals. Recently, the Max Planck Society for the Advancement of Science e.V. has taken the next step and expanded the classic 3R principle to the 4R principle, also taking 'Responsibility' into account. Researchers commit to using their knowledge in order to further promote animal welfare by engaging in public discourse, improving the social structure of housed experimental animals and expanding the knowledge about the experience of pain, intelligence and consciousness in animals [7]. Animal experimentation is not limited to laboratory mice and rats but also includes other vertebrates such as fish, rabbits, cats, dogs, pigs, and others.

Dogs, for instance, are mainly used for toxicology studies. In the European Union, the number of dogs used for any scientific purpose for the first time accounted for 17.711 in 2018, adding up to 25.717, including dogs already in use [8]. By far, the number is exceeded by the United States, with them having used 58.511 dogs for research in 2019 [9]. These numbers clearly demonstrate the need for replacing animal experimentation with meaningful in vitro or in silico methods according to the 3Rs (and 4R concept) principle or at least reducing them to an absolute minimum. This leads to the improvement of the state-of-the-art in vitro methods to reduce the animal numbers used for research and minimise the pain experienced during experiments. These comprise but are not limited to the use of classical cell culture models as well as more advanced methods such as three-dimensional model systems such as tumour spheroids, organoids, organ-on-a-chip technologies, or computer-based models such as prediction methods based on artificial intelligence (AI), as previously applied to diabetes [10], cardiovascular disease [11], tuberculosis [12], and drug discovery [13]. Spheroids pose a model of compact three-dimensional cell aggregates consisting of cells at different states, e.g., proliferating, hypoxic, and quiescent, which are generated on non-adherent surfaces. These do not necessarily represent complex organ architecture on a miniature scale [14]. On the other hand, organoids are three-dimensional models of organ systems reflecting organ microanatomy. Due to their stem-cell-originating nature, organoids are usually indefinitely expandable [15,16]. Modelling different organ systems of various animals will help to replace animal experimentation in accordance with the 3Rs (and 4R concept) principle. This leads to an improved understanding of the biological principles in a broader context, as humans and different species of animals may react differently to various irritants (Figure 1). In-depth knowledge of diverse species and their organs is pivotal for research in a One Health context, taking humans, animals, and the environment into account. Thus, this review deals with the importance of organoids for today's research and provides an overview of different methods for disease modelling and highlights the limitations of organoids, differences between humans and animals and the possible future applications of organoid-based in vitro research.

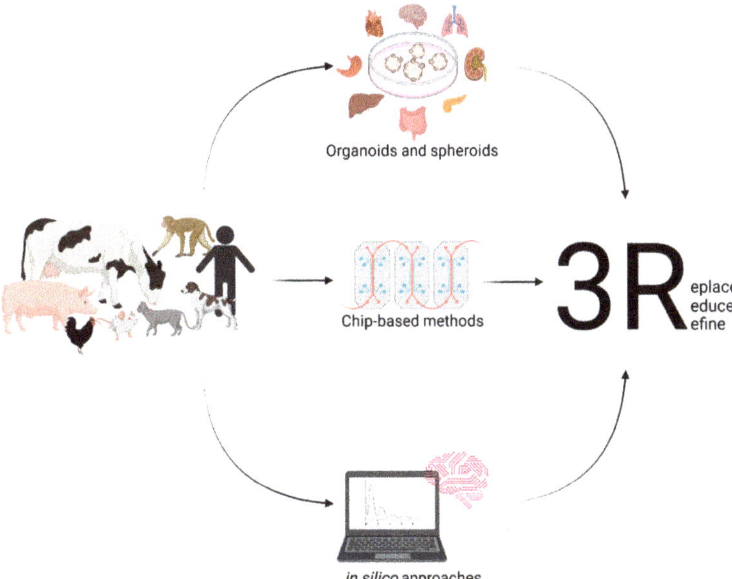

Figure 1. Setting up useful in vitro models from different animals and their various organs and the use of in silico modelling will help to replace the need for animal experimentation.

2. The Importance of Organoids for One Health

The establishment of meaningful in vitro systems to model complex diseases is very important. At the moment, the world is progressing from using classical cell culture models to more sophisticated three-dimensional models to investigate the effects of commensal or pathogenic organisms on certain cells/organs of humans, companion animals as well as farm animals. In humans, many different organs are available as organoid systems, e.g., the brain [17], retina [18], salivary gland [19], thyroid [20], lung [21], blood vessels [22] and the heart [23], mammary gland [24], stomach [25], liver [26], kidney [27], pancreas [28], intestine [29,30], fallopian tube [31], endometrium [32], bladder [33] and the prostate [34]. Many of these can be adapted to cancer organoid cultures, and some have been translated to animal organoid models. There are also very sophisticated air–liquid interface models of patient-derived cancer organoids. One of these models even includes the complex tumour microenvironment with immune cells, making it a very attractive and complex model [35]. A lot of work has been undertaken on organoids from companion animals, including the canine and feline intestine [36–39], the canine and feline liver [40,41], and canine kidney [42], bladder cancer [43], prostate cancer [44], skin [45], and thyroid tissue [46]. These companion animal models are further complemented by organoids derived from farm animals. Among them are primarily intestinal organoids from several species such as pigs, cattle, sheep, horses, and chickens [47], which have recently been reviewed more in-depth elsewhere [48]. In this context, organoids may develop towards a central model connecting the three cornerstones of the One Health concept regarding the physiological and pathophysiological interrelation of human, animal, and environmental health.

Gastrointestinal (GI) diseases do not only affect humans but also constitute a major threat to farm and companion animals and are associated with high costs to healthcare systems and animal owners. Just as in humans, conceivably lethal GI diseases also affect animals. Enteropathogenic viruses and bacteria are frequently responsible for the initiation or further impairment of GI afflictions [49–51]. There are numerous examples of the pathogenic organisms involved in the development of health problems in humans as well as animals. Several reviews have recently highlighted the importance of One Health approaches putting surveillance, monitoring, and treatment options in a broader context compared to studies investigating only one aspect of potentially zoonotic pathogens. Due to the fact that some pathogens can survive in the environment or animal products consumed by humans, the transmission routes should be examined more closely.

Especially, enteric pathogens are a major threat in a zoonotic One Health context, including parasites such as helminths [52], *Giardia duodenalis*, *Blastocystis*, and *Cryptosporidium* spp. [53], as well as bacteria such as *Clostridioides difficile* (*C. difficile*) [54–56], *Bacillus cereus* sensu lato [57], and *Salmonella* [58], which all affect humans as well as animals. Particularly, the widespread *C. difficile* has been well studied, with the faeces of animals contaminating soil and water with *C. difficile* spores, leading to the spread of the disease to other animals. Alike, the spores from infected humans show up in wastewater, highlighting the importance of *C. difficile* for the environment as well as human and veterinary medicine [59]. This is complemented by reports that animals may be important asymptomatic carriers of toxigenic *C. difficile* [60,61]. Additionally, the co-clustering of isolates from cattle and dogs with isolates from human newborns has been documented, indicating the opportunity for inter-species transmission, either directly or indirectly, via contaminated environments [62]. How food intake shapes gut health has also been reviewed many times. Especially, fermented foods have received a lot of attention because of their ability to substantially change gut microbiota composition and therefore influence physiologic as well as pathologic processes [63].

In recent years, intestinal organoids have become increasingly important in research. They do not only represent a more complex system than classical two-dimensional cell cultures, but their three-dimensional nature also allows for the long-term maintenance and differentiation of many different cell types within one dish. Despite their complexity, intestinal organoids bear the advantage of only consisting of one layer of epithelial cells, thus putting the intestinal epithelial lining at the heart of the research. Intestinal organoids are not only valuable models for the investigation of complex diseases, such as IBD [64,65], but also represent a system which makes it possible to propagate pathogens in vitro, which previously could not be cultured, such as *Cryptosporidium* [66]. Beyond that, organoids even open up opportunities for precision medicine, as any effects can be studied in a patient-specific manner. Organoids can be the missing piece in the puzzle of performing research in a One Health context (Figure 2).

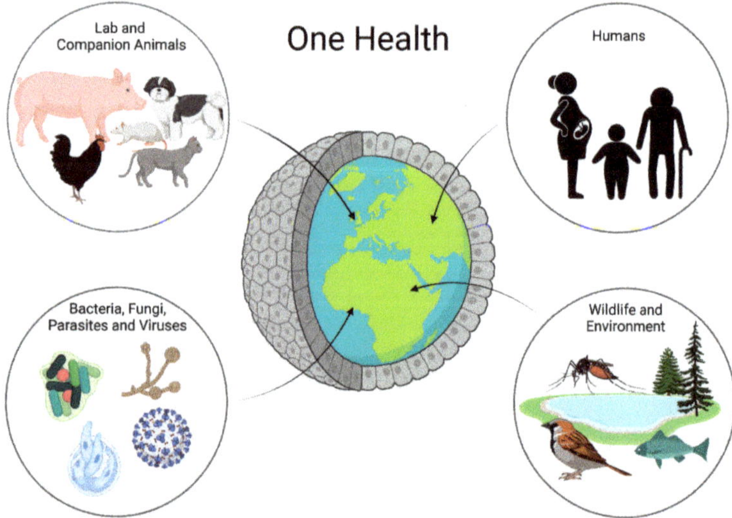

Figure 2. Organoids in One Health Research: Organoids are a possible way to work with all parts of One Health in one platform. Using organoids, one can learn about animal and human health and disease as well as interactions with the environment and bacteria, fungi, parasites, and even viruses.

3. Organoids Modelling the Intestinal Epithelium

The mammalian intestines consist of the small intestine, i.e., duodenum, jejunum and ileum, and the large intestine, i.e., caecum and colon. There are fundamental differences between the small and large intestines, ranging from distinctive cell types over different tissue architectures to different physiological functions as a whole [67,68]. While all sections of the intestine contain certain cell types, such as stem cells, enteroendocrine cells, and goblet cells, and other cell types are only present in specific parts. M-cells, for instance, are only present in the epithelium on top of immune follicles, the intraintestinal lymphatic tissue, also known as gut-associated lymphatic tissue (GALT). There they interact with microbial antigens on their apical cell surface and then present these antigens on the basolateral cell surface to immune cells, thereby initiating an immunologic response [69]. An even more prominent example is Paneth cells in intestinal crypts, where they are intermingled with stem cells and pose an indispensable part of the so-called stem cell niche. These Paneth cells can only be found in crypts of the small intestine but not the colonic epithelium [70]. In 2019, van Es et al. reported that the depletion of Paneth cells from mouse intestines is leading to the adaptation and migration of enteroendocrine cells as well as tuft cells into the crypts in order to supply the stem cell niche with essential growth factors. This may be an alternative also for species in which the existence of Paneth cells has not yet been documented, as is the case for dogs and cats [71,72].

When culturing adult-stem-cell-derived organoids, many of the aforementioned characteristics can be recapitulated in vitro, starting from a single stem cell [30]. Usually, intestinal organoids represent a polarised epithelium of several different cell types, with the basolateral cell surface presented to the outside and the microvilli-bearing apical cell surface oriented towards the lumen side [73]. In 2021, a report highlighted the importance of using organoids from different organisms when it comes to drug toxicity and not simply extrapolating existing results to other species. Anti-cancer drugs have been tested in pig, monkey, and human intestinal organoids and demonstrated differing sensitivities between all three species [74]. Interestingly, Rosselot et al. showed that intestinal organoids even follow a circadian rhythm and that mouse and human organoids react differently to *C. difficile* toxin B depending on their circadian phase, which introduces a whole new level of complexity [75].

Standard intestinal organoids can also be used to model inflammatory bowel diseases. One study shows that human Crohn's Disease (CD) patients have increased interleukin-28A (IL-28A) plasma levels, and organoids were used to model this system and its effects. When they applied IL-28A to human intestinal organoids, their barrier integrity was disrupted in a JAK-STAT-pathway-dependent manner, possibly modelling an important process in CD pathogenesis, as an impaired intestinal barrier is one major aspect of CD. In veterinary science, organoids recently helped to overcome the problem of not being able to propagate serotype I feline coronaviruses (FCoVs). Making this possible now allows for an in-depth functional analysis of the pathogenesis of feline infectious peritonitis and possible treatments [39].

However, to study gastrointestinal diseases using intestinal organoids, many applications depend on the ability to gain access to the apical cell surface on the inside of the organoids, which poses a major hurdle in disease modelling. In order to make the apical cell surface more accessible, several methods have been developed over the last few years:

3.1. Microinjection

Microinjection is a rather laborious method to gain access to the apical cell surface. It may require a lot of training of the experimenter and is not feasible for large-scale screening approaches. However, it is a well suitable method for studying host–microbe interactions. Hill et al. have established a microinjection approach using human intestinal organoids to study the host microbe interactions of non-pathogenic *Escherichia coli* (*E. coli*). Microinjected *E. coli* were able to colonise the intestinal epithelium and establish a stable interaction between microbes and the host cells. This interaction was characterised by pronounced changes in the transcriptomic profile, epithelial proliferation, improved barrier integrity, and many more physiologically relevant adaptations [76]. This study was fundamental to a recent follow-up study by Abuaita and colleagues. They used different Salmonella serovars to find out whether known in vivo immune reactions could be modelled in vitro using intestinal organoids. As expected, different serovars led to different levels of immune responses, with the *Salmonella enterica* serovar Typhi infection leading to the weakest response, which is in accordance with its need to induce a weak host response in order to systemically infect the host. Additionally, many transcriptomic alterations induced by the three tested serovars were noticeable, which again highlights the usefulness of organoids for exploring new signalling pathways targetable in disease treatment and prevention [77]. This method cannot only be used to study bacteria–host interactions but is also applicable for the investigation of small parasitic organisms with the host epithelium, as shown by a model using *Cryptosporidium parvum* microinjection for infection and subsequent oocyst harvest [78]. However, as shown elegantly by the microinjection of *Lactobacilli*, when using pluripotent stem cell-derived intestinal organoids, one has to be cautious since the maturation stage of organoids can be increased using different culture media and can drastically influence the success of *Lactobacillus* colonisation of the organoid epithelium [79].

3.2. Apical-Out Organoids

Another useful method to gain access to the apical surface of the epithelium whilst not disrupting the three-dimensional structure of the organoids is the generation of so-called "apical-out organoids". This method was first described in 2019 in a human enteroid model that appealingly demonstrated the importance of turning organoids inside out, providing the example of two different infection models. The rather simple method relies solely on the fact that organoids reverse their polarity once they are cultured floating in the culture medium without being embedded in an extracellular matrix [73]. A slightly modified version of this method was recently provided as a step-by-step protocol [80]. While *Salmonella* were used again to show their potential to infect the apical cell surface, organoids needed to be in their standard basal-out configuration to be infected by *Listeria monocytogenes* [73]. This study also used insights from research from 1994, which already used a three-dimensional model of canine cells (Madin–Darby canine kidney cells), which indicated an inherent function for beta 1 integrin in cell polarity [81]. Co et al. then demonstrated the importance of beta 1 integrin also for enteroid polarity, as applying a beta 1 integrin blocking antibody showed the same effects as the removal of extracellular matrix and led to organoid polarity reversal [73]. This is just one of many examples where first indications from animal cells give rise to novel approaches in more frequently used model systems, clearly highlighting the importance of interdisciplinary research.

Interestingly, intestinal apical-out organoids have been explored intensely in different animal species but not so much in mouse and human organoids over the last few years. A study using pig organoids analysed their potential to form apical-out organoids and set up functional readouts as fatty acid uptake and barrier integrity analyses [82]. There are several groups working on apical-out organoids for disease modelling in different contexts. For instance, porcine apical-out organoids were employed as an in vitro system to analyse the possibility of infecting organoids with the swine-enteric transmissible gastroenteritis virus (TGEV) and their immune response elicited by this virus [83]. Apart from using sheep gastrointestinal basal-out organoids for investigating the host–parasite interaction of *Teladorsagia circumcincta* with the epithelium, ovine apical-out organoids have also been tested in co-culture with *Salmonella enterica* serovar Typhimurium [84]. Meanwhile, chicken apical-out organoids have also proven to be a valuable tool for analysing different host–pathogen interactions. The protozoan *Eimeria tenella* can infect avian apical-out organoids just as well as the influenza A virus. This study also used *Salmonella enterica* as a bacterial example for avian gastrointestinal infection [85]. This is probably due to *Salmonella* being a facultative anaerobic bacterium relevant for the intestinal epithelium, as using obligate anaerobes such as *Fusobacterium* or *Clostridia* with low oxygen tolerance would not be compatible with the cultivation of apical-out organoids. However, especially these bacterial genera might be of interest for research in the future as they are frequently implicated in gastrointestinal diseases such as colon cancer and ulcerated regions in the human intestine [86] as well as in acute haemorrhagic diarrhoea syndrome (AHDS) in dogs [87]. Specifically, *Clostridia* might bear the risk of being a zoonotic bacterium in the context of One Health, as already outlined above. However, some *Clostridia*, such as *Clostridium hiranonis*, may also exert positive effects on gastrointestinal health. It was reduced in the dysbiosis index of dogs with chronic enteropathy in general [88] and, more specifically, in dogs with IBD [89]. *Clostridium hiranonis* possesses the ability to metabolise bile acids, and the dysregulation of bile acids has been associated with human IBD [90] and in dog enteropathies [91,92], which could potentially be modelled in vitro in the future.

3.3. Organoid-Derived Monolayers

Since handling organoids can be tedious and many standard assays are not adapted to three-dimensional structures, great efforts have been made during the last few years to find a way to reduce the complexity of the organoid system while simultaneously maintaining as many advantages of the organoids as possible. One way to do so is the use of organoid-derived monolayers (ODMs), which serve as a model of an intact intestinal barrier [93]. Classical two-dimensional in vitro models such as the Caco-2 cell system are most frequently used for drug screening and basic research. However, Caco-2 cells are derived from cancer cells and lack some possibly important epithelial enzymes and transporters [94]. Organoid-derived monolayers can be analysed, such as standard two-dimensional cell cultures, and have the advantage that they consist of several different cell types. Additionally, you can prepare them from whatever species you are able to culture organoids from. ODMs can thus be of great help in exploring transepithelial transport of nutrients, damage to the epithelial barrier integrity or similar approaches.

Human intestinal organoid-derived monolayers have been previously used as a model for pharmacokinetics and toxicology. In two-dimensional monolayers, the drug-metabolising enzyme CYP3A4 and several transporters were upregulated compared to Caco-2 cells and intestinal epithelial cells derived from induced pluripotent stem cells and resembled the adult duodenum more closely. These papers also showed the existence of all major differentiated cell types (enterocytes, enteroendocrine cells, goblet cells, and Paneth cells) in these monolayers, while stem cells decreased over time [95,96]. Another study demonstrated the ability of differentiated monolayers to actively transport ions (sodium, potassium, and chloride) and that the hormones serotonin and GLP-1 are produced by epithelial cells [97]. The functional transport of chloride ions has also been shown in porcine organoid-derived monolayers consisting of enterocytes, goblet cells and enteroendocrine cells [98]. Likewise, canine organoids have been used to create transwell-based ODMs that build up a functional barrier that can be used for dog gut research [99]. Aside from functional characteristics, ODMs have also been developed much further as co-culture models with bacteria. Mayorgas et al. used human ODMs as a proxy for the infection with invasive *E. coli* [100]. A slightly more complex system has been introduced by Sasaki et al. [101]. Here, ODMs are produced on transwell inserts. Once these monolayers reach confluence, the transwell chamber is sealed by a butyl rubber plug. This leads to an anaerobic apical chamber, while the bottom chamber, which is in contact with the basolateral cell surface, still has continuous access to oxygen. To test the so-called Intestinal Hemi-Anaerobic Co-culture System (iHACS), the apical chambers were challenged with four different anaerobic bacterial strains (*Bifidobacterium adolescentis*, *Bacteroides fragilis*, *Clostridium butyricum*, and *Akkermansia muciniphila*) and showed the possibility for bacterial survival and propagation over five days of co-culture. This complex example showcases the possibility of using monolayers for bacterial co-culture and possible invasion analyses or co-cultures with commensal bacteria.

4. Limitations

Despite offering outstanding new possibilities for research, e.g., in vitro analysis of physiologic processes, disease modelling and genetic manipulation, organoids also confront researchers with some difficulties and limitations. For example, imaging approaches are more difficult to carry out compared to classical 2D cell culture approaches due to the three-dimensional structure of organoids and the resulting thickness of the sample in whole-mount stainings. However, imaging technology is gradually becoming better, and as confocal laser scanning microscopy (CLSM) is available virtually everywhere, this problem is also becoming smaller. New imaging techniques, such as spinning disk confocal imaging, offer new possibilities, especially for live-cell imaging, as the imaging process itself becomes much faster than in classical CLSM [102]. To overcome the problem of imaging depth, several different tissue-clearing methods have been developed [103,104]. These protocols enable the optical clearing of whole organoids or in vitro 3D tissues for considerably improved clarity and easier imaging of whole-mount samples.

Another difficulty is the batch-to-batch variations of the conditioned media, media supplements, and inhibitors. As organoids require a complex mix of stimulatory and inhibitory components in the medium to simulate the stem cell niche and/or provide the right cues for cell differentiation, all these supplements need to be of high and standardised quality. Chemically synthesised molecules tend not to be a problem as they are of extremely high and pure quality and undergo the appropriate quality checks. However, many labs rely on self-produced conditioned media as supplements for organoid culture media. These conditioned media can substantially vary, depending on the production process, hence skewing the results and hindering reproducibility, even though a report shows that the conditioned media production appears to be reproducible from batch to batch across several different laboratories [105]. To overcome this problem, more cost-intensive, specially designed so-called "surrogate" proteins can be used at defined concentrations [106–108]. Another problem arising from organoid culture media is the variation of media composition between laboratories. While some laboratories still rely on the original culture media [29,30,36,47], certain media are available for driving organoid differentiation while simultaneously ensuring a certain level of stemness in the same dish in human and canine intestinal organoids [37,109] or promoting full differentiation, for example in liver organoids [41,110]. These differences require a highly transparent methodology to ensure reproducibility and highlight the need for standardisation, as outlined by Gabriel et al. [38].

Because organoids are a very complex 3D model, it can be hard to precisely identify the specific factors that provoke the observed changes. Organoids receive various cues from media components and also the extracellular matrix they are grown in that need to be integrated into a physiologic context within the organoid. Therefore, small deviations from standard parameters can provoke drastic changes in the organoids. Matrix proteins are a major part of this dilemma. Matrigel still is the most prominently used extracellular matrix for the cultivation of organoids. However, Matrigel and comparable alternatives are basement membrane extracts derived from Engelbreth–Holm–Swarm sarcomas from mice. Thus, using organoids for research is not necessarily reducing the need for animal experimentation, as large quantities of mice are needed to produce the required extracellular matrix. Additionally, since Matrigel is derived from animals, quality control is rather difficult, no standardised mixture of components is defined, and batch-to-batch variability can be problematic [111]. For the last few years, a lot of money has been invested to produce non-mammalian or even animal-free alternatives to Matrigel. These include but are not limited to peptide-based hydrogels [112], a highly tuneable polysaccharide-based synthetic hydrogel [113], plant-based nanofibrillar cellulose [114,115] and collagen derived from jellyfish [116]. Despite the availability of these mammalian-free matrices, many people have not adopted them in their labs because of time- and cost-intensive procedures.

5. Outlook

Organoids have one major advantage for future research, which is the opportunity to study diseases in a patient-specific manner. Organoids can be established from small biopsies of tissues and expanded in vitro for experimental needs. These can be used for patient-specific drug-screening approaches or the analysis of genetic risk factors for certain diseases [117,118]. However, increasing individuality inevitably leads to less standardised models. In future research, patient-specific models will always have to be analysed with reference to a specific benchmark, i.e., a standardised control sample. Companies such as HUB Organoids in the Netherlands are building large biobanks for human organoids where researchers can apply for licencing agreements in order to use specific healthy or diseased organoids for certain projects [119]. However, for animal research, no such biobank is available, most probably because the research community working with animal-derived organoids is only starting to develop and is still too small.

Organoids may also be used for therapeutic approaches in the future. Kruitwagen et al. showed hepatocyte transplantation in canines with the possibility of curing copper storage disease caused by a mutation of the copper metabolism-domain-containing 1 (COMMD1) gene. Liver organoids were established from COMMD1-deficient dogs, genetically modified to restore COMMD1 function and, subsequently, transplanted back into the dogs of origin. Despite the engraftment percentages being low, the transplanted cells were able to survive for more than two years after transplantation [41]. Sampaziotis et al. used cholangiocyte organoids for direct bile duct regeneration. Importantly, delivering organoids to regenerate damaged bile ducts was demonstrated in mice and humans. While live mice were injected with organoids, normothermic machine perfusion (NMP) was used for human studies, which allows for the physiological perfusion of organs ex vivo. This makes it much easier to control the environmental influences and analyse different parameters. Perfusing these livers with human cholangiocyte organoid cells led to successful engraftment in human bile ducts, demonstrating the proof of principle, that organoid transplantation is feasible in mice as well as humans in the future [120].

Another human/mouse study generated human islet-like organoids to pave the way for diabetes treatment via pancreas islet transplantation. Human induced pluripotent stem cells were differentiated to human islet-like organoids (HILOs) expressing insulin and subsequently transplanted into diabetic mice. These pancreatic island cells could re-establish glucose homeostasis and may be more effective than conventional glucose monitoring and insulin injections as island cells can take on multiple additional roles [121]. In 2020, Meran et al. used organoids from child patients with intestinal failure and expanded them in vitro. They subsequently seeded organoid cells on decellularised small and large intestinal matrices and transplanted these scaffolds into mouse kidney capsules or subcutaneous pockets. These grafts formed luminal structures after transplantation and demonstrated the possibility of re-populating decellularised scaffolds with in vitro expanded cells for transplantation [122]. Similarly, Sugimoto et al. grafted small intestinal organoids onto the surface of the colon. These grafts started to form villus structures and ameliorated the symptoms of small intestinal short bowel syndrome in rats by structurally replacing colon epithelium with small intestinal cells [123].

Scientists are making efforts worldwide to lift organoid technology to the next level, explore new model systems, and generate more meaningful and complex models that mimic in vivo physiology even closer. Recent improvements include organoids with increased complexity, as by Koike et al., who modelled endoderm organogenesis at the foregut–midgut boundary by differentiating human induced pluripotent stem cells. Using this model, they created organoids containing cells from the liver, bile ducts, pancreas, and duodenum organised in one single organoid [124].

Other approaches for combining several organs in one model system mostly go towards using organ-on-a-chip applications. Such a chip incorporates one or many microchannels to connect the chip with a capillary system. This allows for the injection of fluids in a controlled manner that also supports directed flow of a medium, as, for instance, the intestine also experiences in vivo. Chip technologies can also be upgraded with microsensors and pose an extremely complex system [125]. The advances in organoid technology, microfabrication, cell engineering, and imaging technologies have led organ-on-a-chip to become an innovative technology capable of reproducing physiological cell behaviours in vitro [126]. However, the use of species other than mice and humans for chip-based technologies is very limited, with only two reports. The combination of multiple interconnected organ-on-a-chip systems in a single platform is now bringing this technology to the next level that aims to emulate an entire biological entity that is seldom limited to a single organ termed "body-on-a-chip" [127,128]. Despite these new advancements, there is still a lot to learn from organoids themselves and together with organoids, organ-on-a-chip technologies will take science a step further to replace animal experimentation. The comparison of in vitro organ models from various species will also guide new ways to explore the interconnection of humans, animals, and the environment in the context of One Health and help to explore new treatment strategies for various diseases.

6. Conclusions

Organoids are a promising tool for modern research. The continuous developments of new technologies, co-cultures, and organoid manipulation techniques lead to constant advancement in the field and open up new possibilities for treatments. Organoids of the liver, pancreas, stomach, and intestine are currently the in vitro method of choice for gastrointestinal research. Learning from mouse and human studies, many organoid systems have been adapted to other species. People are just beginning to explore these organoids and their differences from well-characterised models. Animal organoids pose a valuable in vitro method to model and study diseases, test environmental irritants on different organ systems of various species and develop new therapeutics. Keeping this in mind, organoids are becoming increasingly important in regard to the 3Rs (and 4R concept) and One Health research.

Author Contributions: Conceptualisation, G.C., B.P. and I.A.B.; Writing—Original Draft Preparation, G.C. and B.P.; Writing—Review & Editing, G.C., B.P. and I.A.B.; Visualisation, G.C.; Supervision, I.A.B.; Funding Acquisition, G.C. and I.A.B. All authors have read and agreed to the published version of the manuscript.

Funding: G.C. was funded by the Austrian Academy of Sciences (ÖAW), DOC fellowship grant number 26349. The APC was funded in part by Vetmeduni. Open Access Funding by the University of Veterinary Medicine Vienna.

Institutional Review Board Statement: Not Applicable.

Informed Consent Statement: Not Applicable.

Data Availability Statement: Not Applicable.

Acknowledgments: All figures were created with Biorender.com.

Conflicts of Interest: The authors declare no conflict of interest.

References

1. Anholt, M.; Barkema, H. What Is One Health? *Can Vet. J.* **2021**, *62*, 641–644. [PubMed]
2. Nerpel, A.; Yang, L.; Sorger, J.; Käsbohrer, A.; Walzer, C.; Desvars-Larrive, A. SARS-ANI: A Global Open Access Dataset of Reported SARS-CoV-2 Events in Animals. *Sci Data* **2022**, *9*, 438. [CrossRef] [PubMed]
3. Amman, F.; Markt, R.; Endler, L.; Hupfauf, S.; Agerer, B.; Schedl, A.; Richter, L.; Zechmeister, M.; Bicher, M.; Heiler, G.; et al. Viral Variant-Resolved Wastewater Surveillance of SARS-CoV-2 at National Scale. *Nat. Biotechnol.* **2022**. [CrossRef] [PubMed]
4. Russell, W.; Burch, R. The Principles of Humane Experimental Technique. In *John Hopkins Bloomberg School of Public Health*; 1959. Available online: https://caat.jhsph.edu/principles/the-principles-of-humane-experimental-technique (accessed on 9 September 2022).
5. Office of the Law Revision Counsel. *Title 7 of the Code of Laws of the United States of America, Chapter 54: Transportation, Sale, and Handling of Certain Animals*; US Congress. Available online: http://uscode.house.gov/view.xhtml?path=/prelim@title7/chapter54&edition=prelim (accessed on 9 September 2022).

6. European Parliament; European Council. *Directive 2010/63/EU*; European Union, 2010. Available online: https://norecopa.no/legislation/eu-directive-201063 (accessed on 9 September 2022).
7. Max Planck Society for the Advancement of Science, e.V. Reduce, Refine, Replace–Responsibility. Available online: https://www.mpg.de/10973438/4rs (accessed on 4 July 2022).
8. European Commission. *Summary Report on the Statistics on the Use of Animals for Scientific Purposes in the Member States of the European Union and Norway in 2018*; Brussels, 2021. Available online: https://www.google.com.hk/url?sa=t&rct=j&q=&esrc=s&source=web&cd=&ved=2ahUKEwj2nInjn536AhVCumMGHXqaCcEQFnoECA0QAQ&url=https%3A%2F%2Fec.europa.eu%2Fenvironment%2Fchemicals%2Flab_animals%2Fpdf%2FSWD_%2520part_A_and_B.pdf&usg=AOvVaw0opxqbviYJ0WzPyGdJUFKo (accessed on 9 September 2022).
9. United States Department of Agriculture. *Annual Report Animal Usage by Fiscal Year*. Available online: https://Www.Aphis.Usda.Gov/Animal_welfare/Downloads/Reports/Fy19-Summary-Report-Column-F.Pdf (accessed on 4 July 2022).
10. Ellahham, S. Artificial Intelligence: The Future for Diabetes Care. *Am. J. Med.* **2020**, *133*, 895–900. [CrossRef] [PubMed]
11. Mohd Faizal, A.S.; Thevarajah, T.M.; Khor, S.M.; Chang, S.W. A Review of Risk Prediction Models in Cardiovascular Disease: Conventional Approach vs. Artificial Intelligent Approach. *Comput Methods Programs Biomed* **2021**, *207*, 106190. [CrossRef] [PubMed]
12. Peetluk, L.S.; Ridolfi, F.M.; Rebeiro, P.F.; Liu, D.; Rolla, V.C.; Sterling, T.R. Systematic Review of Prediction Models for Pulmonary Tuberculosis Treatment Outcomes in Adults. *BMJ Open* **2021**, *11*, e044687. [CrossRef] [PubMed]
13. Jiménez-Luna, J.; Grisoni, F.; Weskamp, N.; Schneider, G. Artificial Intelligence in Drug Discovery: Recent Advances and Future Perspectives. *Expert Opin. Drug Discov.* **2021**, *16*, 949–959. [CrossRef] [PubMed]
14. Nath, S.; Devi, G.R. Three-Dimensional Culture Systems in Cancer Research: Focus on Tumor Spheroid Model. *Pharmacol. Ther.* **2016**, *163*, 94–108. [CrossRef]
15. Lancaster, M.A.; Knoblich, J.A. Organogenesis in a Dish: Modeling Development and Disease Using Organoid Technologies. *Science* **2014**, *345*, 1247125. [CrossRef]
16. Willyard, C. Rise of the Organoids. *Nature* **2015**, *523*, 1–10.
17. Lancaster, M.A.; Renner, M.; Martin, C.A.; Wenzel, D.; Bicknell, L.S.; Hurles, M.E.; Homfray, T.; Penninger, J.M.; Jackson, A.P.; Knoblich, J.A. Cerebral Organoids Model Human Brain Development and Microcephaly. *Nature* **2013**, *501*, 373–379. [CrossRef] [PubMed]
18. Cowan, C.S.; Renner, M.; de Gennaro, M.; Gross-Scherf, B.; Goldblum, D.; Hou, Y.; Munz, M.; Rodrigues, T.M.; Krol, J.; Szikra, T.; et al. Cell Types of the Human Retina and Its Organoids at Single-Cell Resolution. *Cell* **2020**, *182*, 1623–1640. [CrossRef] [PubMed]
19. Jeong, S.Y.; Choi, W.H.; Jeon, S.G.; Lee, S.; Park, J.M.; Park, M.; Lee, H.; Lew, H.; Yoo, J. Establishment of Functional Epithelial Organoids from Human Lacrimal Glands. *Stem Cell Res. Ther.* **2021**, *12*, 247. [CrossRef] [PubMed]
20. Ogundipe, V.M.L.; Groen, A.H.; Hosper, N.; Nagle, P.W.K.; Hess, J.; Faber, H.; Jellema, A.L.; Baanstra, M.; Links, T.P.; Unger, K.; et al. Generation and Differentiation of Adult Tissue-Derived Human Thyroid Organoids. *Stem Cell Rep.* **2021**, *16*, 913–925. [CrossRef] [PubMed]
21. Sachs, N.; Papaspyropoulos, A.; Zomer-van Ommen, D.D.; Heo, I.; Böttinger, L.; Klay, D.; Weeber, F.; Huelsz-Prince, G.; Iakobachvili, N.; Amatngalim, G.D.; et al. Long-term Expanding Human Airway Organoids for Disease Modeling. *EMBO J.* **2019**, *38*, e100300. [CrossRef]
22. Wimmer, R.A.; Leopoldi, A.; Aichinger, M.; Wick, N.; Hantusch, B.; Novatchkova, M.; Taubenschmid, J.; Hämmerle, M.; Esk, C.; Bagley, J.A.; et al. Human Blood Vessel Organoids as a Model of Diabetic Vasculopathy. *Nature* **2019**, *565*, 505–510. [CrossRef] [PubMed]
23. Lewis-Israeli, Y.R.; Wasserman, A.H.; Gabalski, M.A.; Volmert, B.D.; Ming, Y.; Ball, K.A.; Yang, W.; Zou, J.; Ni, G.; Pajares, N.; et al. Self-Assembling Human Heart Organoids for the Modeling of Cardiac Development and Congenital Heart Disease. *Nat. Commun.* **2021**, *12*, 5142. [CrossRef]
24. Dekkers, J.F.; van Vliet, E.J.; Sachs, N.; Rosenbluth, J.M.; Kopper, O.; Rebel, H.G.; Wehrens, E.J.; Piani, C.; Visvader, J.E.; Verissimo, C.S.; et al. Long-Term Culture, Genetic Manipulation and Xenotransplantation of Human Normal and Breast Cancer Organoids. *Nat. Protoc.* **2021**, *16*, 1936–1965. [CrossRef]
25. Bartfeld, S.; Bayram, T.; van de Wetering, M.; Huch, M.; Begthel, H.; Kujala, P.; Vries, R.; Peters, P.J.; Clevers, H. In Vitro Expansion of Human Gastric Epithelial Stem Cells and Their Responses to Bacterial Infection. *Gastroenterology* **2015**, *148*, 126–136.e6. [CrossRef] [PubMed]
26. Huch, M.; Gehart, H.; van Boxtel, R.; Hamer, K.; Blokzijl, F.; Verstegen, M.M.A.; Ellis, E.; van Wenum, M.; Fuchs, S.A.; de Ligt, J.; et al. Long-Term Culture of Genome-Stable Bipotent Stem Cells from Adult Human Liver. *Cell* **2015**, *160*, 299–312. [CrossRef] [PubMed]
27. Takasato, M.; Er, P.X.; Chiu, H.S.; Maier, B.; Baillie, G.J.; Ferguson, C.; Parton, R.G.; Wolvetang, E.J.; Roost, M.S.; de Sousa Lopes, S.M.C.; et al. Kidney Organoids from Human IPS Cells Contain Multiple Lineages and Model Human Nephrogenesis. *Nature* **2015**, *526*, 564–568. [CrossRef] [PubMed]
28. Broutier, L.; Andersson-Rolf, A.; Hindley, C.J.; Boj, S.F.; Clevers, H.; Koo, B.K.; Huch, M. Culture and Establishment of Self-Renewing Human and Mouse Adult Liver and Pancreas 3D Organoids and Their Genetic Manipulation. *Nat. Protoc.* **2016**, *11*, 1724–1743. [CrossRef] [PubMed]

29. Sato, T.; Stange, D.E.; Ferrante, M.; Vries, R.G.J.; van Es, J.H.; van den Brink, S.; van Houdt, W.J.; Pronk, A.; van Gorp, J.; Siersema, P.D.; et al. Long-Term Expansion of Epithelial Organoids from Human Colon, Adenoma, Adenocarcinoma, and Barrett's Epithelium. *Gastroenterology* **2011**, *141*, 1762–1772. [CrossRef] [PubMed]

30. Sato, T.; Vries, R.G.; Snippert, H.J.; van de Wetering, M.; Barker, N.; Stange, D.E.; van Es, J.H.; Abo, A.; Kujala, P.; Peters, P.J.; et al. Single Lgr5 Stem Cells Build Crypt-Villus Structures in Vitro without a Mesenchymal Niche. *Nature* **2009**, *459*, 262–265. [CrossRef]

31. Chang, Y.H.; Chu, T.Y.; Ding, D.C. Human Fallopian Tube Epithelial Cells Exhibit Stemness Features, Self-Renewal Capacity, and Wnt-Related Organoid Formation. *J Biomed Sci* **2020**, *27*, 32. [CrossRef] [PubMed]

32. Boretto, M.; Cox, B.; Noben, M.; Hendriks, N.; Fassbender, A.; Roose, H.; Amant, F.; Timmerman, D.; Tomassetti, C.; Vanhie, A.; et al. Development of Organoids from Mouse and Human Endometrium Showing Endometrial Epithelium Physiology and Long-Term Expandability. *Development (Cambridge)* **2017**, *144*, 1775–1786. [CrossRef] [PubMed]

33. Kim, E.; Choi, S.; Kang, B.; Kong, J.; Kim, Y.; Yoon, W.H.; Lee, H.-R.; Kim, S.; Kim, H.-M.; Lee, H.; et al. Creation of Bladder Assembloids Mimicking Tissue Regeneration and Cancer. *Nature* **2020**, *588*, 664–669. [CrossRef]

34. Drost, J.; Karthaus, W.R.; Gao, D.; Driehuis, E.; Sawyers, C.L.; Chen, Y.; Clevers, H. Organoid Culture Systems for Prostate Epithelial and Cancer Tissue. *Nat. Protoc.* **2016**, *11*, 347–358. [CrossRef]

35. Neal, J.T.; Li, X.; Zhu, J.; Giangarra, V.; Grzeskowiak, C.L.; Ju, J.; Liu, I.H.; Chiou, S.H.; Salahudeen, A.A.; Smith, A.R.; et al. Organoid Modeling of the Tumor Immune Microenvironment. *Cell* **2018**, *175*, 1972–1988. [CrossRef] [PubMed]

36. Chandra, L.; Borcherding, D.C.; Kingsbury, D.; Atherly, T.; Ambrosini, Y.M.; Bourgois-Mochel, A.; Yuan, W.; Kimber, M.; Qi, Y.; Wang, Q.; et al. Derivation of Adult Canine Intestinal Organoids for Translational Research in Gastroenterology. *BMC Biol.* **2019**, *17*, 1–21. [CrossRef] [PubMed]

37. Kramer, N.; Pratscher, B.; Meneses, A.M.C.; Tschulenk, W.; Walter, I.; Swoboda, A.; Kruitwagen, H.S.; Schneeberger, K.; Penning, L.C.; Spee, B.; et al. Generation of Differentiating and Long-Living Intestinal Organoids Reflecting the Cellular Diversity of Canine Intestine. *Cells* **2020**, *9*, 822. [CrossRef] [PubMed]

38. Gabriel, V.; Zdyrski, C.; Sahoo, D.K.; Dao, K.; Bourgois-Mochel, A.; Kopper, J.; Zeng, X.L.; Estes, M.K.; Mochel, J.P.; Allenspach, K. Standardization and Maintenance of 3D Canine Hepatic and Intestinal Organoid Cultures for Use in Biomedical Research. *J. Vis. Exp.* **2022**, *2022*. [CrossRef]

39. Tekes, G.; Ehmann, R.; Boulant, S.; Stanifer, M.L. Development of Feline Ileum- and Colon-Derived Organoids and Their Potential Use to Support Feline Coronavirus Infection. *Cells* **2020**, *9*, 2085. [CrossRef]

40. Kruitwagen, H.S.; Oosterhoff, L.A.; Vernooij, I.G.W.H.; Schrall, I.M.; van Wolferen, M.E.; Bannink, F.; Roesch, C.; van Uden, L.; Molenaar, M.R.; Helms, J.B.; et al. Long-Term Adult Feline Liver Organoid Cultures for Disease Modeling of Hepatic Steatosis. *Stem Cell Rep.* **2017**, *8*, 822–830. [CrossRef]

41. Kruitwagen, H.S.; Oosterhoff, L.A.; van Wolferen, M.E.; Chen, C.; Nantasanti Assawarachan, S.; Schneeberger, K.; Kummeling, A.; van Straten, G.; Akkerdaas, I.C.; Vinke, C.R.; et al. Long-Term Survival of Transplanted Autologous Canine Liver Organoids in a COMMD1-Deficient Dog Model of Metabolic Liver Disease. *Cells* **2020**, *9*, 410. [CrossRef]

42. Chen, T.C.; Neupane, M.; Chien, S.J.; Chuang, F.R.; Crawford, R.B.; Kaminski, N.E.; Chang, C.C. Characterization of Adult Canine Kidney Epithelial Stem Cells That Give Rise to Dome-Forming Tubular Cells. *Stem Cells Dev.* **2019**, *28*, 1424–1433. [CrossRef]

43. Elbadawy, M.; Usui, T.; Mori, T.; Tsunedomi, R.; Hazama, S.; Nabeta, R.; Uchide, T.; Fukushima, R.; Yoshida, T.; Shibutani, M.; et al. Establishment of a Novel Experimental Model for Muscle-Invasive Bladder Cancer Using a Dog Bladder Cancer Organoid Culture. *Cancer Sci.* **2019**, *110*, 2806–2821. [CrossRef]

44. Usui, T.; Sakurai, M.; Nishikawa, S.; Umata, K.; Nemoto, Y.; Haraguchi, T.; Itamoto, K.; Mizuno, T.; Noguchi, S.; Mori, T.; et al. Establishment of a Dog Primary Prostate Cancer Organoid Using the Urine Cancer Stem Cells. *Cancer Sci.* **2017**, *108*, 2383–2392. [CrossRef]

45. Wiener, D.J.; Basak, O.; Asra, P.; Boonekamp, K.E.; Kretzschmar, K.; Papaspyropoulos, A.; Clevers, H. Establishment and Characterization of a Canine Keratinocyte Organoid Culture System. *Vet. Dermatol.* **2018**, *29*, 375-e126. [CrossRef]

46. Jankovic, J.; Dettwiler, M.; Fernández, M.G.; Tièche, E.; Hahn, K.; April-Monn, S.; Dettmer, M.S.; Kessler, M.; Rottenberg, S.; Campos, M. Validation of Immunohistochemistry for Canine Proteins Involved in Thyroid Iodine Uptake and Their Expression in Canine Follicular Cell Thyroid Carcinomas (FTCs) and FTC-Derived Organoids. *Vet. Pathol.* **2021**, *58*, 1172–1180. [CrossRef]

47. Powell, R.H.; Behnke, M.S. WRN Conditioned Media Is Sufficient for in Vitro Propagation of Intestinal Organoids from Large Farm and Small Companion Animals. *Biol. Open* **2017**, *6*, 698–705. [CrossRef] [PubMed]

48. Kawasaki, M.; Goyama, T.; Tachibana, Y.; Nagao, I.; Ambrosini, Y.M. Farm and Companion Animal Organoid Models in Translational Research: A Powerful Tool to Bridge the Gap Between Mice and Humans. *Front. Med. Technol.* **2022**, *4*, 895379. [CrossRef]

49. Hoffmann, A.R.; Proctor, L.M.; Surette, M.G.; Suchodolski, J.S. The Microbiome: The Trillions of Microorganisms That Maintain Health and Cause Disease in Humans and Companion Animals. *Vet. Pathol.* **2016**, *53*, 10–21. [CrossRef] [PubMed]

50. Honneffer, J.B.; Minamoto, Y.; Suchodolski, J.S. Microbiota Alterations in Acute and Chronic Gastrointestinal Inflammation of Cats and Dogs. *World J. Gastroenterol.* **2014**, *20*, 16489–16497. [CrossRef]

51. Mondo, E.; Marliani, G.; Accorsi, P.A.; Cocchi, M.; di Leone, A. Role of Gut Microbiota in Dog and Cat's Health and Diseases. *Open Vet. J.* **2019**, *9*, 253–258. [CrossRef] [PubMed]

52. Sack, A.; Palanisamy, G.; Manuel, M.; Paulsamy, C.; Rose, A.; Kaliappan, S.P.; Ward, H.; Walson, J.L.; Halliday, K.E.; Rao Ajjampur, S.S. A One Health Approach to Defining Animal and Human Helminth Exposure Risks in a Tribal Village in Southern India. *Am. J. Trop. Med. Hyg.* **2021**, *105*, 196–203. [CrossRef] [PubMed]

53. Galán-Puchades, M.T.; Trelis, M.; Sáez-Durán, S.; Cifre, S.; Gosálvez, C.; Sanxis-Furió, J.; Pascual, J.; Bueno-Marí, R.; Franco, S.; Peracho, V.; et al. One Health Approach to Zoonotic Parasites: Molecular Detection of Intestinal Protozoans in an Urban Population of Norway Rats, Rattus Norvegicus, in Barcelona, Spain. *Pathogens* **2021**, *10*, 311. [CrossRef]

54. Usui, M. Prevalence of Clostridioides Difficile in Animals and Its Relationship with Human Infections. *Jpn. J. Chemother.* **2020**, *68*, 557–562.

55. Usui, M. One Health Approach to Clostridioides Difficile in Japan. *J. Infect. Chemother.* **2020**, *26*, 643–650. [CrossRef]

56. Finsterwalder, S.K.; Loncaric, I.; Cabal, A.; Szostak, M.P.; Barf, L.M.; Marz, M.; Allerberger, F.; Burgener, I.A.; Tichy, A.; Feßler, A.T.; et al. Dogs as Carriers of Virulent and Resistant Genotypes of Clostridioides Difficile. *Zoonoses Public Health* **2022**, *69*, 673–681. [CrossRef]

57. Ehling-Schulz, M.; Lereclus, D.; Koehler, T.M. The Bacillus Cereus Group: Bacillus Species with Pathogenic Potential. *Microbiol. Spectr.* **2019**, *7*. [CrossRef] [PubMed]

58. Stevens, M.P.; Humphrey, T.J.; Maskell, D.J. Molecular Insights into Farm Animal and Zoonotic Salmonella Infections. *Philos. Trans. R. Soc. B Biol. Sci.* **2009**, *364*. [CrossRef] [PubMed]

59. Rodriguez Diaz, C.; Seyboldt, C.; Rupnik, M. Non-Human C. Difficile Reservoirs and Sources: Animals, Food, Environment. *Adv. Exp. Med. Biol.* **2018**, *1050*, 227–243. [PubMed]

60. McLure, A.; Clements, A.C.A.; Kirk, M.; Glass, K. Modelling Diverse Sources of Clostridium Difficile in the Community: Importance of Animals, Infants and Asymptomatic Carriers. *Epidemiol. Infect.* **2019**, *147*, E152. [CrossRef] [PubMed]

61. Otten, A.M.; Reid-Smith, R.J.; Fazil, A.; Weese, J.S. Disease Transmission Model for Community-Associated Clostridium Difficile Infection. *Epidemiol. Infect.* **2010**, *138*, 907–914. [CrossRef] [PubMed]

62. Redding, L.E.; Tu, V.; Abbas, A.; Alvarez, M.; Zackular, J.P.; Gu, C.; Bushman, F.D.; Kelly, D.J.; Barnhart, D.; Lee, J.J.; et al. Genetic and Phenotypic Characteristics of Clostridium (Clostridioides) Difficile from Canine, Bovine, and Pediatric Populations. *Anaerobe* **2022**, *74*, 102539. [CrossRef]

63. Bell, V.; Ferrão, J.; Pimentel, L.; Pintado, M.; Fernandes, T. One Health, Fermented Foods, and Gut Microbiota. *Foods* **2018**, *7*, 195. [CrossRef]

64. Schulte, L.; Hohwieler, M.; Müller, M.; Klaus, J. Intestinal Organoids as a Novel Complementary Model to Dissect Inflammatory Bowel Disease. *Stem Cells Int.* **2019**, *2019*, 1–15. [CrossRef]

65. Kopper, J.J.; Iennarella-Servantez, C.; Jergens, A.E.; Sahoo, D.K.; Guillot, E.; Bourgois-Mochel, A.; Martinez, M.N.; Allenspach, K.; Mochel, J.P. Harnessing the Biology of Canine Intestinal Organoids to Heighten Understanding of Inflammatory Bowel Disease Pathogenesis and Accelerate Drug Discovery: A One Health Approach. *Front. Toxicol.* **2021**, *3*. [CrossRef]

66. Gunasekera, S.; Zahedi, A.; O'dea, M.; King, B.; Monis, P.; Thierry, B.; Carr, J.M.; Ryan, U. Organoids and Bioengineered Intestinal Models: Potential Solutions to the Cryptosporidium Culturing Dilemma. *Microorganisms* **2020**, *8*, 715. [CrossRef]

67. Bowcutt, R.; Forman, R.; Glymenaki, M.; Carding, S.R.; Else, K.J.; Cruickshank, S.M. Heterogeneity across the Murine Small and Large Intestine. *World J. Gastroenterol.* **2014**, *20*, 15216. [CrossRef] [PubMed]

68. Cramer, J.M.; Thompson, T.; Geskin, A.; Laframboise, W.; Lagasse, E. Distinct Human Stem Cell Populations in Small and Large Intestine. *PLoS ONE* **2015**, *10*, e0118792. [CrossRef] [PubMed]

69. Kobayashi, N.; Takahashi, D.; Takano, S.; Kimura, S.; Hase, K. The Roles of Peyer's Patches and Microfold Cells in the Gut Immune System: Relevance to Autoimmune Diseases. *Front. Immunol.* **2019**, *10*, 2345. [CrossRef] [PubMed]

70. Burclaff, J.; Bliton, R.J.; Breau, K.A.; Ok, M.T.; Gomez-Martinez, I.; Ranek, J.S.; Bhatt, A.P.; Purvis, J.E.; Woosley, J.T.; Magness, S.T. A Proximal-to-Distal Survey of Healthy Adult Human Small Intestine and Colon Epithelium by Single-Cell Transcriptomics. *CMGH* **2022**, *13*, 1554–1589. [CrossRef]

71. Ghoos, Y.; Vantrappen, G. The Cytochemical Localization of Lysozyme in Paneth Cell Granules. *Histochem. J.* **1971**, *3*, 175–178. [CrossRef]

72. Sheahan, D.G.; Jervis, H.R. Comparative Histochemistry of Gastrointestinal Mucosubstances. *Am. J. Anat.* **1976**, *146*, 103–131. [CrossRef]

73. Co, J.Y.; Margalef-Català, M.; Li, X.; Mah, A.T.; Kuo, C.J.; Monack, D.M.; Amieva, M.R. Controlling Epithelial Polarity: A Human Enteroid Model for Host-Pathogen Interactions. *Cell Rep.* **2019**, *26*, 2509–2520.e4. [CrossRef]

74. Li, H.; Wang, Y.; Zhang, M.; Wang, H.; Cui, A.; Zhao, J.; Ji, W.; Chen, Y.G. Establishment of Porcine and Monkey Colonic Organoids for Drug Toxicity Study. *Cell Regen.* **2021**, *10*, 32. [CrossRef]

75. Rosselot, A.E.; Park, M.; Kim, M.; Matsu-Ura, T.; Wu, G.; Flores, D.E.; Subramanian, K.R.; Lee, S.; Sundaram, N.; Broda, T.R.; et al. Ontogeny and Function of the Circadian Clock in Intestinal Organoids. *EMBO J* **2022**, *41*, e106973. [CrossRef]

76. Hill, D.R.; Huang, S.; Nagy, M.S.; Yadagiri, V.K.; Fields, C.; Mukherjee, D.; Bons, B.; Dedhia, P.H.; Chin, A.M.; Tsai, Y.H.; et al. Bacterial Colonization Stimulates a Complex Physiological Response in the Immature Human Intestinal Epithelium. *Elife* **2017**, *6*, e29132. [CrossRef]

77. Abuaita, B.H.; Lawrence, A.L.E.; Berger, R.P.; Hill, D.R.; Huang, S.; Yadagiri, V.K.; Bons, B.; Fields, C.; Wobus, C.E.; Spence, J.R.; et al. Comparative Transcriptional Profiling of the Early Host Response to Infection by Typhoidal and Non-Typhoidal Salmonella Serovars in Human Intestinal Organoids. *PLoS Pathog.* **2021**, *17*, e1009987. [CrossRef] [PubMed]

78. Dutta, D.; Heo, I.; O'connor, R. Studying Cryptosporidium Infection in 3D Tissue-Derived Human Organoid Culture Systems by Microinjection. *J. Vis. Exp.* **2019**, *2019*, e59610. [CrossRef] [PubMed]

79. Son, Y.S.; Ki, S.J.; Thanavel, R.; Kim, J.J.; Lee, M.O.; Kim, J.; Jung, C.R.; Han, T.S.; Cho, H.S.; Ryu, C.M.; et al. Maturation of Human Intestinal Organoids in Vitro Facilitates Colonization by Commensal Lactobacilli by Reinforcing the Mucus Layer. *FASEB J.* **2020**, *34*, 9899–9910. [CrossRef]

80. Stroulios, G.; Stahl, M.; Elstone, F.; Chang, W.; Louis, S.; Eaves, A.; Simmini, S.; Conder, R.K. Culture Methods to Study Apical-Specific Interactions Using Intestinal Organoid Models. *J. Vis. Exp.* **2021**, *2021*, e62330. [CrossRef] [PubMed]

81. Ojakian, G.K.; Schwimmer, R. Regulation of Epithelial Cell Surface Polarity Reversal by B1 Integrins. *J. Cell Sci.* **1994**, *107 Pt 3*, 561–576.e6. [CrossRef]

82. Joo, S.S.; Gu, B.H.; Park, Y.J.; Rim, C.Y.; Kim, M.J.; Kim, S.H.; Cho, J.H.; Kim, H.B.; Kim, M. Porcine Intestinal Apical-Out Organoid Model for Gut Function Study. *Animals* **2022**, *12*, 372. [CrossRef]

83. Li, Y.; Yang, N.; Chen, J.; Huang, X.; Zhang, N.; Yang, S.; Liu, G.; Liu, G. Next-Generation Porcine Intestinal Organoids: An Apical-Out Organoid Model for Swine Enteric Virus Infection and Immune Response Investigations. *J. Virol.* **2020**, *94*, e01006–e01020. [CrossRef]

84. Smith, D.; Price, D.R.G.; Burrells, A.; Faber, M.N.; Hildersley, K.A.; Chintoan-Uta, C.; Chapuis, A.F.; Stevens, M.; Stevenson, K.; Burgess, S.T.G.; et al. The Development of Ovine Gastric and Intestinal Organoids for Studying Ruminant Host-Pathogen Interactions. *Front Cell Infect Microbiol.* **2021**, *11*. [CrossRef]

85. Nash, T.J.; Morris, K.M.; Mabbott, N.A.; Vervelde, L. Inside-out Chicken Enteroids with Leukocyte Component as a Model to Study Host–Pathogen Interactions. *Commun. Biol.* **2021**, *4*, 377. [CrossRef]

86. Bullman, S.; Pedamallu, C.S.; Sicinska, E.; Clancy, T.E.; Zhang, X.; Cai, D.; Neuberg, D.; Huang, K.; Guevara, F.; Nelson, T.; et al. Analysis of Fusobacterium Persistence and Antibiotic Response in Colorectal Cancer. *Science* **2017**, *358*, 1443–1448. [CrossRef]

87. Unterer, S.; Busch, K. Acute Hemorrhagic Diarrhea Syndrome in Dogs. *Vet. Clin. N. Am. -Small Anim. Pract.* **2021**, *51*, 79–92. [CrossRef] [PubMed]

88. AlShawaqfeh, M.K.; Wajid, B.; Minamoto, Y.; Markel, M.; Lidbury, J.A.; Steiner, J.M.; Serpedin, E.; Suchodolski, J.S. A Dysbiosis Index to Assess Microbial Changes in Fecal Samples of Dogs with Chronic Inflammatory Enteropathy. *FEMS Microbiol. Ecol.* **2017**, *93*, 1–8. [CrossRef] [PubMed]

89. Vázquez-Baeza, Y.; Hyde, E.R.; Suchodolski, J.S.; Knight, R. Dog and Human Inflammatory Bowel Disease Rely on Overlapping yet Distinct Dysbiosis Networks. *Nat. Microbiol.* **2016**, *1*, 16177. [CrossRef]

90. Duboc, H.; Rajca, S.; Rainteau, D.; Benarous, D.; Maubert, M.A.; Quervain, E.; Thomas, G.; Barbu, V.; Humbert, L.; Despras, G.; et al. Connecting Dysbiosis, Bile-Acid Dysmetabolism and Gut Inflammation in Inflammatory Bowel Diseases. *Gut* **2013**, *62*, 531–539. [CrossRef]

91. Kent, A.C.C.; Cross, G.; Taylor, D.R.; Sherwood, R.A.; Watson, P.J. Measurement of Serum 7α-Hydroxy-4-Cholesten-3-One as a Marker of Bile Acid Malabsorption in Dogs with Chronic Diarrhoea: A Pilot Study. *Vet. Rec. Open* **2016**, *3*. [CrossRef] [PubMed]

92. Guard, B.C.; Suchodolski, J.S. Horse Species Symposium: Canine Intestinal Microbiology and Metagenomics: From Phylogeny to Function. *J. Anim. Sci.* **2016**, *94*, 2247–2261. [CrossRef]

93. Weiß, F.; Holthaus, D.; Kraft, M.; Klotz, C.; Schneemann, M.; Schulzke, J.D.; Krug, S.M. Human Duodenal Organoid-Derived Monolayers Serve as a Suitable Barrier Model for Duodenal Tissue. *Ann. N. Y. Acad. Sci.* **2022**. [CrossRef]

94. Sun, H.; Chow, E.C.Y.; Liu, S.; Du, Y.; Pang, K.S. The Caco-2 Cell Monolayer: Usefulness and Limitations. *Expert Opin. Drug Metab. Toxicol.* **2008**, *4*, 395–411. [CrossRef] [PubMed]

95. Yamashita, T.; Inui, T.; Yokota, J.; Kawakami, K.; Morinaga, G.; Takatani, M.; Hirayama, D.; Nomoto, R.; Ito, K.; Cui, Y.; et al. Monolayer Platform Using Human Biopsy-Derived Duodenal Organoids for Pharmaceutical Research. *Mol. Ther. Methods Clin. Dev.* **2021**, *22*, 263–278. [CrossRef] [PubMed]

96. Takahashi, Y.; Noguchi, M.; Inoue, Y.; Sato, S.; Shimizu, M.; Kojima, H.; Okabe, T.; Kiyono, H.; Yamauchi, Y.; Sato, R. Organoid-Derived Intestinal Epithelial Cells Are a Suitable Model for Preclinical Toxicology and Pharmacokinetic Studies. *iScience* **2022**, *25*, 104542. [CrossRef]

97. Kozuka, K.; He, Y.; Koo-McCoy, S.; Kumaraswamy, P.; Nie, B.; Shaw, K.; Chan, P.; Leadbetter, M.; He, L.; Lewis, J.G.; et al. Development and Characterization of a Human and Mouse Intestinal Epithelial Cell Monolayer Platform. *Stem Cell Rep.* **2017**, *9*, 1976–1990. [CrossRef]

98. Hoffmann, P.; Schnepel, N.; Langeheine, M.; Künnemann, K.; Grassl, G.A.; Brehm, R.; Seeger, B.; Mazzuoli-Weber, G.; Breves, G. Intestinal Organoid-Based 2D Monolayers Mimic Physiological and Pathophysiological Properties of the Pig Intestine. *PLoS ONE* **2021**, *16*, e0256143. [CrossRef] [PubMed]

99. Ambrosini, Y.M.; Park, Y.; Jergens, A.E.; Shin, W.; Min, S.; Atherly, T.; Borcherding, D.C.; Jang, J.; Allenspach, K.; Mochel, J.P.; et al. Recapitulation of the Accessible Interface of Biopsy-Derived Canine Intestinal Organoids to Study Epithelial-Luminal Interactions. *PLoS ONE* **2020**, *15*, e0231423. [CrossRef] [PubMed]

100. Mayorgas, A.; Dotti, I.; Martínez-Picola, M.; Esteller, M.; Bonet-Rossinyol, Q.; Ricart, E.; Salas, A.; Martínez-Medina, M. A Novel Strategy to Study the Invasive Capability of Adherent-Invasive Escherichia Coli by Using Human Primary Organoid-Derived Epithelial Monolayers. *Front. Immunol.* **2021**, *12*. [CrossRef] [PubMed]

101. Sasaki, N.; Miyamoto, K.; Maslowski, K.M.; Ohno, H.; Kanai, T.; Sato, T. Development of a Scalable Coculture System for Gut Anaerobes and Human Colon Epithelium. *Gastroenterology* **2020**, *159*, 388–390. [CrossRef]

102. Bayguinov, P.O.; Oakley, D.M.; Shih, C.C.; Geanon, D.J.; Joens, M.S.; Fitzpatrick, J.A.J. Modern Laser Scanning Confocal Microscopy. *Curr. Protoc. Cytom.* **2018**, *85*, e39. [CrossRef] [PubMed]
103. Chen, Y.Y.; Silva, P.N.; Syed, A.M.; Sindhwani, S.; Rocheleau, J.V.; Chan, W.C.W. Clarifying Intact 3D Tissues on a Microfluidic Chip for High-Throughput Structural Analysis. *Proc. Natl. Acad. Sci. USA* **2016**, *113*, 14915–14920. [CrossRef] [PubMed]
104. van Ineveld, R.L.; Ariese, H.C.R.; Wehrens, E.J.; Dekkers, J.F.; Rios, A.C. Single-Cell Resolution Three-Dimensional Imaging of Intact Organoids. *J. Vis. Exp.* **2020**, *2020*. [CrossRef] [PubMed]
105. VanDussen, K.L.; Sonnek, N.M.; Stappenbeck, T.S. L-WRN Conditioned Medium for Gastrointestinal Epithelial Stem Cell Culture Shows Replicable Batch-to-Batch Activity Levels across Multiple Research Teams. *Stem Cell Res.* **2019**, *37*. [CrossRef] [PubMed]
106. Janda, C.Y.; Dang, L.T.; You, C.; Chang, J.; de Lau, W.; Zhong, Z.A.; Yan, K.S.; Marecic, O.; Siepe, D.; Li, X.; et al. Surrogate Wnt Agonists That Phenocopy Canonical Wnt and β-Catenin Signalling. *Nature* **2017**, *545*, 234–237. [CrossRef] [PubMed]
107. Miao, Y.; Ha, A.; de Lau, W.; Yuki, K.; Santos, A.J.M.; You, C.; Geurts, M.H.; Puschhof, J.; Pleguezuelos-Manzano, C.; Peng, W.C.; et al. Next-Generation Surrogate Wnts Support Organoid Growth and Deconvolute Frizzled Pleiotropy In Vivo. *Cell Stem Cell* **2020**, *27*, 840–851.e6. [CrossRef] [PubMed]
108. Luca, V.C.; Miao, Y.; Li, X.; Hollander, M.J.; Kuo, C.J.; Christopher Garcia, K. Surrogate R-Spondins for Tissue-Specific Potentiation of Wnt Signaling. *PLoS ONE* **2020**, *15*, e0226928. [CrossRef] [PubMed]
109. Fujii, M.; Matano, M.; Toshimitsu, K.; Takano, A.; Mikami, Y.; Nishikori, S.; Sugimoto, S.; Sato, T. Human Intestinal Organoids Maintain Self-Renewal Capacity and Cellular Diversity in Niche-Inspired Culture Condition. *Cell Stem Cell* **2018**, *23*, 787–793.e6. [CrossRef]
110. Nantasanti, S.; Spee, B.; Kruitwagen, H.S.; Chen, C.; Geijsen, N.; Oosterhoff, L.A.; van Wolferen, M.E.; Pelaez, N.; Fieten, H.; Wubbolts, R.W.; et al. Disease Modeling and Gene Therapy of Copper Storage Disease in Canine Hepatic Organoids. *Stem Cell Rep.* **2015**, *5*, 895–907. [CrossRef] [PubMed]
111. Aisenbrey, E.A.; Murphy, W.L. Synthetic Alternatives to Matrigel. *Nat. Rev. Mater.* **2020**, *5*, 539–551. [CrossRef] [PubMed]
112. Raphael, B.; Khalil, T.; Workman, V.L.; Smith, A.; Brown, C.P.; Streuli, C.; Saiani, A.; Domingos, M. 3D Cell Bioprinting of Self-Assembling Peptide-Based Hydrogels. *Mater. Lett.* **2017**, *190*, 103–106. [CrossRef]
113. Cherne, M.D.; Sidar, B.; Sebrell, T.A.; Sanchez, H.S.; Heaton, K.; Kassama, F.J.; Roe, M.M.; Gentry, A.B.; Chang, C.B.; Walk, S.T.; et al. A Synthetic Hydrogel, VitroGel®ORGANOID-3, Improves Immune Cell-Epithelial Interactions in a Tissue Chip Co-Culture Model of Human Gastric Organoids and Dendritic Cells. *Front Pharmacol.* **2021**, *12*, 2279. [CrossRef] [PubMed]
114. Bhattacharya, M.; Malinen, M.M.; Lauren, P.; Lou, Y.R.; Kuisma, S.W.; Kanninen, L.; Lille, M.; Corlu, A.; Guguen-Guillouzo, C.; Ikkala, O.; et al. Nanofibrillar Cellulose Hydrogel Promotes Three-Dimensional Liver Cell Culture. *Proc. J. Control. Release* **2012**, *164*, 291–298.
115. Krüger, M.; Oosterhoff, L.A.; van Wolferen, M.E.; Schiele, S.A.; Walther, A.; Geijsen, N.; de Laporte, L.; van der Laan, L.J.W.; Kock, L.M.; Spee, B. Cellulose Nanofibril Hydrogel Promotes Hepatic Differentiation of Human Liver Organoids. *Adv. Healthc. Mater.* **2020**, *9*, 1901658. [CrossRef] [PubMed]
116. Flaig, I.; Radenković, M.; Najman, S.; Pröhl, A.; Jung, O.; Barbeck, M. In Vivo Analysis of the Biocompatibility and Immune Response of Jellyfish Collagen Scaffolds and Its Suitability for Bone Regeneration. *Int. J. Mol. Sci.* **2020**, *21*, 4518. [CrossRef] [PubMed]
117. Kim, M.; Mun, H.; Sung, C.O.; Cho, E.J.; Jeon, H.J.; Chun, S.M.; Jung, D.J.; Shin, T.H.; Jeong, G.S.; Kim, D.K.; et al. Patient-Derived Lung Cancer Organoids as in Vitro Cancer Models for Therapeutic Screening. *Nat. Commun.* **2019**, *10*, 3991. [CrossRef] [PubMed]
118. Driehuis, E.; Kretzschmar, K.; Clevers, H. Establishment of Patient-Derived Cancer Organoids for Drug-Screening Applications. *Nat. Protoc.* **2020**, *15*, 3380–3409. [CrossRef] [PubMed]
119. HUB Organoids A Patient in the Lab®. Available online: https://www.huborganoids.nl (accessed on 15 July 2022).
120. Sampaziotis, F.; Muraro, D.; Tysoe, O.C.; Sawiak, S.; Beach, T.E.; Godfrey, E.M.; Upponi, S.S.; Brevini, T.; Wesley, B.T.; Garcia-Bernardo, J.; et al. Cholangiocyte Organoids Can Repair Bile Ducts after Transplantation in the Human Liver. *Science* **2021**, *371*, 839–846. [CrossRef] [PubMed]
121. Yoshihara, E.; O'Connor, C.; Gasser, E.; Wei, Z.; Oh, T.G.; Tseng, T.W.; Wang, D.; Cayabyab, F.; Dai, Y.; Yu, R.T.; et al. Immune-Evasive Human Islet-like Organoids Ameliorate Diabetes. *Nature* **2020**, *586*, 606–611. [CrossRef] [PubMed]
122. Meran, L.; Massie, I.; Campinoti, S.; Weston, A.E.; Gaifulina, R.; Tullie, L.; Faull, P.; Orford, M.; Kucharska, A.; Baulies, A.; et al. Engineering Transplantable Jejunal Mucosal Grafts Using Patient-Derived Organoids from Children with Intestinal Failure. *Nat. Med.* **2020**, *26*, 1593–1601. [CrossRef] [PubMed]
123. Sugimoto, S.; Kobayashi, E.; Fujii, M.; Ohta, Y.; Arai, K.; Matano, M.; Ishikawa, K.; Miyamoto, K.; Toshimitsu, K.; Takahashi, S.; et al. An Organoid-Based Organ-Repurposing Approach to Treat Short Bowel Syndrome. *Nature* **2021**, *592*, 99–104. [CrossRef] [PubMed]
124. Koike, H.; Iwasawa, K.; Ouchi, R.; Maezawa, M.; Giesbrecht, K.; Saiki, N.; Ferguson, A.; Kimura, M.; Thompson, W.L.; Wells, J.M.; et al. Modelling Human Hepato-Biliary-Pancreatic Organogenesis from the Foregut–Midgut Boundary. *Nature* **2019**, *574*, 112–116. [CrossRef] [PubMed]
125. Pattanayak, P.; Singh, S.K.; Gulati, M.; Vishwas, S.; Kapoor, B.; Chellappan, D.K.; Anand, K.; Gupta, G.; Jha, N.K.; Gupta, P.K.; et al. Microfluidic Chips: Recent Advances, Critical Strategies in Design, Applications and Future Perspectives. *Microfluid Nanofluidics* **2021**, *25*, 99. [CrossRef] [PubMed]

126. Clapp, N.; Amour, A.; Rowan, W.C.; Candarlioglu, P.L. Organ-on-Chip Applications in Drug Discovery: An End User Perspective. *Biochem. Soc. Trans.* **2021**, *49*, 1881–1890. [CrossRef] [PubMed]
127. Picollet-D'hahan, N.; Zuchowska, A.; Lemeunier, I.; le Gac, S. Multiorgan-on-a-Chip: A Systemic Approach To Model and Decipher Inter-Organ Communication. *Trends Biotechnol.* **2021**, *39*, 788–810. [CrossRef]
128. Ronaldson-Bouchard, K.; Teles, D.; Yeager, K.; Tavakol, D.N.; Zhao, Y.; Chramiec, A.; Tagore, S.; Summers, M.; Stylianos, S.; Tamargo, M.; et al. A Multi-Organ Chip with Matured Tissue Niches Linked by Vascular Flow. *Nat. Biomed. Eng.* **2022**, *6*, 351–371. [CrossRef] [PubMed]

Article

Can Chromoendoscopy Improve the Early Diagnosis of Gastric Carcinoma in Dogs?

Marcus Vinicius Candido [1,*], Pernilla Syrjä [2], Susanne Kilpinen [1], Søren Meisner [3], Mohsen Hanifeh [1] and Thomas Spillmann [1]

1 Department of Equine and Small Animal Medicine, Faculty of Veterinary Medicine, P.O. Box 57, Helsinki University, 00014 Helsinki, Finland
2 Department of Veterinary Biosciences, Veterinary Pathology, Faculty of Veterinary Medicine, P.O. Box 66, Helsinki University, 00014 Helsinki, Finland
3 Digestive Disease Center, Bispebjerg Hospital, University of Copenhagen, 2400 Copenhagen, Denmark
* Correspondence: marcus.candido@helsinki.fi

Simple Summary: Currently, canine gastric carcinoma is mainly diagnosed in its late, incurable phase, and strategies for early diagnosis are lacking. In human medicine, chromoendoscopic (CE) methods such as staining the gastric mucosal surface with indigo carmine (IC), and narrow band imaging (NBI), have improved the diagnosis of precancerous gastric mucosal changes and early gastric carcinoma. This study aimed at investigating whether IC-CE and NBI-CE can improve the diagnostic yield of endoscopy in dogs. Belgian Shepherd dogs are predisposed to gastric carcinoma; thus, 30 dogs of the breed served as the study population. As a result, the study revealed that especially the combination of standard white light endoscopy (WLE) with NBI-CE allows better recognition of gastric mucosal structural changes than WLE alone. However, CE assessment templates used to predict the type of mucosal change in humans were not applicable in dogs. The value of the study lies in providing evidence that CE can improve the diagnosis of precancerous changes and early gastric carcinoma in dogs. However, current image assessment templates from human medicine need major adjustments to comprehend canine gastric mucosal conditions.

Abstract: Chromoendoscopy has improved the early diagnosis of gastric cancer in humans but its usefulness in dogs is unknown. This study aimed at assessing whether adding narrow band imaging (NBI) or indigo carmine (IC) chromoendoscopy (CE) can improve the diagnostic yield of standard white light endoscopy (WLE). We compared the real-time findings of canine WLE, NBI-CE, and IC-CE and corresponding histology reports with endoscopic mucosal pattern assessment templates used in human medicine. Belgian Shepherd dogs are predisposed to gastric carcinoma. Therefore, 30 dogs of this breed served as the study population. According to histology, 17/30 dogs had mucosal changes (mucous metaplasia, glandular dysplasia, and gastric carcinoma). Diagnostic yield was best when targeted biopsies were taken with WLE and NBI-CE combined (15/17 cases). WLE alone positively identified only 8/17 cases and missed a gastric carcinoma in 3/6 cases. CE assessment templates based on macroscopic mucosal patterns, broadly used in human medicine, were not readily applicable in dogs. In conclusion, the study provides evidence that using CE in dogs has the potential to improve the diagnosis of precancerous gastric mucosal pathology and early gastric carcinoma. However, current image assessment templates from human medicine need major adjustments to the patterns of canine gastric mucosa.

Keywords: dog; gastric carcinoma; mucous metaplasia; glandular dysplasia; endoscopy; chromoendoscopy; narrow band imaging; indigo carmine; cancer

Citation: Candido, M.V.; Syrjä, P.; Kilpinen, S.; Meisner, S.; Hanifeh, M.; Spillmann, T. Can Chromoendoscopy Improve the Early Diagnosis of Gastric Carcinoma in Dogs? *Animals* **2022**, *12*, 2253. https://doi.org/10.3390/ani12172253

Academic Editors: Aarti Kathrani and Romy M. Heilmann

Received: 27 July 2022
Accepted: 29 August 2022
Published: 31 August 2022

Publisher's Note: MDPI stays neutral with regard to jurisdictional claims in published maps and institutional affiliations.

1. Introduction

Canine gastric carcinoma is the most common gastric tumor in dogs older than 8–10 years. Currently, the tumor is diagnosed mostly at an advanced stage with a 70–90%

rate of metastasis [1]. Non-polypoid tumors have a rate of invasion and metastasis of 100% and polypoid tumors of 21% [2]. Treatment of choice is tumor resection with wide margins and removal of local lymph nodes, but the median survival time after surgery (178 days; range 1–1902 days) is rather short and depends on the rate of local invasion and metastasis [3]. Current research looks for possibilities to advance the early diagnosis of gastric carcinoma, to improve the therapeutic outcome.

Gastroscopy and mucosal biopsy sampling for histology are established as the standard procedure to detect and differentiate gastric mucosal pathology in dogs [4]. In humans, gross mucosal changes are easily identified under standard white light endoscopy (WLE), but more subtle, early changes may be overlooked, and random biopsy might fail to sample from pathologically changed mucosa, leading to possible misdiagnosis [5]. Recent endoscopic studies in the Belgian Shepherd dog breed types Tervuren and Groenendael revealed an association of gastric mucosal structural changes such as mucous metaplasia and glandular dysplasia with gastric carcinoma. Like in humans, these pathological gastric mucosal changes should be considered preneoplastic. Overlooking those changes when using WLE could have harmful consequences for the patient [6]. Attempts in human medicine to overcome the limitations of WLE in visualizing subtle macroscopic mucosal changes have led to the development of chromoendoscopy (CE). CE allows an enhancement of mucosal surface patterns, color, and vascularization, inducing a paradigm shift from random to targeted biopsies in patients with subtle gastric mucosal changes [7]. Examples of such novel imaging technologies are narrow band imaging (NBI), Storz professional image enhancement system (SPIES), and flexible spectral imaging color enhancement, all using specifically developed endoscopy equipment [8–10]. Studies in humans revealed that, for example, NBI allows the identification of gastric mucosal patterns with high predictive value for the final histologic diagnosis of normal mucosa (83%), intestinal metaplasia (84%), and dysplasia (95%) [11]. To achieve such image improvement, NBI technology uses the wavelengths of blue and green light and further digital processing.

Another, and less expensive way to improve the visualization of gastric mucosal structures, allowing for targeted instead of random biopsies in humans, has been the use of in-vivo staining methods, by spraying dyes on the mucosal surface during standard WLE [12]. Indigo carmine (IC) works by enhancing the contours of mucosal groves and has been one of the more commonly used CE dyes. IC-CE increases the detection of dysplastic lesions in humans by three to five-fold as compared to using WLE alone [13].

In veterinary gastroenterology, the possibilities to detect subtle changes through targeted biopsies have been limited to the use of WLE. It remains rather challenging to predict the possible mucosal pathology and to detect preneoplastic processes and malignancies early and accurately, especially when they are still subtle and confined to the gastric mucosa [1]. Despite the increased availability of CE equipment in veterinary medicine, there have been only very few studies exploring the possibilities of CE for canine and feline patients [9,14]. It remains so far uncertain whether CE techniques can improve the diagnostic yield of endoscopy in dogs.

This study was designed to assess the possibilities of WLE, NBI-CE, and IC-CE in visualizing gastric mucosal structural changes such as mucous metaplasia, glandular dysplasia, and gastric carcinoma in dogs. For this, we compared histologic reports with real-time gastroscopy findings, and images saved using WLE and two CE methods taken during a prospective clinical trial in Belgian Shepherd dogs. The primary objective of this study was to assess whether real-time findings noted during examination with WLE, NBI-CE, and IC-CE, or a template-based interpretation of recorded images of mucosal patterns in NBI-CE, or IC-CE images, can predict the histologically confirmed presence of mucosal changes of concern (mucous metaplasia, glandular dysplasia, and gastric carcinoma). The second objective was to assess whether targeted biopsies directed by WLE in combination with NBI-CE or IC-CE have a better diagnostic yield at histology than an approach using non-targeted biopsies during WLE alone.

2. Materials and Methods

2.1. Study Population and Ethical Licensing

WLE, NBI-CE, and IC-CE of the stomach were performed successively during the same endoscopic procedure in 30 Belgian Shepherd dogs. Here, 19 Tervuren, eight Groenendael, and three Malinois attended a prospective clinical trial on the diagnosis of gastric mucosal changes, namely mucous metaplasia, glandular dysplasia, and carcinoma [3]. Prior to endoscopy, all dogs fasted for 12–16 h, then received orally a mixture of 20% acetylcysteine (3 mL, Aurum Pharmaceuticals Ltd., Essex, UK) and 0.526% simethicone (15 mL, Endoparactol, Hormosan Pharma GmbH, Frankfurt am Main, Germany) to improve the visibility of the gastric mucosa [15]. Endoscopy was performed under general anesthesia with two different anesthesia protocols as previously described [16]. WLE, NBI-CE, and IC-CE images were saved in a local image bank during the same study for an external blinded assessment by a medical doctor experienced in gastroscopy in humans (S.M.). The study protocol was approved by the Viikki Campus Research Ethics Committee, University of Helsinki (Statement 5/2015).

2.2. Endoscopic Equipment and Image Acquisition

WLE and NBI-CE were performed successively by experienced endoscopists (M.C., T.S.) with a first-generation NBI 180 system (Olympus Europa SE & Co. KG, Hamburg, Germany) consisting of an endoscope (CF180AL), light source (CLV-180), image processor (CV-180), carbon dioxide insufflation unit (UCR), and an irrigation pump (OFP-2) for washing debris from the gastric mucosa during the endoscopy.

IC-CE was performed right after NBI-CE with the same equipment but under white light and by inserting a spray catheter (PW-5V-1, Olympus Europe) through the endoscope's working channel. IC stock solution (indigo carmine 1%, Life partners Europe, Bagnolet, France) was diluted in sterile water to a 0.2% working solution. To avoid excessive dye accumulation, the working solution was sprayed sparingly onto the entire gastric mucosal surface.

During the WLE, NBI-CE, and IC-CE procedures, images were taken at full gastric insufflation from standardized viewpoints and captured with MediCap USB200 (Medi-Capture Inc., Plymouth Meeting, PA, USA). The files were finally stored in the Microsoft Outlook cloud of the University of Helsinki, Finland. The standard viewpoints for taking images during WLE, NBI-CE, and IC-CE were (1) gastric body, (2) fundus and cardia, (3) angular fold (incisura) with fundus and antrum (two chamber view), and (4) pylorus. The gastric mucosa was then examined closely in search of diffuse or focal changes. The location and distribution of changes were recorded when the pictures were taken, and were numbered in a schematic gastric map (Figure 1). The changes were listed in a numbered table along with a short macroscopic description.

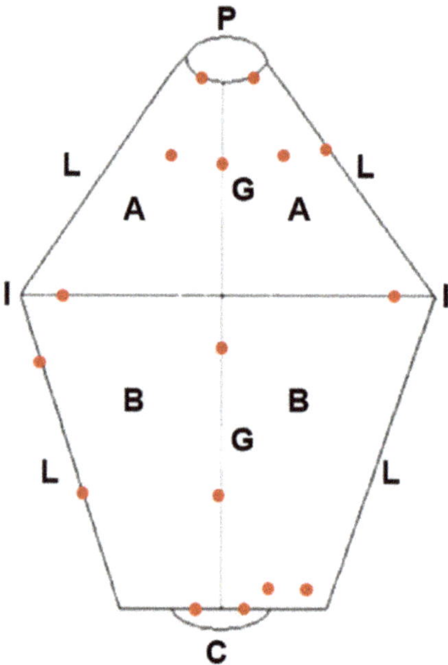

Figure 1. Schematic map of the stomach showing standard biopsy sites (red dots) used for non-targeted gastric mucosal sampling (modified from Simone et al. [17]). A: antrum; B: gastric body; G: greater curvature; L: lesser curvature; P: pylorus; I: incisura angularis (angular fold); C: cardia.

2.3. Targeted and Non-Targeted Biopsy Sampling and Histologic Assessment of Gastric Mucosal Microstructure

For histology, two separate sets of gastric mucosal samples were taken to compare the diagnostic outcome of biopsies as indicated by WLE alone, or by the combination of WLE and CE. When the endoscopic appearance was unremarkable, a non-targeted biopsy routine was applied. In short, two samples each were taken from cardia, fundus, greater and lesser curvature of the gastric body, incisura, and pylorus, and one from each quadrant of the antrum (clockwise at 12, 3, 6, and 9 o'clock) (Figure 1).

Biopsy Set 1 included targeted biopsies from the sites presenting mucosal changes visible after the three subsequent endoscopic techniques: (1) WLE, (2) NBI-CE, and (3) IC-CE. The samples were identified according to the gastric region, or a short description of the targeted change (e.g., ulcer, mass, unusual texture). In the absence of visible mucosal changes, non-targeted sampling was applied (Figure 1). After that, the second set of samples *(Biopsy Set 2)* was taken either from the predefined locations shown in Figure 1, or from the mucosal changes recorded in those regions during WLE, and labeled according to the gastric region sampled.

All mucosal samples were collected with disposable biopsy forceps (FB-210U, Olympus Europe) and placed onto wood chips, then immersed (upside-down) in 10% formalin for fixation, in separate vials. Each vial was labeled as part of the Biopsy Set 1 or 2, as well as the gastric region, and/or respective change (Biopsy Set 1). Therefore, at least four separate vials (and up to 6 in Biopsy Set 1), were used for the gastric regions (1) cardia + fundus, (2) body, (3) incisura, and (4) antrum + pylorus. All biopsy specimens were collected after IC dye-spraying and IC-CE image recording to avoid interference of blood from biopsy sites with the IC-CE imaging.

For histology, the labeled samples were paraffin-embedded and cut into sections of 4 μm. The sections were routinely stained with hematoxylin-eosin, but special staining techniques such as periodic acid-Schiff reaction were applied in selected cases at the pathologist's discretion. Immuno-histochemical staining for epithelial cells was performed with anti-cytokeratin AE1/AE antibodies (anti-human Ck AE1/AE3, mouse monoclonal M3515, Dako Agilent, Santa Clara, CA, USA). Antigens were retrieved with 0.01 M citrate buffer at pH 6 and heated for 20 min at 99 °C, then revealed following the instructions for UltraVision Detection System HRP/DAB kit (Thermo Fisher Scientific, Waltham, MA, USA). Biopsy Set 1 was used for the diagnostic workup, and Biopsy Set 2 was examined in a blinded fashion at the end of the study for comparison of results from biopsies directed by the combination of WLE and CE, or by WLE alone.

The histological slides of both sample sets were examined according to recommendations by the World Small Animal Veterinary Association's International Gastrointestinal Standardization Group [18]. The slides were also examined for histopathologic structural changes, including mucous metaplasia, glandular dysplasia, and type of gastric carcinoma [19,20]. The presence and severity of mucosal inflammation (intraepithelial lymphocytes, and lymphoplasmacytic, eosinophilic or neutrophilic infiltration into lamina propria, was scored as normal = 0, mild = 1, moderate = 2 or severe = 3 [21].

2.4. Real-Time Assessment of Endoscopic Findings during WLE, NBI-CE, and IC-CE

Every endoscopic procedure started with standard WLE, and the gastric mucosa was subjectively assessed in real time as unremarkable, diffusely changed, or focally changed. Diffuse and focal changes were described in the patient record and drawn in a printout of a schematic map of the stomach (Figure 1) that was added to the patient's records. In addition, standard images were taken as described above. Once this examination was finalized, the NBI setting of the processor was activated and the endoscopic examination was repeated in the same fashion as for WLE. Gastric mucosal changes that differed from WLE findings (e.g., focal changes) were recorded in the same fashion as for WLE. As last, IC-CE was performed by spraying IC on the gastric mucosa surface and the same assessment steps were repeated.

Results of the real-time assessment of WLE, NBI-CE, and IC-CE findings were compared with the histological findings of targeted and non-targeted biopsies to estimate the sensitivity, specificity, and accuracy of every single endoscopic method, or the combination of WLE with NBI-CE and IC-CE for predicting mucous metaplasia, glandular dysplasia, and gastric carcinoma.

2.5. Template-Based NBI-CE and IC-CE Image Assessment of Gastric Mucosal Surface Patterns

During each endoscopic examination, two separate sets of images were captured, first using NBI-CE and then IC-CE. The images saved from each patient were reviewed by the endoscopist (M.C.) to select a subset of images that corresponded to localizations that had also been biopsied for histology. Among the images available, those with the best quality (focus, resolution, color) were selected. The images included focal changes, as well as pictures taken from standard viewpoints in case no macroscopic changes were seen in a dog during NBI- or IC-CE. The files were named according to the localization of the image (e.g., pylorus, body distal greater curvature, etc.). These digital images were submitted for evaluation by a human gastroenterologist experienced in NBI- and IC-CE (S.M.). The sets of images taken with NBI-CE and with IC-CE were sent in separate lots two months apart to minimize possible identification of the cases by recollection. The external examiner was unaware of the histopathological diagnoses.

To evaluate the NBI-CE images, the examiner applied the template validated and widely used in humans to identify changes related to intestinal metaplasia, dysplasia, and carcinoma. The original template comprises standard attributes for mucosal pattern, mucosal color, and vascular pattern [22].

The mucosal patterns detected in NBI-CE images of the dogs were classified either as normal round, normal mosaic, and polygonal, enlarged mosaic and polygonal, or tubulovillous (Table 1), conforming to the template applied in human gastroenterology.

Table 1. Examples of mucosal pattern categories recognized in narrow band imaging chromoendoscopy (NBI-CE) of the gastric mucosa of dogs.

Gastric Mucosal Pattern Category	NBI-CE Example Image
Normal round	
Normal mosaic and polygonal	
Enlarged mosaic and polygonal	
Tubulo-villous	

For the assessment of IC-CE images, the mucosal patterns were graded as normal or suspicious according to the examiner's (S.M.) expertise in human gastroscopy because there have been no templates for IC-CE assessment in human medicine. Examples of such categories are shown in Table 2.

Table 2. Examples of mucosal categories recognized in indigo carmine chromoendoscopy (IC-CE) of the gastric mucosa of dogs.

Gastric Mucosal Pattern Category	IC-CE Example Image
Normal	
Suspicious	

2.6. Statistical Analysis

Descriptive statistics were used to assess the ability of WLE, NBI-CE, and IC-CE to select the biopsy site with the most severe gastric mucosal changes.

NBI-CE images graded as normal mucosal patterns (normal round, mosaic, and polygonal) were regarded as suggesting the absence of disease, thus were counted and analyzed as negative test results. Abnormal patterns (enlarged mosaic and tubulo-villous) were regarded as indicating the presence of gastric mucosal pathology (positive for metaplasia, dysplasia, or carcinoma). For IC-CE, images suggesting normal mucosal structure were assessed as negative for gastric mucosal structural pathology. The images regarded as suspicious for gastric mucosal pathology were assessed as positive. The ability of NBI-CE or IC-CE to predict gastric mucosal changes was assessed by calculating sensitivity, specificity, and diagnostic accuracy.

3. Results

3.1. Dogs and Endoscopic Findings

Endoscopic examinations were performed in 30 Belgian Shepherd dogs belonging to the breed varieties Tervuren ($n = 19$), Groenendael ($n = 8$), and Malinois ($n = 3$). Patient signalment and the diagnostically most important endoscopic findings of the gastric mucosa are summarized in Table 3. Based on real-time endoscopic assessment, seven dogs had unremarkable gastric mucosa and 23 dogs had changes that were diffuse in 14 dogs, focal (non-ulcerative) in three dogs, and ulcerative in six dogs.

Conformity between the real-time findings from WLE and the expert assessments of still images taken during NBI- and IC-CE (as illustrated in Figure 2) was overall moderate. Regarding the presence (positive) or absence of changes of concern (negative), WLE and NBI-CE agreed in 18/30 dogs and disagreed in 12/30 dogs. Concordance between WLE and IC-CE was seen in 26/30 dogs and disparity in only 4/30. The blinded expert assessments of NBI- and IC-CE images were compatible in 18/30 and divergent in 12/30 dogs.

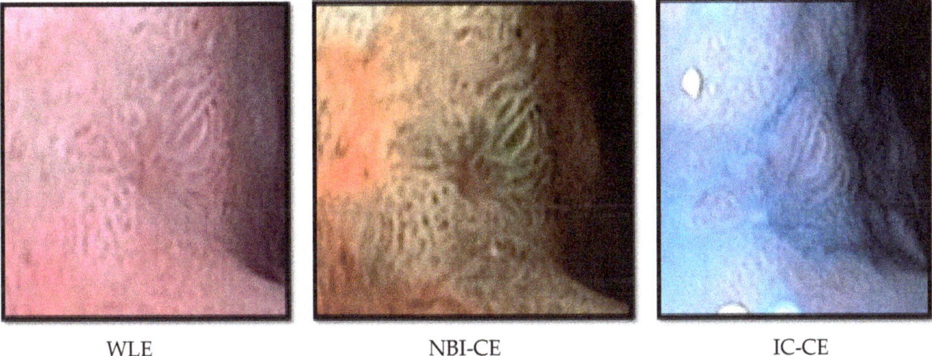

WLE   NBI-CE   IC-CE

Figure 2. Sample images of the same focal change in Dog 6, as seen at regular white light endoscopy (WLE), narrow band imaging chromoendoscopy (NBI-CE) and indigo carmine chromoendoscopy (IC-CE). Histology revealed mucous metaplasia and glandular dysplasia.

Table 3. Signalment of 30 Belgian shepherd dogs, findings during endoscopy, and histology from gastric mucosal biopsies: (1) sampling directed by white-light endoscopy (WLE) followed by chromoendoscopy (CE), and (2) biopsies taken as indicated by WLE alone.

Dog Nr.	Dog's Study ID	Breed Type	Age in Years	Sex	Body Weight (kg)	Distribution of the Most Relevant Endoscopic Finding	Type of In-flammation	Mucous Metaplasia		Glandular Dysplasia		Carcinoma	
								Biopsy Set 1	Biopsy Set 2	Biopsy Set 1	Biopsy Set 2	Biopsy Set 1	Biopsy Set 2
1	641	Tervuren	10.8	M	26.1	Diffuse Abnormal folding surrounded by irregular mucosal surface in corpus: distal greater curvature	LP	Yes	Yes	Yes	No	Yes	No
2	642	Tervuren	9.9	M	22.7	Diffuse Very prominent blood vessels in proximal antrum; abnormal folding/ thicker mucosa in corpus: distal greater curvature	LP, EO, NP	No	No	No	No	No	No
3	644	Tervuren	10.1	FS	18.5	Diffuse Major blood vessels visible in most of corpus, especially along greater curvature; marked mucosal folding in pylorus	LP, EO, NP	No	No	No	No	No	No
4	645	Tervuren	11.6	FS	25.3	Focal Sessile polyp 2 mm × 5 mm	LP, EO, NP	No	No	No	No	No	No
5	646	Groenendael	11.2	M	23.0	Diffuse Persistent folds after full insufflation in corpus: distal greater curvature	LP, NP	No	No	No	Yes	No	No
6	647	Tervuren	10.1	FS	23.4	Diffuse Abnormal folding in corpus: distal greater curvature	LP, NP	Yes	Yes	Yes	Yes	Yes	Yes
7	649	Tervuren	7.9	F	20.5	Diffuse Irregular mucosal surface in corpus: distal greater curvature	LP, EO, NP	No	No	No	No	No	No

Table 3. *Cont.*

Dog Nr.	Dog's Study ID	Breed Type	Age in Years	Sex	Body Weight (kg)	Distribution of the Most Relevant Endoscopic Finding	Type of Inflammation	Mucous Metaplasia Biopsy Set 1	Mucous Metaplasia Biopsy Set 2	Glandular Dysplasia Biopsy Set 1	Glandular Dysplasia Biopsy Set 2	Carcinoma Biopsy Set 1	Carcinoma Biopsy Set 2
8	650	Groenendael	10.6	M	23.0	Unremarkable	LP, NP	No	No	No	No	No	No
9	702	Malinois	11.1	MC	31.6	Focal Irregular mucosal surface in corpus: distal greater curvature	LP, EO	Yes	No	No	No	No	No
10	703	Tervuren	9.0	M	21.3	Diffuse More than 20 discolored patches 5 mm all over antrum	LP, EO	No	No	No	Yes	No	No
11	652	Tervuren	10.1	MC	30.7	Ulcerative More than 30 erosions 2 mm in proximal greater curvature	LP, EO	No	No *	Yes	No	No	No
12	653	Malinois	6.7	FS	36.6	Diffuse Hyperemic spots around cardia (lymphoid follicles)	LP, EO	No	No	No	No	No	No
13	654	Tervuren	7.6	M	29.3	Ulcerative Linear ulcers in corpus: distal lesser curvature	LP, EO, NP	No	No	No	No	No	No
14	655	Groenendael	9.1	FS	22.2	Diffuse Hyperemic spots around cardia	LP, EO	No	No	No	No	No	No
15	658	Tervuren	7.2	M	27.8	Ulcerative Massive elevation in mucosa, two large crater-like ulcers in corpus	LP, EO, NP	Yes	No	Yes	No	Yes	No
16	664	Tervuren	10.4	FS	22.6	Focal Tiny polyp in pylorus	LP, EO, NP	No	No	Yes	No	No	No

Table 3. *Cont.*

Dog Nr.	Dog's Study ID	Breed Type	Age in Years	Sex	Body Weight (kg)	Distribution of the Most Relevant Endoscopic Finding	Type of In-flammation	Mucous Metaplasia		Glandular Dysplasia		Carcinoma	
								Biopsy Set 1	Biopsy Set 2	Biopsy Set 1	Biopsy Set 2	Biopsy Set 1	Biopsy Set 2
17	677	Groenendael	8.3	F	17.4	Ulcerative Long ulcer crater in corpus, closer to lesser curvature; gross (mural) thickening of the incisura	LP, EO, NP	No	No	Yes	Yes	Yes	Yes
18	689	Malinois	8.8	FS	22.2	Unremarkable	LP, EO, NP	No	No	Yes	Yes	No	No
19	687	Groenendael	9.7	FS	27.7	Unremarkable	LP, EO, NP	No	No	Yes	Yes	No	No
20	686	Tervuren	5.6	F	21.5	Unremarkable	LP, EO, NP	No	No	No	No	No	No
21	679	Tervuren	9.0	FS	17.1	Diffuse Visible blood vessels on gastric body	LP, EO, NP	No	No	No	No	No	No
22	722	Tervuren	11.2	M	29.8	Diffuse Greater curvature: persisting folds in distal corpus, irregular surface in antrum	LP, EO, NP	Yes	Yes	Yes	No	No	No
23	721	Tervuren	11.2	FS	23.2	Ulcerative Three small ulcers along lesser curvature and thickened mucosa in distal greater curvature of corpus; hyperemic antrum greater curvature	LP, EO, NP	No	Yes	Yes	Yes	Yes	Yes
24	680	Tervuren	5.5	F	22.7	Diffuse Distal greater curvature of corpus: very marked persistent folding with irregular surface and multifocal hyperemia, linear depression	LP, NP	Yes	No	Yes	Yes	No	No

Table 3. *Cont.*

Dog Nr.	Dog's Study ID	Breed Type	Age in Years	Sex	Body Weight (kg)	Distribution of the Most Relevant Endoscopic Finding	Type of In-flammation	Mucous Metaplasia		Glandular Dysplasia		Carcinoma	
								Biopsy Set 1	Biopsy Set 2	Biopsy Set 1	Biopsy Set 2	Biopsy Set 1	Biopsy Set 2
25	750	Tervuren	10.1	FS	25.5	Diffuse Corncob texture along the left side of the corpus	LP, EO	**Yes**	**No**	Yes	Yes	No	No
26	749	Groenendael	8.3	MC	27.2	Unremarkable	LP	No	No	No	No	No	No
27	747	Tervuren	8.2	F	29.5	Unremarkable	LP, EO	Yes	Yes	No	No	No	No
28	724	Groenendael	7.6	F	24.9	Diffuse Corpus: elevated mucosa with multifocal, depressed darker spots along proximal greater curvature; visible small vessels in distal lesser curvature	LP, EO	No	No	No	No	No	No
29	711	Groenendael	8.3	FS	24.6	Ulcerative Elongated, small transversal, non-bleeding crater in transition between corpus and antrum, left of greater curvature	LP, EO	No	No	**Yes**	**No**	**Yes**	**No**
30	719	Tervuren	11.3	FS	18.2	Unremarkable	LP, EO	No	No	No	No	No	No

Legend: M—male; MC—male castrated; F—female, FS—female spayed; LP—lymphoplasmacytic; EO—eosinophilic; NP—neutrophilic infiltrations in lamina propria. Histologies that differ between the two biopsy sets are highlighted in a gray background, the presence of changes of concern are highlighted in bold. * Intestinal metaplasia (AB-PAS positive).

3.2. Histology of Targeted and Non-Targeted Gastric Mucosal Biopsies

Histology was performed in two sample sets. Biopsy Set 1 contained biopsies directed by the combination of WLE and CE. Biopsy Set 2 comprised biopsies taken as dictated by WLE alone. The results of Biopsy Set 1 were used for diagnostic purposes and assessed immediately. Biopsy Set 2 was withheld after processing: the histologic slides were assessed in a blinded fashion at the end of the study and their results were compared with those of the targeted biopsies.

Targeted biopsies were taken from the focal or diffuse changes visible in WLE, NBI-CE, and/or IC-CE. However, in 5/30 dogs (Dogs 18, 20, 26, 27, 30) no obvious changes were seen in WLE, NBI-CE, or IC-CE. In these dogs, non-targeted biopsies were taken for both sample sets from predefined locations as presented in Figure 1.

Based on the histologic assessment of biopsy set 1 (CE-directed, mostly targeted biopsies), none of the dogs showed entirely normal gastric mucosa and the diagnostically most important histologic findings were mucosal inflammation of different types in 30/30 dogs (Table 3). While 16/30 dogs displayed submucosal and mucosal inflammation of different severity, 15/30 dogs had also mucosal structural changes such as mucous metaplasia ($n = 8$), glandular dysplasia ($n = 13$), and/or gastric carcinoma ($n = 6$). As summarized in Table 3, the parallel occurrence of two or more types of mucosal change was seen in 9/15 dogs. Three dogs had concurrent mucous metaplasia, glandular dysplasia, and carcinoma (Dogs 1, 6, 15); three had dysplasia and carcinoma (Dogs 17, 23, 29); and the other three (Dogs 22, 24, 25) had metaplasia and dysplasia. One of them (Dog 24) was diagnosed with gastric carcinoma one year later [6].

When comparing the most severe histological changes detected in biopsy set 1 with those in biopsy set 2 of individual dogs, a complete agreement was seen in 18/30 dogs but disagreements in 12/30 dogs (Table 3). From the disagreements in 12 dogs, biopsy set 1 revealed more severe mucosal structural changes than biopsy set 2 in 9/12 dogs. Misdiagnosis based on non-targeted biopsies occurred in three of the dogs with carcinoma (one misdiagnosed as mucous metaplasia (dog 1) and two misdiagnosed as inflammatory (dogs 15, 29)), and in three dogs with glandular dysplasia (two misdiagnosed as inflammation (dogs 11, 16) and one as mucous metaplasia (dog 22)). Dog 11 was the only patient where intestinal, but not mucous, metaplasia was reported from non-targeted, set-2 biopsies. In three dogs (dogs 15, 24, 25), mucous metaplasia was seen in targeted (set 1) but not in non-targeted (set 2) biopsies. However, there were 2/12 dogs with disagreeing results in which non-targeted Set-2 biopsies revealed glandular dysplasia but set-1 biopsies were regarded as inflammatory without mucosal structural changes (dogs 5, 10). In dog 23, glandular dysplasia and gastric carcinoma were confirmed in both biopsy sets but only non-targeted biopsies revealed mucous metaplasia. In all five dogs that had non-targeted biopsies at both biopsy sets (dogs 18, 20, 26, 27, 30), the histology findings agreed.

Combining both sample sets revealed no mucosal changes in 13/30 dogs, but 17/30 dogs had either mucous metaplasia, glandular dysplasia, or gastric carcinoma as the most severe histologic finding. From 17 dogs with mucosal pathology, nine were diagnosed in targeted biopsies only, six in both targeted and non-targeted biopsies, and two in non-targeted biopsies only. When linking histology with the endoscopic methods, the diagnostic yield of mucosal structural pathology was best when targeted biopsies were taken under WLE combined with NBI-CE (15/17 cases). WLE alone detected these lesions only in 8/17 cases and missed a gastric carcinoma in 3/6 cases (50%).

3.3. Template-Based Assessment of Images Taken during NBI-CE

For an independent image assessment, 143 NBI-CE images were selected and sent to the external evaluator (S.M.), who was blinded to the histologic findings. Of these, 27 images were used only for anatomical orientation and were not included in the template-based assessment. Seventeen other images had insufficient image quality and another 12 images that failed to match the template were excluded by the examiner. Thus, a total of 87 NBI-CE images were assessed according to the given NBI-CE assessment (Table 4).

However, it was not possible to apply the human NBI template for assessing mucosal color and vascular pattern to the gastric mucosa of dogs, because the images neither showed color changes typical for gastric pathologies in humans nor were the vascular patterns relatable.

Table 4. Most severe histopathological diagnoses of biopsy sites and categories of gastric mucosal patterns of dogs identified in 87 endoscopic still images obtained during NBI-CE. 4.1. 2 × 2 table for histopathological diagnosis and NBI-CE mucosal patterns.

| | Type of Mucosal Pattern at NBI-CE | | | |
| | Normal | | Abnormal | |
Histopathological Diagnosis	**Round** ($n = 44$)	**Mosaic and Polygonal** ($n = 21$)	**Enlarged Mosaic and Polygonal** ($n = 12$)	**Tubulo-Villous** ($n = 10$)
Normal ($n = 24$)	14	7	3	0
Inflammation ($n = 32$)	19	6	4	3
Mucous metaplasia ($n = 2$)	1	1	0	0
Glandular dysplasia ($n = 16$)	8	4	4	0
Carcinoma ($n = 13$)	2	3	1	7

(4.1)

| **NBI-CE Mucosal Pattern** | Histopathological Diagnosis | |
	Normal/Inflammation ($n = 56$)	**Meta-/Dysplasia/Carcinoma ($n = 31$)**
Normal ($n = 65$)	46	19
Abnormal ($n = 22$)	10	12

The mucosal patterns of 12 NBI-CE images of 10 dogs that did not match the template were therefore reported separately. Of these images, three were assessed as representing active or healing ulcers, one of which was not confirmed by histology (dog 3). The other two were confirmedly ulcerative, with one being inflammatory (dog 24) and the other an adenocarcinoma of intestinal type (dog 23). Dog 23 also showed marked differences between two types of gastric mucosa with bluish color, which was however not confirmed as intestinal metaplasia, as described in humans [23]. Small hyperplastic areas were suspected in two dogs, of which histology revealed normal mucosa with mild acute hemorrhage in one case (dog 10) and, in the other case, moderate lymphoplasmacytic infiltration, multifocal hyperemia, mild hemorrhage, fibrosis in the lamina propria, and mild superficial edema (dog 9). In two dogs, the mucosa was assessed as normal despite differing from the template (dogs 20 and 30). Both dogs had chronic diffuse gastritis with an eosinophilic component and different degrees of diffuse or multifocal fibrosis. In two dogs, the endoscopic appearance suggested local blood flow disorders such as ischemia (dog 11) or hypervascularity (dog 25). While dog 11 had histologically a mild diffuse chronic eosinophilic gastritis with mild fibrosis and mucous metaplasia in the angular fold, dog 25 had a severe diffuse chronic atrophic gastritis, with mild multifocal eosinophilic and neutrophilic components. In dog 16, a small adenoma in the pylorus area was suspected. Histology revealed intact epithelium and diffuse severe lymphoplasmacytic infiltration with a moderate neutrophilic and eosinophilic component. The gland structure was in part moderately thinned and replaced by fibrosis.

The application of the human template for the evaluation of gastric NBI-CE images as normal (negative) or abnormal (positive) showed a sensitivity of 39%, a specificity of 82% and a diagnostic accuracy of 67%.

3.4. Template-Based Assessment of Images Taken during IC-CE

When the IC dye was sprayed on the gastric mucosa during IC-CE, the perception of surface details of the mucosa improved, but only when the amount of dye was not excessive, causing it to pool over the gastric mucosal surface. In cases with extensive IC dye accumulation, suction and flushing were necessary to remove some of the dye and improve mucosal texture recognition.

For image assessment, 142 IC-CE images were selected and sent to the external evaluator (S.M.) two months after the submission of NBI-CE images, to reduce the risk of bias. Again, the evaluator was blinded to the histologic findings and excluded four of the images from assessment, two used for anatomical orientation and two due to insufficient quality. Thus, 138 images were assessed according to the respective mucosal appearance with IC-CE and categorized as normal, suspicious, or abnormal. Table 5 summarizes the gastric mucosal patterns found in the IC-CE images and their corresponding histopathological diagnoses.

Table 5. Most severe histopathological diagnosis of biopsy sites and categories of gastric mucosal assessment of dogs identified in 138 endoscopic still images taken during IC-CE. 5.1. 2 × 2 table for histopathological diagnosis and IC-CE mucosal assessment.

	IC-CE Mucosal Appearance		
Histopathological Diagnosis	**Normal** ($n = 109$)	**Suspicious** ($n = 20$)	**Abnormal** ($n = 9$)
Normal ($n = 46$)	40	6	0
Inflammation ($n = 50$)	40	8	2
Mucous metaplasia ($n = 1$)	1	0	0
Glandular dysplasia ($n = 23$)	20	2	1
Carcinoma ($n = 18$)	8	4	6

(5.1)

IC-CE Mucosal Appearance	**Histopathological Diagnosis**	
	Normal/Inflammation ($n = 96$)	**Meta-/Dysplasia/Carcinoma** ($n = 42$)
Normal ($n = 109$)	80	29
Suspicious/Abnormal ($n = 29$)	16	13

For the 138 IC-CE images, histological diagnoses were as follows: normal gastric mucosa ($n = 46$), inflammation of different type and degree ($n = 50$), mucous metaplasia ($n = 1$), glandular dysplasia ($n = 23$), and carcinoma ($n = 18$). The assessment of gastric IC-CE images as normal (negative) or suspicious/abnormal (positive) for pathological changes in histology had a sensitivity of 31%, specificity of 83%, and diagnostic accuracy of 67%.

4. Discussion

We found a higher diagnostic yield of biopsies taken during endoscopic procedures combining WLE with NBI-CE or IC-CE than from biopsies taken under WLE alone. The histology of both biopsy sets agreed in 60% of the dogs (18/30), but in other 30% (9/30) the gastric mucosal pathologies were more often diagnosed in targeted biopsies from WLE in combination with NBI-CE or IC-CE than in biopsies taken as indicated by WLE alone. The most prominent difference was that the diagnosis of gastric carcinoma was missed by biopsies taken under WLE. Independent from the biopsy set, histology revealed no mucosal changes in 13/30 dogs and changes in 17/30 dogs. Gastric mucosal structural changes were detected in 9/17 dogs (53%) using targeted biopsies guided by the combination of WLE and CE. In contrast, WLE biopsies found such histologic changes in only 2/17 dogs (12%). An agreement of mucosal structural pathology in both biopsy sets was seen in 6/17 dogs (35%). These results give some evidence that targeted biopsies guided by WLE in combination with CE can improve the diagnostic yield at histology, in comparison to WLE alone or non-targeted biopsies.

Chromoendoscopic procedures have gained wide application in human medicine but studies about their usefulness in veterinary medicine are scarce [6,13]. In people, improved image quality and light manipulation have led to major improvements in early gastric cancer diagnosis [24] and staging [25]. Our first objective was to assess whether real-time findings of WLE combined with two chromoendoscopic methods (NBI-CE and IC-CE) allow an improved biopsy sampling with a higher diagnostic yield at histology than using WLE alone. The second objective was to investigate whether templates used in humans for NBI-CE or IC-CE image assessments can predict gastric mucosal changes in dogs. However, the image assessment based on the NBI-CE template and expertise in IC-CE from human medicine revealed differing results. The estimated sensitivities, specificities, and accuracies for gastric mucosal changes of concern (gastric metaplasia, dysplasia, or carcinoma) showed that the human template is insufficient to match histologic confirmation or exclusion of gastric mucosal pathology in dogs.

Several aspects need to be discussed that might have influenced the outcome of the study. These include the investigated dog population, the study design, and some limitations that need to be taken into consideration.

Gastric carcinoma accounts for only 1% of all neoplastic diseases in dogs, which means it is rather rare [26]. Belgian Shepherd dogs have been considered the best study population because especially the breed types Tervuren and Groenendael are highly predisposed not only for gastric carcinoma but also for mucous metaplasia and glandular dysplasia [27], both considered as preneoplastic changes of the gastric mucosa in humans and dogs [28]. Previous studies in humans used different prospective study designs comparing NBI-CE with dye-based CE (e.g., Lugol's solution, methylene blue, IC) in the gastrointestinal tract (esophagus to colon). They included (1) a randomized order of endoscopy methods requiring two successive procedures in the same patient [29,30], (2) comparison of two groups of patients being randomly allocated to either NBI or dye-based CE [31,32], or (3) performing WLE followed by dye-based CE and NBI-CE, or NBI-CE followed by dye-based CE [33,34]. We decided to perform first WLE followed by NBI-CE and successively IC-CE for three reasons. First, pet dog owners would not consent otherwise, as there would be no ethical justification for two consecutive endoscopies just for research purposes. Second, we expected a case number too small to allow for the statistical comparison of two patient groups, one undergoing WLE and NBI-CE and the other WLE and IC-CE. Third, IC-CE was chosen to be the last procedure after NBI-CE to avoid lengthy washing of the gastric mucosa for dye removal and to prevent any interference of IC-CE with NBI-CE. This allowed for all three endoscopic procedures to be performed during one general anesthesia and obtaining endoscopic biopsies at the end of the procedure because these lead to microhemorrhage that could otherwise interfere with the CE results. One limitation of this approach is that the same endoscopist performed all three consecutive endoscopic examinations, which can cause a bias towards the assessment of the gastric mucosal patterns and the decision about sampling sites for the gastric mucosal biopsies. To address this bias, targeted biopsies were taken based on the findings of WLE combined with both CE methods, followed by biopsies that were non-targeted or directed by the initial records from WLE at the start of the procedure. In addition, NBI-CE and IC-CE images were sent two months apart to an external examiner (S.M.) who was blinded to both patient data and histologic diagnosis.

Our study showed an improved diagnostic yield when combining WLE with CE. This is consistent to some degree with data reported in human medicine. In one prospective study in humans, the diagnostic accuracy of WLE for extensive gastric intestinal metaplasia was 60% in a cohort of 25 patients. In comparison, the real-time diagnostic accuracy of high-resolution light-NBI in 35 patients was 87%. Those authors concluded that more than 90% of individuals at risk for gastric cancer could be identified without the need for biopsies [35]. In contrast, a recently published, open label, randomized, controlled tandem trial in 4523 Japanese patients revealed that the overall sensitivity of primary endoscopy to detect early gastric cancer in humans was only 75% when using second generation NBI-CE, therefore not increasing the detection rate over conventional WLE [36]. These conflicting experiences in human medicine should be taken into consideration when interpreting the results of our study. It may seem obvious that a definitive diagnosis of precancerous or early cancerous gastric mucosal pathology cannot be done without histology. However, improved visualization of changed gastric mucosal patterns by adding NBI-CE or IC-CE to WLE can lead to improvement by directing biopsies to target the areas with gastric mucosal structural changes. The addition of CE methods to WLE might minimize the risk of inadvertently sampling from areas without gastric mucosal pathology, thus achieving an inaccurate diagnosis with possibly detrimental consequences for the affected dog.

In 3/6 dogs, the carcinoma was undiagnosed in the biopsy set 2, referring to WLE-directed sampling. Although two of those dogs had very subtle, initial lesions (dogs 1 and 29), another case (dog 15) was diagnosed with a mass-like ulcerated carcinoma. The most plausible reason for the failure to diagnose the advanced case could be explained by insufficient sampling, missing the actual tumor. It is known that the diagnosis of gastric carcinoma sometimes requires full-thickness biopsies [1] as the invasive tumor cells are often found only deeper into the gastric wall, under the superficial mucosal layer. For the more subtle, early cases, it might have been expected that CE could facilitate selecting suspected biopsy sites, but there can also be other causes for the misdiagnosis. The WLE biopsy set 2 was collected last when blood from the previous sampling could have impaired exact visualization of the intended site, which could have been relevant with small changes.

Some issues affecting image quality and assessment are worthy of mention. We used the first-generation, 180-series NBI technology, missing the magnification function that can be enabled in the latest, more costly NBI equipment. In general, that restricted the ability to achieve ideal focus and final resolution of the still images submitted for assessment, which were paired with the histopathological diagnoses. The exact biopsy site was therefore considerably smaller than the area of the image assessed. On the other hand, the intensity of the light was often insufficient during real-time endoscopic examination when NBI function was on, which only allowed the adequate examination of restricted areas at a time, at close-up, and yet with limited resolution. In addition, the regular (white) light was insufficient to assess the mucosa after it had been sprayed with the dark-blue IC.

CE requires excellent mucosal preparation and image quality was affected by the smallest amounts of food and debris (soil, hair) even after adequate fasting. Similarly, despite the mucolytic and antifoaming agents administered to improve mucosal preparation, remaining strains of mucus or abundant fluid and foam (saliva) often required extensive flushing using the irrigation pump and cycles of suction to allow for detailed visualization. Even small amounts of bile-tinged duodenal mucus were a nuisance, as bile shows a misleading reddish hue in NBI-CE. Also, blood from ulcerated sites not only interfered with local mucosal imaging but also impaired the IC-CE staining. During the biopsy sampling steps, the combination of the dark-blue IC and bleeding from biopsy sites, with the limited illumination provided by the equipment, often required additional wash steps to carry on with the extensive sampling protocol.

In our experience, IC-CE does improve the perception of mucosal topography and texture, making it easier to identify suspected areas even without a close-up view during the initial evaluation of the stomach. Dye-spraying is much more affordable than NBI technologies, requiring only simple equipment and a couple of minutes to apply. It effectively enhances the texture of what is already present with WLE, grossly and superficially. On the other hand, NBI has the potential to show more subtle, inconspicuous, and deeper mucosal details.

Our findings suggest that CE in dogs has the potential to improve the diagnosis of precancerous gastric mucosal pathology and early gastric carcinoma. However, the differences between the gastric mucosa of humans and dogs are such that a specific template must be developed for clinical application in canines. Future research should apply next-generation NBI equipment enabling magnification (e.g., Olympus 190/200 series) in search of early changes related to gastric carcinoma. Tervuren and Groenendael Belgian Shepherd dogs aged between five and six years, with the highest risk to be positive for precancerous structural changes of the gastric mucosa such as mucous metaplasia and glandular dysplasia [6,26], should be targeted in the next phase of the study, with focal imaging in exact correspondence with targeted biopsy sites.

Author Contributions: Conceptualization, T.S.; methodology, T.S. and M.V.C.; formal analysis, T.S. and M.V.C.; investigation, T.S., M.V.C., S.K., P.S. and S.M.; data curation, T.S., M.V.C. and M.H.; writing—original draft preparation, T.S. and M.V.C.; writing—review and editing, T.S., M.V.C., S.K., M.H., P.S. and S.M.; visualization, T.S. and M.V.C.; supervision, T.S. and S.K.; project administration, T.S.; funding acquisition, T.S. and M.V.C. All authors have read and agreed to the published version of the manuscript.

Funding: This research was supported by a grant for doctoral studies (201568/2015–2) by the Brazilian Research Council (CNPq). Additional funds for research material costs were granted by the Finnish Veterinary Research Foundation (ETTS) and the Finnish Foundation of Veterinary Research (SELS). None of the funding bodies were involved in the design, collection of samples, analyses, interpretation, or writing of the manuscript.

Institutional Review Board Statement: The study protocol was approved by the University of Helsinki—Viikki Campus Research Ethics Committee (Statement 5/2015).

Informed Consent Statement: Informed consent was obtained from all subjects involved in the study.

Data Availability Statement: Not applicable.

Acknowledgments: The authors would like to thank everyone directly or indirectly involved in the procedures or supporting this research work. Special thanks to the Finnish Belgian Shepherd dog association (SBPKY) for their support in publicizing our research effort and outcomes. Open access funding provided by University of Helsinki.

Conflicts of Interest: The authors declare no conflict of interest. SM was, during the study period, employed by Olympus Europa SE & Co. KG as an external consultant, but the study was not sponsored by the enterprise. SM also worked as a consultant in gastrointestinal endoscopy for Ambu A/S and SATC CENTER, House of Research, Odense University Hospital, Denmark, as well as Coloplast A/S, where he also acted as a primary investigator in research and development of clinical studies, having received personal payments.

References

1. Hugen, S.; Thomas, R.E.; German, A.J.; Burgener, I.A.; Mandigers, P.J.J. Gastric carcinoma in canines and humans, a review. *Vet. Comp. Oncol.* **2017**, *15*, 692–705. [CrossRef] [PubMed]
2. Saito, T.; Nibe, K.; Chambers, J.K.; Uneyama, M.; Nakashima, K.; Ohno, K.; Tsujimoto, H.; Uchida, K.; Nakayama, H. A histopathological study on spontaneous gastrointestinal epithelial tumors in dogs. *J. Toxicol. Pathol.* **2020**, *33*, 105–113. [CrossRef]
3. Abrams, B.; Wavreille, V.A.; Husbands, B.D.; Matz, B.M.; Massari, F.; Liptak, J.M.; Cray, M.T.; de Mello Souza, C.H.; Wustefeld-Janssens, B.G.; Oblak, M.L.; et al. Perioperative complications and outcome after surgery for treatment of gastric carcinoma in dogs: A Veterinary Society of Surgical Oncology retrospective study of 40 cases (2004–2018). *Vet. Surg.* **2019**, *48*, 923–932. [CrossRef] [PubMed]
4. Spillmann, T.; Candido, M. Stomach. In *BSAVA Manual of Canine and Feline Gastroenterology*, 3rd ed.; Hall, E.J., Williams, D.A., Kathrani, A., Eds.; BSAVA: Gloucester, UK, 2020; pp. 117–197.
5. Coda, S.; Thillainayagam, A.V. State of the art in advanced endoscopic imaging for the detection and evaluation of dysplasia and early cancer of the gastrointestinal tract. *Clin. Exp. Gastroenterol.* **2014**, *7*, 133–150. [CrossRef] [PubMed]
6. Cândido, M.V.; Syrjä, P.; Hanifeh, M.; Lepajõe, J.; Salla, K.; Kilpinen, S.; Noble, P.J.; Spillmann, T. Gastric mucosal pathology in Belgian Shepherd dogs with and without clinical signs of gastric disease. *Acta Vet. Scand.* **2021**, *63*, 7. [CrossRef] [PubMed]
7. Ngamruengphong, S.; Abe, S.; Oda, I. Endoscopic management of early gastric adenocarcinoma and preinvasive gastric lesions. *Surg. Clin. N. Am.* **2017**, *97*, 371–385. [CrossRef]
8. Pimentel-Nunes, P.; Libânio, D.; Lage, J.; Abrantes, D.; Coimbra, M.; Esposito, G.; Hormozdi, D.; Pepper, M.; Drasovean, S.; White, J.R.; et al. A multicenter prospective study of the real-time use of narrow-band imaging in the diagnosis of premalignant gastric conditions and lesions. *Endoscopy* **2016**, *48*, 723–730. [CrossRef]
9. Salavati, M.; Perez-Accino, J.; Tan, Y.L.; Liuti, T.; Smith, S.; Morrison, L.; Salavati Schmitz, S. Correlation of minimally invasive imaging techniques to assess intestinal mucosal perfusion with established markers of chronic inflammatory enteropathy in dogs. *J. Vet. Intern. Med.* **2021**, *35*, 162–171. [CrossRef]

10. Yokoyama, T.; Miyahara, R.; Funasaka, K.; Furukawa, K.; Yamamura, T.; Ohno, E.; Nakamura, M.; Kawashima, H.; Watanabe, O.; Hirooka, Y.; et al. The utility of ultrathin endoscopy with flexible spectral imaging color enhancement for early gastric cancer. *Nagoya J. Med. Sci.* **2019**, *81*, 241–248.

11. Capelle, L.G.; Haringsma, J.; de Vries, A.C.; Steyerberg, E.W.; Biermann, K.; van Dekken, H.; Kuipers, E.J. Narrow band imaging for the detection of gastric intestinal metaplasia and dysplasia during surveillance endoscopy. *Dig. Dis. Sci.* **2010**, *55*, 3442–3448. [CrossRef]

12. Gershman, G.; Ament, M. Chromoendoscopy. In *Practical Pedriatric Gastrointestinal Endoscopy*; Gershman, G., Ament, M., Eds.; John Wiley & Sons: Chichester, UK, 2008; pp. 181–193.

13. Uedo, N.; Yao, K. Endoluminal diagnosis of early gastric cancer and its precursors: Bridging the gap between endoscopy and pathology. *Adv. Exp. Med. Biol.* **2016**, *908*, 293–316.

14. Spillmann, T.; Willard, M.D.; Ruhnke, I.; Suchodolski, J.S.; Steiner, J.M. Feasibility of endoscopic retrograde cholangiopancreatography in healthy cats. *Vet. Radiol. Ultrasound* **2014**, *55*, 85–91. [CrossRef]

15. Basford, P.J.; Brown, J.; Gadeke, L.; Fogg, C.; Haysom-Newport, B.; Ogollah, R.; Bhattacharyya, R.; Longcroft-Wheaton, G.; Thursby-Pelham, F.; Neale, J.R.; et al. A randomized controlled trial of pre-procedure simethicone and N-acetylcysteine to improve mucosal visibility during gastroscopy NICEVIS. *Endosc. Int. Open* **2016**, *4*, E1197–E1202. [CrossRef]

16. Salla, K.M.; Lepajoe, J.; Candido, M.V.; Spillmann, T.; Casoni, D. Comparison of the effects of methadone and butorphanol combined with acepromazine for canine gastroduodenoscopy. *Vet. Anaesth. Analg.* **2020**, *47*, 748–756. [CrossRef]

17. Simone, A.; Casadei, A.; De Vergori, E.; Morgagni, P.; Saragoni, L.; Ricci, E. Rescue endoscopy to identify site of gastric dysplasia or carcinoma found at random biopsies. *Dig. Liver Dis.* **2011**, *43*, 721–725. [CrossRef]

18. Washabau, R.J.; Day, M.J.; Willard, M.D.; Hall, E.J.; Jergens, A.E.; Mansell, J.; Minami, T.; Bilzer, T.W. Endoscopic, biopsy, and histopathologic guidelines for the evaluation of gastrointestinal inflammation in companion animals. *J. Vet. Intern. Med.* **2010**, *24*, 10–26.

19. Amorim, I.; Taulescu, M.A.; Day, M.J.; Catoi, C.; Reis, C.A.; Carneiro, F.; Gärtner, F. Canine gastric pathology: A review. *J. Comp. Pathol.* **2016**, *154*, 9–37. [CrossRef]

20. Plattner, B.; Hostetter, J.M.; Uzal, F.A. Alimentary system and peritoneum. In *Jubb, Kennedy and Palmer's Pathology of Domestic Animals*, 6th ed.; Maxie, M.G., Ed.; Elsevier: St. Louis, MO, USA, 2016; Volume 2, p. 47.

21. Jergens, A.E.; Evans, R.B.; Ackermann, M.; Hostetter, J.; Willard, M.; Mansell, J.; Bilzer, T.; Wilcock, B.; Washabau, R.; Hall, E.J.; et al. Design of a simplified histopathologic model for gastrointestinal inflammation in dogs. *Vet. Pathol.* **2014**, *51*, 946–950. [CrossRef]

22. Pimentel-Nunes, P.; Dinis-Ribeiro, M.; Soares, J.B.; Marcos-Pinto, R.; Santos, C.; Rolanda, C.; Bastos, R.P.; Areia, M.; Afonso, L.; Bergman, J.; et al. A multicenter validation of an endoscopic classification with narrow band imaging for gastric precancerous and cancerous lesions. *Endoscopy* **2012**, *44*, 236–246. [CrossRef]

23. Uedo, N.; Ishihara, R.; Iishi, H.; Yamamoto, S.; Yamada, T.; Imanaka, K.; Takeuchi, Y.; Higashino, K.; Ishiguro, S.; Tatsuta, M. A new method of diagnosing gastric intestinal metaplasia: Narrow-band imaging with magnifying endoscopy. *Endoscopy* **2006**, *38*, 819–824. [CrossRef]

24. Waddingham, W.; Graham, D.; Banks, M.; Jansen, M. The evolving role of endoscopy in the diagnosis of premalignant gastric lesions. *F1000Research* **2018**, *7*, 715. [CrossRef] [PubMed]

25. Rugge, M.; Fassan, M.; Pizzi, M.; Farinati, F.; Sturniolo, G.C.; Plebani, M.; Graham, D.Y. Operative link for gastritis assessment vs. operative link on intestinal metaplasia assessment. *World J. Gastroenterol.* **2011**, *17*, 4596–4601. [CrossRef] [PubMed]

26. Seim-Wikse, T.; Jörundsson, E.; Nødtvedt, A.; Grotmol, T.; Bjornvad, C.R.; Kristensen, A.T.; Skancke, E. Breed predisposition to canine gastric carcinoma—A study based on the Norwegian canine cancer register. *Acta Vet. Scand.* **2013**, *55*, 25. [CrossRef] [PubMed]

27. Candido, M.V.; Syrjä, P.; Kilpinen, S.; Spillmann, T. Canine breeds associated with gastric carcinoma, metaplasia and dysplasia diagnosed by histopathology of endoscopic biopsy samples. *Acta Vet. Scand.* **2018**, *60*, 37. [CrossRef]

28. Munday, J.S.; Löhr, C.V.; Kiupel, M. Tumors of the alimentary tract. In *Tumors in Domestic Animals*, 5th ed.; Meuten, D.J., Ed.; John Wiley and Sons Inc.: Ames, IA, USA, 2020; pp. 499–600.

29. Hoffman, A.; Kiesslich, R.; Bender, A.; Neurath, M.F.; Nafe, B.; Herrmann, G.; Jung, M. Acetic acid-guided biopsies after magnifying endoscopy compared with random biopsies in the detection of Barrett's esophagus: A prospective randomized trial with crossover design. *Gastrointest Endosc.* **2006**, *64*, 1–8. [CrossRef]

30. Kara, M.A.; Peters, F.P.; Rosmolen, W.D.; Krishnadath, K.K.; Ten Kate, F.J.; Fockens, P.; Bergman, J.J. High-resolution endoscopy plus chromoendoscopy or narrow-band imaging in Barrett's esophagus: A prospective randomized crossover study. *Endoscopy* **2005**, *37*, 929–936. [CrossRef]

31. Gruner, M.; Denis, A.; Masliah, C.; Amil, M.; Metivier-Cesbron, E.; Luet, D.; Kaasis, M.; Coron, E.; Le Rhun, M.; Lecleire, S.; et al. Narrow-band imaging versus Lugol chromoendoscopy for esophageal squamous cell cancer screening in normal endoscopic practice: Randomized controlled trial. *Endoscopy* **2021**, *53*, 674–682. [CrossRef]

32. Bisschops, R.; Bessissow, T.; Joseph, J.A.; Baert, F.; Ferrante, M.; Ballet, V.; Willekens, H.; Demedts, I.; Geboes, K.; De Hertogh, G.; et al. Chromoendoscopy versus narrow band imaging in UC: A prospective randomised controlled trial. *Gut* **2018**, *67*, 1087–1094. [CrossRef]

33. Ishioka, M.; Chino, A.; Ide, D.; Saito, S.; Igarashi, M.; Nagasaki, T.; Akiyoshi, T.; Nagayama, S.; Fukunaga, Y.; Ueno, M.; et al. Adding narrow-band imaging to chromoendoscopy for the evaluation of tumor response to neoadjuvant therapy in rectal cancer. *Dis. Colon Rectum* **2021**, *64*, 53–59. [CrossRef]
34. Pennachi, C.M.P.S.; Moura, D.T.H.; Amorim, R.B.P.; Guedes, H.G.; Kumbhari, V.; Moura, E.G.H. Lugol's iodine chromoendoscopy versus narrow band image enhanced endoscopy for the detection of esophageal cancer in patients with stenosis secondary to caustic/corrosive agent ingestion. *Arq. Gastroenterol.* **2017**, *54*, 250–254. [CrossRef]
35. Lage, J.; Pimentel-Nunes, P.; Figueiredo, P.C.; Libanio, D.; Ribeiro, I.; Jacome, M.; Afonso, L.; Dinis-Ribeiro, M. Light-NBI to identify high-risk phenotypes for gastric adenocarcinoma: Do we still need biopsies? *Scand. J. Gastroenterol.* **2016**, *51*, 501–506. [CrossRef]
36. Yoshida, N.; Doyama, H.; Yano, T.; Horimatsu, T.; Uedo, N.; Yamamoto, Y.; Kakushima, N.; Kanzaki, H.; Hori, S.; Yao, K.; et al. Early gastric cancer detection in high-risk patients: A multicentre randomised controlled trial on the effect of second-generation narrow band imaging. *Gut* **2021**, *70*, 67–75. [CrossRef]

Article

Intestinal S100/Calgranulin Expression in Cats with Chronic Inflammatory Enteropathy and Intestinal Lymphoma

Denise S. Riggers [1], Corinne Gurtner [2], Martina Protschka [3], Denny Böttcher [4], Wolf von Bomhard [5], Gottfried Alber [3], Karsten Winter [6], Joerg M. Steiner [7] and Romy M. Heilmann [1,*]

[1] Small Animals Department, College of Veterinary Medicine, Leipzig University, 04103 Leipzig, Germany
[2] Institute of Animal Pathology, Department of Infectious Diseases and Pathobiology (DIP), Vetsuisse Faculty, University of Bern, 3012 Bern, Switzerland
[3] Institute of Immunology/Molecular Pathogenesis, Center for Biotechnology and Biomedicine, College of Veterinary Medicine, Leipzig University, 04103 Leipzig, Germany
[4] Institute of Veterinary Pathology, College of Veterinary Medicine, Leipzig University, 04103 Leipzig, Germany
[5] Veterinary Pathology Center, 80689 Munich, Germany
[6] Institute of Anatomy, Medical Faculty, Leipzig University, 04103 Leipzig, Germany
[7] Gastrointestinal Laboratory, College of Veterinary Medicine and Biomedical Sciences, Texas A&M University, College Station, TX 77843, USA
* Correspondence: romy.heilmann@kleintierklinik.uni-leipzig.de

Citation: Riggers, D.S.; Gurtner, C.; Protschka, M.; Böttcher, D.; von Bomhard, W.; Alber, G.; Winter, K.; Steiner, J.M.; Heilmann, R.M. Intestinal S100/Calgranulin Expression in Cats with Chronic Inflammatory Enteropathy and Intestinal Lymphoma. *Animals* **2022**, *12*, 2044. https://doi.org/10.3390/ani12162044

Academic Editor: Edward J. Hall

Received: 15 July 2022
Accepted: 9 August 2022
Published: 11 August 2022

Publisher's Note: MDPI stays neutral with regard to jurisdictional claims in published maps and institutional affiliations.

Simple Summary: Intestinal diseases in cats are complicated diseases in which intestinal inflammation is difficult to distinguish from lymphoma, which is a neoplasm. In this study, the expression of the proteins S100A8/A9 and S100A12 (also called calgranulins) in the intestine is investigated in both diseases and for potential correlations with microscopically visible changes in the intestine or the clinical severity of the disease. Only small differences were seen between healthy and diseased animals, and there were no differences between cats with intestinal inflammation and lymphoma. However, several correlations of cells staining positive for calgranulins and inflammatory changes at the microscopic level and clinical disease severity were shown. This indicates that calgranulins play a role in both gastrointestinal lymphoma and inflammation and would support the recent theory that these two diseases might not be separate disease entities but instead are related. Further insights into the role of the calgranulins in these feline diseases will lead to a better understanding of the disease pathogenesis and, thus, potentially novel diagnostics and treatment avenues.

Abstract: Diagnosing chronic inflammatory enteropathies (CIE) in cats and differentiation from intestinal lymphoma (IL) using currently available diagnostics is challenging. Intestinally expressed S100/calgranulins, measured in fecal samples, appear to be useful non-invasive biomarkers for canine CIE but have not been evaluated in cats. We hypothesized S100/calgranulins to play a role in the pathogenesis of feline chronic enteropathies (FCE) and to correlate with clinical and/or histologic disease severity. This retrospective case-control study included patient data and gastrointestinal (GI) tissues from 16 cats with CIE, 8 cats with IL, and 16 controls with no clinical signs of GI disease. GI tissue biopsies were immunohistochemically stained using polyclonal α-S100A8/A9 and α-S100A12 antibodies. S100A8/A9$^+$ and S100A12$^+$ cells were detected in all GI segments, with few significant differences between CIE, IL, and controls and no difference between diseased groups. Segmental inflammatory lesions were moderately to strongly correlated with increased S100/calgranulin-positive cell counts. Clinical disease severity correlated with S100A12$^+$ cell counts in cats with IL ($\rho = 0.69$, $p = 0.042$) and more severe diarrhea with colonic lamina propria S100A12$^+$ cells with CIE ($\rho = 0.78$, $p = 0.021$) and duodenal S100A8/A9$^+$ cells with IL ($\rho = 0.71$, $p = 0.032$). These findings suggest a role of the S100/calgranulins in the pathogenesis of the spectrum of FCE, including CIE and IL.

Keywords: calgranulins; calgranulin C; calprotectin; chronic enteropathy; chronic inflammatory enteropathy; feline; food-responsive enteropathy; intestinal lymphoma; small-cell lymphoma; S100A8/A9; S100A12

1. Introduction

Feline chronic enteropathy (FCE) is one of the most common disorders in elderly cats, with an increasing prevalence over the last years [1,2]. While there are many studies on chronic enteropathies in dogs, much less is known about FCE [3–5]. Currently, FCE is defined as the presence of gastrointestinal (GI) clinical signs for ≥3 weeks and exclusion of other GI conditions (e.g., endoparasitic infections) and extra-GI diseases (e.g., hyperthyroidism) [3,6]. The most common types of FCE are chronic inflammatory enteropathy (CIE) and small-cell alimentary lymphoma (SCL) [6].

The etiopathogenesis of CIE has not been fully elucidated but appears to result from a complex interplay between environmental factors (e.g., food antigens, intestinal microbiome), a dysregulated immune response, and genetic predisposition [4,6]. The subclassification of CIE used in most investigations is retrospectively based on the clinical response to treatment into food-responsive enteropathy (FRE) and immunosuppressant-responsive enteropathy (IRE), which cannot necessarily be differentiated by histology [6,7]. Older literature also includes the CIE subclass antibiotic-responsive enteropathy (ARE), the existence of which in cats is debated and more likely presents an idiopathic or secondary small intestinal dysbiosis [8].

The three main types of feline GI lymphoma are low-grade SCL, intermediate-to-large-cell lymphoma (LCL), and large granular lymphoma (LGL). These subtypes vary in histologic appearance, treatment, and prognosis, whereby SCL accounts for approximately 75% of all GI lymphoma cases in cats. SCL is characterized by a well-differentiated population of small lymphocytes with low mitotic rates and usually slow clinical progression (median survival time [MST]: 19–29 months) [9]. In contrast, LGLs have an immature and less differentiated cell population, higher mitotic rates, and faster progression, leading to more rapid clinical courses (MST: 7–10 months) [10,11].

Currently, the subclassification of feline CIE as either FRE or IRE and the differentiation from diffuse infiltrative neoplasia (particularly SCL) is challenging [12]. SCL and CIE largely overlap concerning patient signalment (age, breed), clinical signs (weight loss, vomiting, anorexia, diarrhea), and the results of non-invasive diagnostics (routine blood work, abdominal ultrasonography) [5,6,13–19]. Histopathology is currently the most reliable diagnostic to document the presence of FCE [6]. The most common histologic type of feline CIE is lymphoplasmacytic enteritis (LPE), with the inflammatory infiltrate localized to the lamina propria (and in some cases, the intestinal epithelium) and often accompanied by architectural lesions of tissue. In SCL cases, epithelial, mucosal, or transmural infiltration with well-differentiated neoplastic lymphocytes is often accompanied by lymphoplasmacytic inflammation [20–23]. However, diagnosing LPE alone is not equivalent to confirming CIE and can also be associated with food hypersensitivity, intestinal parasites, and hyperthyroidism [1], which need to be excluded as differentials. Often, CIE and SCL cannot be distinguished based on histology alone, and advanced diagnostics, including immunohistochemistry (IHC) and polymerase chain reaction (PCR)-based methods, are needed [24,25]. A recent study also showed promise for the diagnostic utility of histology-guided mass spectrometry [26].

The mainstay of treatment for feline IRE and SCL cases involves immunosuppression or chemotherapy with corticosteroids (e.g., prednisolone), calcineurin inhibitors (cyclosporine), and/or alkylating agents (chlorambucil) [1,4,6], all of which can have significant adverse effects. More targeted pathway-specific treatment options would be desirable in the hands of the small animal practitioner. Thus, further evaluating the inflammatory pathways involved in FCE will improve our understanding of these conditions and potentially open new avenues to individualized therapeutic intervention.

Calprotectin, the S100A8/A9 protein complex, belongs to the damage-associated molecular pattern (DAMP) molecules of the innate immune response and is a fecal biomarker of intestinal inflammation in dogs [27–32]. It is expressed primarily by neutrophils and activated macrophages (MΦ) and also in epithelial cells [33–35] and presents a ligand for Toll-like receptor (TLR)-4, a key molecule in acute and chronic inflammatory processes [36]. The S100A12 protein, also referred to as calgranulin C, is also a DAMP molecule [33,34] and is primarily released from activated polymorphonuclear cells [33–35]. S100A12 can bind to several innate immune receptors, including the receptor for advanced glycation end products (RAGE) [37], with a central role in innate and acquired immune responses [27,38]. Concentrations of S100A8/A9 and S100A12 in fecal samples are increased in dogs with CIE and correlate with the severity of clinical signs, endoscopic lesions, histologic changes, and disease subclassification [28–31]. Similar to dogs, the intestinal mucosal cellular infiltrates in cats with CIE are predominated by lymphocytes, plasma cells, and eosinophils, whereas neutrophils seem to represent only a small proportion of cells [3,4,39,40]. The role of MΦ, particularly of the MΦ1/MΦ2 polarization, remains unclear in canine CIE [41,42] and has not been evaluated in feline CIE. Preliminary data in cats suggest that fecal calprotectin [43] and S100A12 [44] are also significantly increased with chronic GI inflammation, but the source of these S100/calgranulins in the GI tract in FCE remains unknown.

Our central hypothesis was that the S100/calgranulin proteins, S100A8/A9 and S100A12, play a role in the pathogenesis of FCE, particularly CIE, and that the expression of S100A8/A9 and S100A12 in GI tissue biopsy specimens is therefore increased (due to primary or tumor-associated inflammation) and reflects the number and/or activity of intestinal mucosal polymorphonuclear cells in cats with feline alimentary lymphoma and/or CIE compared to healthy controls. Further, we hypothesized that the expression of S100A8/A9 and S100A12 correlates with the severity of histologic lesions and clinical disease as assessed by the feline chronic enteropathy activity index (FCEAI [5]) score. Thus, our study aimed to evaluate S100A8/A9 and S100A12 expression in GI tissue biopsies from cats with FCE and healthy controls.

2. Materials and Methods

2.1. Ethics Approval

Ethics approval was not required for the cases enrolled in this study at the University of Leipzig College of Veterinary Medicine (UL-CVM), as patient data and archived sample materials were retrospectively included, and owners provided their written consent for the use of data and surplus materials from routine diagnostics on the standard patient admission form of the UL-CVM Department for Small Animals.

2.2. Study Population and Routine Diagnostics

Patient medical records of all cats that underwent GI endoscopy between 01/2011 and 01/2019 ($n = 71$) for various differentials and of all cats that had surgical or endoscopic GI biopsies for suspicion of CE ($n = 51$) at the UL-CVM Department for Small Animals were searched ($n = 122$) to identify all cats with a diagnosis of CIE or intestinal lymphoma, confirmed by histopathological examination, between January 2011 and July 2021. Complete medical records of cats with differentials CIE or IL ($n = 53$) were reviewed by one of the investigators (DR), and cats were selected for inclusion in the study based on (i) the presence of clinical signs for a minimum of 3 weeks prior to routine diagnostic evaluation and sample collection, (ii) exclusion of other underlying GI (e.g., endoparasites), extra-GI, or non-lymphoid neoplastic conditions, (iii) not receiving any medication that may interfere with calgranulin expression (e.g., corticosteroids [45] or non-steroidal anti-inflammatory drugs at the time of first biopsy), and (iv) a final diagnosis of either CIE or IL by a board-certified veterinary pathologist (Figure 1).

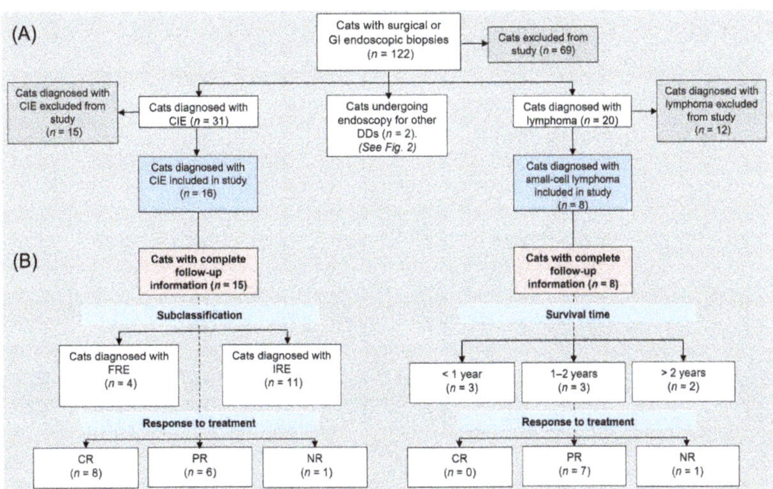

Figure 1. Flowchart of (**A**) all cats with chronic enteropathies included in the study and (**B**) those cats with FCE and complete follow-up information. Abbreviations: n = number; CIE = chronic inflammatory enteropathy; FRE = food-responsive enteropathy, IRE = immunosuppressant-responsive enteropathy; GI = gastrointestinal; DD = differential diagnosis; LCL = large-cell lymphoma; SCL = small-cell lymphoma; CR = complete remission; PR = partial remission; NR = no response. State of remission was based on clinical signs, partial remission in lymphoma cases was defined as "stable disease".

The possibility of other conditions was evaluated based on a minimum database, including a complete blood count, serum biochemistry profile, fecal examination by flotation (n = 23), and *Giardia* spp. antigen ELISA (n = 20), and abdominal ultrasonography. Sonographic findings were evaluated based on the presence of increased total GI wall thickness (>2.5 mm for the duodenum and jejunum and >3.2 mm for the ileum; level of increase graded as normal, mildly increased, or markedly increased [46]), loss of GI wall layering, increased thickness of the muscularis layer (>0.3 mm for the duodenum, >0.4 mm for the jejunum, and >0.9 mm for the ileum; graded as normal, mildly increased, or markedly increased [46]), enlarged mesenteric lymph nodes, and the presence of (scant) ascites [47].

If deemed to be indicated by the attending clinician, serum feline-specific pancreatic lipase (fPLI, measured as Spec fPL) (n = 20), cobalamin (n = 23), folate (n = 16), fructosamine (n = 18), total thyroxine (T4; n = 17), and feline trypsin-like immunoreactivity (fTLI) (n = 6) concentrations, urine analysis (n = 10), retrovirus (FeLV/FIV) status (n = 18), bile acid stimulation test (n = 3), and thoracic radiography or computed tomography scan (n = 22) were performed; 3 cats were tested for *Tritrichomonas foetus*. Further diagnostic evaluation of the cats included an upper (n = 6) or combined upper and lower (n = 13) GI endoscopy with mucosal biopsies or exploratory laparotomy with surgical (full-thickness) GI biopsies (n = 5). Endoscopy was repeated in 2 cats after 29 and 5 months, respectively.

The FCEAI score [5] was retrospectively calculated for all cats diagnosed with CIE or GI lymphoma to assess the clinical disease severity at the time of diagnosis. This 9-parameter scoring system considers the following parameters: GI endoscopic lesions, abnormal serum total protein (TP) concentration, ALT/ALP activity, phosphorous concentration, attitude/activity, and the presence of GI clinical signs, including vomiting, diarrhea, weight loss, and/or hyporexia. Clinical signs were graded from 0 to 3 depending on the severity (0 = normal, 1 = mild, 2 = moderate, 3 = severe), the remaining variables were dichotomously scored (0 = normal, 1 = increased (TP, ALT, ALP) or decreased (albumin, phosphorus)) [5]. Because no ranges have been established for interpreting the disease severity based on the cumulative FCEAI score, the following categories were considered:

mild (FCEAI score of 0–5), moderate (FCEAI score of 6–12), and severe clinical disease (FCEAI score of 13–19).

A standardized questionnaire was used to obtain patient follow-up information from the owners and referring or attending veterinarians, asking for information regarding clinical signs, treatment, and individual outcomes. Based on this follow-up information, cats in the CIE group were subclassified as either FRE or IRE, and the survival times and response to treatment were analyzed.

For establishing a control group (Figure 2), medical records were also searched for cats without GI disease (*n* = 1) based on histopathology results as determined by a board-certified veterinary pathologist (WvB). Data and GI tissues of cats biopsied between January 2019–June 2021, archived at the Texas A&M University Gastrointestinal Laboratory and not currently enrolled in other studies, were also reviewed (*n* = 36), but none of these cats were considered as not having histologic evidence of GI disease. Tissue specimens from cats (*n* = 17) that were euthanized or died because of a non-GI-related cause at the UL-CVM Department for Small Animals or sent for autopsy to the UL-CVM Institute of Veterinary Pathology between April 2021–January 2022 with no prior history of GI signs or disease were obtained and examined by a nationally accredited veterinary pathologist (DB). Because none of the cats were considered free of any GI lesions in all segments of the GI tract, those full-thickness GI segments and areas with no histologic evidence of inflammation were included as control tissues.

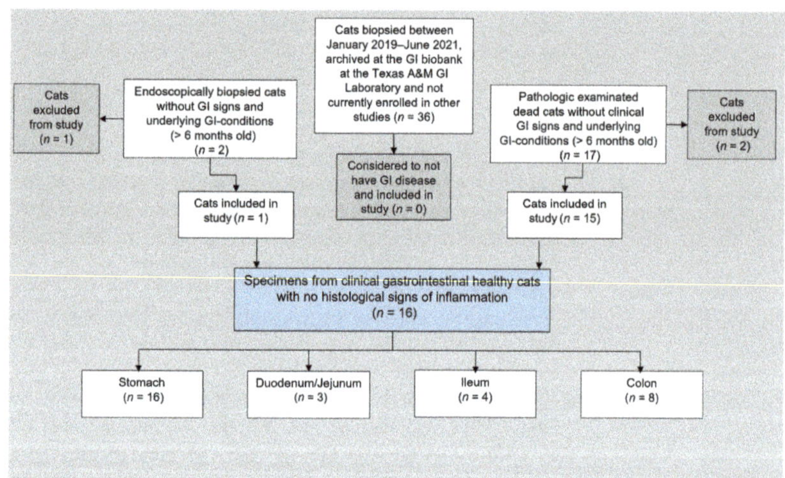

Figure 2. Flowchart of all cats included in the control group. Note that none of the cats was considered to be completely free of GI disease; thus, full-thickness tissue specimens of GI segments and areas with no histologic lesions were included as control tissues. GI = gastrointestinal.

2.3. Tissue Sample Collection and Routine Evaluation

A minimum of 5 endoscopic or between 1 and 3 full-thickness tissue biopsies per section and animal were obtained from the stomach (*n* = 35), duodenum/proximal jejunum (*n* = 27), ileum (*n* = 14), and/or colon (*n* = 18). These biopsies were used for routine histopathology and to evaluate the intestinal mucosal expression of S100A8/A9 and S100A12. Two cats diagnosed with GI lymphoma were re-biopsied when clinical signs recurred or worsened 29 and 5 months after the first diagnostic investigation of the cat; these biopsies were entered in the data analysis as individual tissue samples.

The tissues were submitted for routine histologic evaluation (WvB, DB) using the criteria of the WSAVA Gastrointestinal Standardization grading system [39] to assess morphologic lesions and inflammatory changes (including the severity of neutrophilic and macrophage infiltration) in the stomach, duodenum/proximal jejunum, ileum, and colon each on a 4-point scale (0 = normal, 1 = mild lesions, 2 = moderate lesions, and 3 = severe lesions). Cumulative lesion scores (i.e., the sum of individual lesion scores) were calculated. IHC analysis for CD3 (T cell) and CD20 positivity (B cell) was performed for all animals diagnosed with CIE or GI lymphoma (WvB, DB). For 18 of the cats, CD-staining was semi-quantitatively and spatially analyzed (DB) based on the criteria of Freiche et al. (2021) [48] (Supplementary Figure S1) with specific evaluation of villus atrophy, intraepithelial nests/plaques of lymphoid cells, lymphocytic cryptitis, cell counts and T-lymphocyte gradient in the lamina propria, infiltration of the intestinal epithelium, submucosa and subserosa/serosa with lymphoid cells, and percentage of CD3- and CD20-positive staining cells in the lamina propria and epithelium. Additionally, PCR for antigen receptor rearrangement (PARR) was performed on tissue biopsies from 10 cats at the Department of Pathobiology, University of Veterinary Medicine Vienna, Vienna, Austria.

2.4. IHC for Tissue S100A8/A9 and S100A12 Protein–Protocol Adjustment for Feline Tissues

Polyclonal rabbit antibodies generated against canine S100A8/A9 [49,50] and canine S100A12 protein [51–53] were used for IHC analysis of S100A8/A9 and S100A12 expression. Cross-reactivity of these polyclonal antibodies with feline S100A8/A9 and feline S100A12 has been demonstrated in immunodiffusion experiments and xenospecies immunoassays employing these antibodies or antiserum [49,51,54,55]. Surplus materials provided by the autopsy service at the Institute of Veterinary Pathology at UL-CVM were used as IHC controls. Feline brain tissue served as a negative control, whereas feline spleen was used as a positive control (Figure 3).

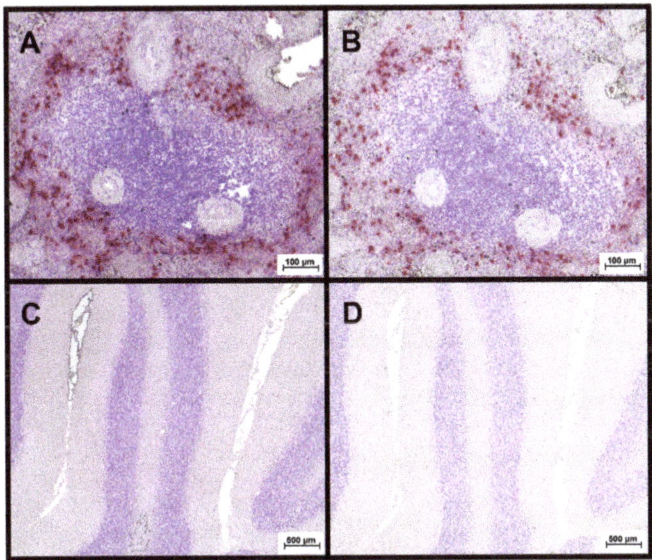

Figure 3. Evaluation of S100A8/A9- and S100A12-specific staining in feline tissues using polyclonal anti-canine S100A8/A9 and anti-S100A12 antibodies [35]. Large numbers of cells staining positive (red) for (**A**) S100A8/A9 and (**B**) S100A12 were detected in feline splenic tissue (used as positive control). In contrast, feline cerebral cortex (negative control) showed no cells staining positive for (**C**) S100A8/A9 nor (**D**) S100A12.

All tissue specimens were fixed in neutral buffered formalin (10%) for a minimum of 24 h at room temperature (approximately 23 °C) and embedded in paraffin wax. The blocks containing the tissue biopsies were then cut into 2 µm or 3 µm thick sections for IHC and other analyses (e.g., CD3 or CD20 staining).

For S100A8/A9 and S100A12 staining, a modification of the IHC protocol originally established for tissues from dogs [35] was used. Briefly, following deparaffinization and rehydration with xylene and ethanol, each of the biopsy sections underwent antigen retrieval by heat treatment (95 °C) in 0.01 M citrate buffer (pH 6.0). Slides were then washed in phosphate-buffered saline (PBS), and nonspecific binding sites were blocked in PBS supplemented with 4% bovine serum albumin (BSA). After incubating the slides with the primary antibodies (purified polyclonal anti-canine S100A8/A9 [49] at 0.2 µg/mL and polyclonal anti-canine S100A12 [53] at 0.25 µg/mL in PBS with 4% BSA; normal rabbit serum at 0.2 µg/mL in PBS supplemented with 0.025% (v/v) Tween-20 (PBST) served as negative staining control), the secondary antibody (goat anti-rabbit alkaline phosphatase-labeled IgG, Dianova, Hamburg, Germany; diluted at 1 µg/mL in PBST) was added. Fast-red (Sigma-Aldrich, St. Louis, MO, USA) was used as a substrate, visualizing positive staining as a red product, and hematoxylin was added for counterstaining. After mounting the slides using Fluoromount™ aqueous mounting media (Sigma), standard light microscopy (AX70 microscope, Olympus, Centre Valley, PA, USA) was used to assess the quality of staining in the tissue biopsies.

2.5. S100A8/A9 and S100A12 IHC Analysis of Feline GI Tissue Biopsies

Tissue samples from 26 cats with FCE, of which 16 cats were diagnosed with CIE and 8 cats with alimentary lymphoma (2 cats biopsied twice), and 16 healthy controls were used for this study. All tissues were stained as described (Supplementary Figure S2).

All slides were converted to digital data using 3D Histech Pannoramic Scan II (3D Histech Ltd., Budapest, Hungary), and the images were analyzed using a free-ware histology slide viewing and evaluation software (SlideViewer v2.3, 3D Histech, Budapest, Hungary). A board-certified veterinary pathologist (CG) and trainee (DR) independently evaluated the number of positive-staining cells in all GI biopsies (blinded to the identity, clinical, clinicopathologic, and histopathologic results); these results were then combined, and a consensus was reached for any deviating cell counts that were obtained.

Mucosal S100A8/A9 and S100A12 positivity was assessed separately for the 2 different mucosal compartments: (i) the epithelial layer and (ii) the lamina propria. S100A8/A9$^+$ and S100A12$^+$ cell counts were determined for each segment (stomach, duodenum/proximal jejunum, ileum, and colon) and compartment (epithelium and lamina propria) (Figure 4) by counting all cells staining positive for S100A8/A9 or S100A12 in 10 defined areas (regions of interest that did not include intravascular cells, 0.01 mm^2 each) under high-power fields (Figure 5). The mean number of positive cells identified in all fields was calculated and used for statistical analysis.

2.6. Statistical Analyses

Commercially available software packages (JMP® v13.0, SAS Institute, Cary, NC, USA and GraphPad Prism® v9.0, GraphPad Software, San Diego, CA, USA) were used for all statistical analyses. Based on the results of a similar study [35] in dogs, a sample size of 10 per group was expected to be sufficient. Normality of the data distribution was tested using a Shapiro–Wilk test, and the summary statistics were reported as medians and interquartile ranges (IQR) or ranges (for continuous data) or counts and percentages (for categorical data). Two-group or multiple-group comparisons were performed using a Wilcoxon rank-sum and Kruskal–Wallis test, respectively. A likelihood ratio test was performed to test for associations of categorical variables and correlation analyses by calculating a Spearman correlation coefficient (ρ). Statistical significance was set at $p < 0.05$, and a Holm-Bonferroni correction was applied for multiple correlations with the number of categories or subcategories considered.

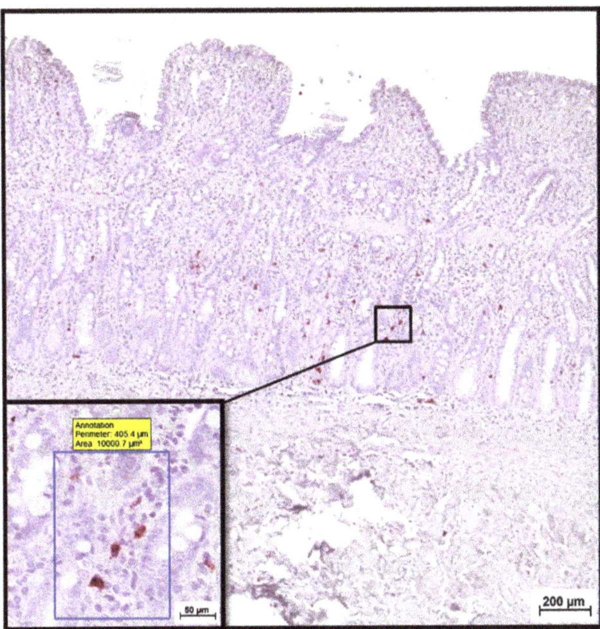

Figure 4. IHC for tissue S100/calgranulin protein expression. Shown is an example for selecting 1 of the 10 defined areas within the lamina propria (0.01 mm^2 each, examined for each mucosal compartment and GI segment) of a duodenal endoscopic biopsy obtained from a cat with CIE.

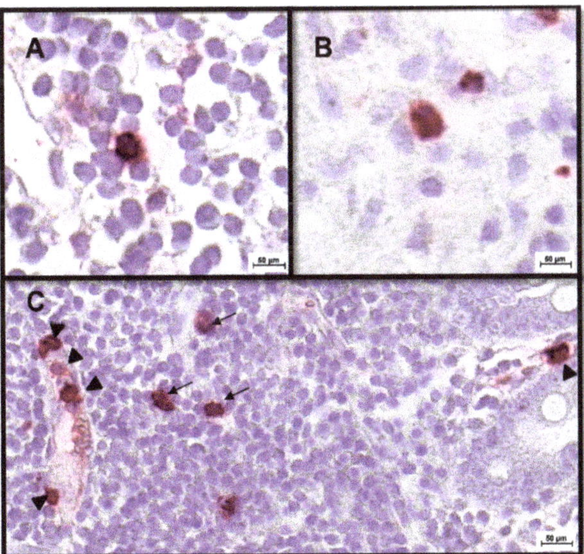

Figure 5. S100A8/A9- and S100A12-IHC of feline GI tissue biopsies. Numbers of cells staining positive for (**A**) S100A8/A9 and (**B**) S100A12 were evaluated. As demonstrated in (**C**), only infiltrating or resident cells staining positive for S100/calgranulins (arrows), but not intravascular cells (arrowheads), were counted and included in the analyses.

3. Results

3.1. Patient Clinical Data

Cats with GI lymphoma were significantly older than those diagnosed with CIE (p = 0.039; Supplementary Table S1) or healthy controls (p = 0.016; age available for 14 healthy controls), with no difference between CIE and healthy control cats (p = 0.901). No differences were seen in the sex distribution (p = 0.340) or body weights (p = 0.368) between both disease groups of cats. Most cats in both groups were Domestic shorthair cats. Disease duration prior to presentation for diagnostic investigation and survival time after diagnosis did not differ between both groups (p = 0.254 and p = 0.364). Retrovirus status was negative in all cats that were tested. A total of 21 cats (13 cats with CIE, 7 IL cases, and 1 control cat) were tested for *Giardia* spp. and 3 cats (2 cats with CIE and 1 IL case) for *Tritrichomonas foetus*, with all being negative for either. Diagnostic imaging was pursued in all cats, with abdominal and thoracic radiographs performed in 15 cats with CIE and 7 cats with IL, and abdominal ultrasound with still images and video sequences available for re-review for all 24 cats. Sonographic findings did not differ significantly between cats with CIE and those with lymphoma, but compared to the latter, CIE cats tended to less frequently have sonographic evidence of ascites (25% vs. 60%; p = 0.075). Cats with CIE had significantly lower FCEAI scores than cats with lymphoma (p = 0.011), with less severe hyporexia (p = 0.039) and weight loss (p = 0.014; Supplementary Table S1).

3.2. Clinicopathologic Parameters

Hypocobalaminemia was significantly more frequently detected in cats with lymphoma compared to cats with CIE (88% vs. 44%, p = 0.045). No significant differences were seen in serum TP, albumin, globulin, total calcium, blood urea nitrogen (BUN), inorganic phosphorus, the prevalence of anemia (19% vs. 25%) and leukocytosis (38% vs. 50%), and serum fPL and fructosamine concentrations (all p > 0.05). Total T4 was within the reference interval in all cats, and serum fTLI concentrations (all > 16.4 μg/L) were determined in 2 cats with CIE and 4 IL cases.

3.3. Endoscopy, Histopathology, and Clonality Testing

Endoscopy and sampling of biopsies were performed and documented (images and reports) in 19 cases (esophagogastroduodenoscopy combined with ileocolonoscopy in 13 cases and esophagogastroduodenoscopy alone in 6 cats; endoscopy was repeated after 29 and 5 months in 2 cats with IL) and surgical biopsies were obtained in 5 cats. For the control group, samples were obtained from 15 cats during autopsy, and 1 cat underwent esophagogastroduodenoscopy (Figure 2). Endoscopic lesions were detected in 11/13 cats (85%) with CIE and in 6/6 cats (100%) with IL (Supplementary Table S1). Linear hyperemic lesions were seen endoscopically in 1 cat, each with CIE and lymphoma.

Inflammatory lesion scores in the stomach, duodenum/proximal jejunum, and ileum were significantly lower in cats with CIE than in those with lymphoma (p = 0.026, p < 0.0001, and p = 0.019), which was not seen for the colon, and there were no differences in morphologic lesion scores between both patient groups. PARR showed polyclonality for TCRγ in 2 cats and bi- or oligoclonality (with a polyclonal background) in 3 cats with CIE, whereas mono-, bi-, or oligoclonality was detected in all 5 cats diagnosed with IL (and a polyclonal background in 2 of those cats). Helicobacter-like organisms (HLO) were present in gastric biopsies from 9 cats (47%), including 6 cats with gastric inflammation and 3 cats with normal gastric histology. The presence of HLO was not linked to gastric mucosal S100/calgranulin-positive cell counts or endoscopic lesions but was associated with less severe histologic lesions (p = 0.012) due to lower numbers of lymphocytes and plasma cells (p = 0.024).

3.4. Follow-Up and Response to Treatment

Long-term follow-up information was available from 15 cats (94%) with CIE and all 8 lymphoma cases (Figure 1). Response to treatment differed significantly between both disease groups ($p = 0.011$). In the CIE group, 8 cats (53%) achieved complete remission, 6 cats (40%) partially responded, and 1 cat (7%) had no clinical response. Four cats (27%) were diagnosed with FRE, of which 2 cats received a commercial novel protein (monoprotein) and 2 cats an easily digestible GI diet. The remaining 11 cats were classified as IRE based on the feeding of a GI or monoprotein diet (5 cats), hydrolyzed protein diet (1 cat), or both tried sequentially (4 cats), yielding only a partial or no clinical response but significant improvement or resolution of clinical signs on anti-inflammatory or immunosuppressive (prednisolone) treatment. Dietary intervention was not possible as part of the diagnostic evaluation in 1 cat classified as IRE. Additional treatments included an antiemetic (2 cats), prebiotic and/or probiotic (3 cats), supplementation of cobalamin (6 cats), pancreatic enzyme replacement therapy (1 cat), and antimicrobial therapy (1 cat).

Most cats (88%) with lymphoma had improved clinical signs/stable disease on chemotherapy, with 1 cat (12%) experiencing progressive disease despite chemotherapy and 1 cat with an initial response relapsing after chemotherapy withdrawal. Chlorambucil was the most frequently administered chemotherapy drug, either as monotherapy (in 2 cats) or in combination with prednisolone (5 cats); 1 cat received a combination of prednisolone and cyclophosphamide. Progressive disease warranted including cyclophosphamide or doxorubicin (1 cat each) into the chemotherapy protocol. Additional treatments included a hydrolyzed protein diet (4 cats), an antiemetic (5 cats), supplementation of cobalamin (4 cats), pancreatic enzymes replacement therapy (2 cats), and antimicrobial therapy (1 cat). GI disease-associated death was less likely in cats with CIE (50%) than IL (100%; $p = 0.012$) with a relative risk (RR) of 0.5 (95% CI: 0.25–1.0). Survival times were longest in cats with FRE (median: 47 months) and differed significantly from cats with IL (median: 18 months, $p = 0.027$), but there was no significant difference compared to IRE (median: 24 months, $p = 0.101$; Figure 6). FCEAI scores at the time of diagnosis did not differ between FRE and IRE cats ($p = 0.025$).

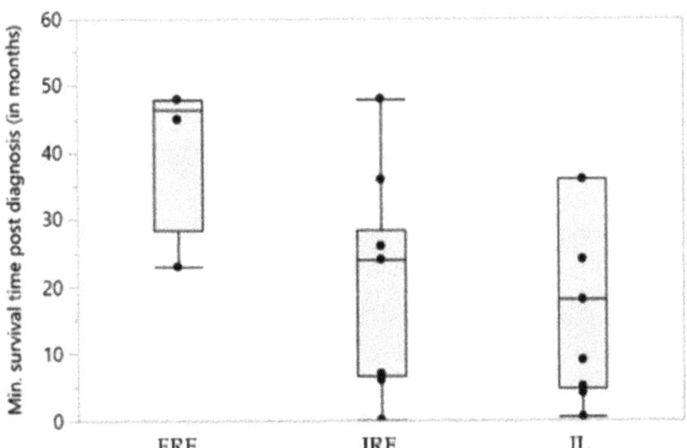

Figure 6. Survival times in cats with chronic enteropathy in this study. Cats diagnosed with chronic inflammatory enteropathies (CIE) survived longer (median: 24 months) after diagnosis than those with intestinal lymphoma (IL) (median: 18 months), but this difference was not significant. Within the CIE group, cats classified as food-responsive enteropathy (FRE) survived longer (median: 47 months) than those cats with immunosuppressant-responsive enteropathy (IRE) (median: 24 months), with no significant difference between both groups but significantly longer survival with FRE compared to IL ($p = 0.027$).

3.5. Gastrointestinal S100/Calgranulin Expression

Cats with FCE generally had slightly higher gastrointestinal mucosal S100A8/A9$^+$ and S100A12$^+$ cell counts than controls, albeit most comparisons were not statistically significant (Tables 1 and 2). The numbers of S100A8/A9$^+$ (Figure 7) and S100A12$^+$ (Figure 8) cells in any of the GI segments evaluated were not significantly different between cats with

CIE and cats with IL (all $p > 0.05$; Tables 1 and 2) nor between cats with FRE or IRE (all $p > 0.05$), except for higher maximum numbers of S100 A8/A9$^+$ cells in the gastric lamina propria in cats with FRE (median: 9, IQR: 3–13) than in IRE cats (median: 1, IQR: 0–7; $p = 0.048$). Strong correlations were detected between S100A12$^+$ cell counts and the severity of lymphoplasmacytic infiltration, eosinophilic component, and the resulting inflammatory and cumulative histologic lesion scores in the colon (Table 3) of cats with CIE, whereas a similar pattern was detected as a trend for correlations with S100A8/A9$^+$ cell counts. S100A8/A9$^+$ cell counts were moderately positively correlated with the severity of follicular hyperplasia in the stomach (Table 3), whereas an inverse trend for a correlation with gastric S100A12$^+$ cell counts was seen for intraepithelial lymphocytes. Finally, particularly S100A8/A9$^+$ counts showed a trend for a positive correlation with ileal neutrophil and macrophage counts but an inverse relationship with the ileal cumulative histopathology score. No significant correlations or trends for associations with histologic lesions were seen for duodenal/proximal jejunal S100/calgranulin-positive cells.

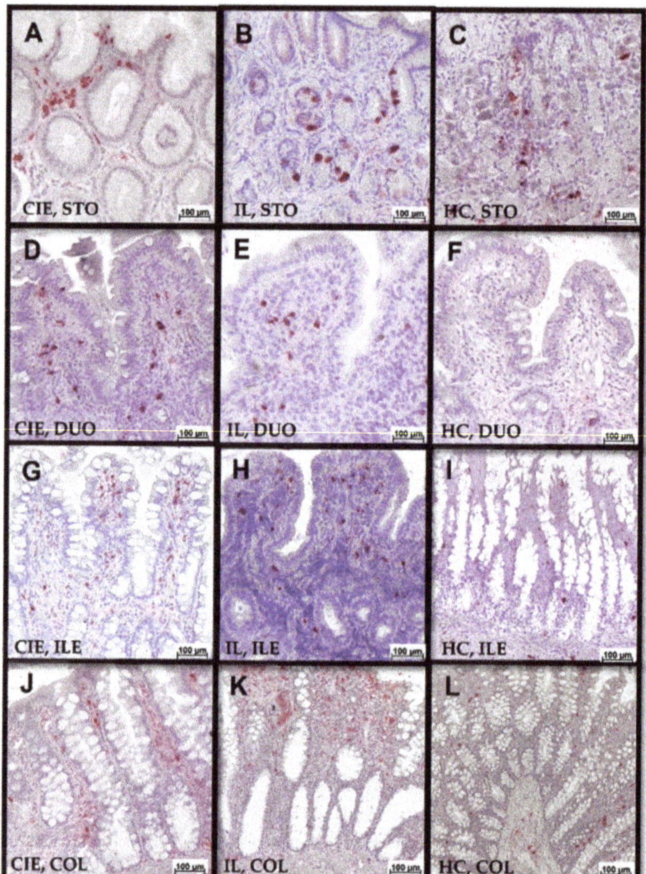

Figure 7. S100A8/A9-IHC of gastrointestinal tissue biopsies with varying numbers of intraepithelial and lamina propria S100A8/A9$^+$ cells. Left to right: chronic inflammatory enteropathy (CIE), intestinal lymphoma (IL), GI healthy control group (HC); stomach (STO, **A–C**), duodenum (DUO, **D–F**), ileum (ILE, **G–I**), colon (COL, **J–L**). Note positively stained epithelial cells in (**B**).

Table 1. Gastrointestinal tissue S100A8/A9 positivity in the cats included in the study ($n = 40$).

Histologic Lesion	n	Epithelial S100A8/A9 Positivity (Cell Counts)			Lamina Propria S100A8/A9 Positivity (Cell Counts)			Submucosal S100A8/A9 Positivity (Cell Counts)		
		Range in All Tissue Biopsies	Median (Range)	p	Range in All Tissue Biopsies	Median (Range)	p	Range in All Tissue Biopsies	Median (Range)	p
Stomach	34									
normal	19	0–1	0 (0) [A]		0–10	1.5 (0–5)		—	—	
inflammation	13	0–10	0 (0–4) [B]	0.011	0–13	1.5 (0.5–3)	0.326	—	—	—
lymphoma	2	0–3	1 (0–1) [B]		0–3	0.5 (0.5–1)		—	—	
Duodenum/Jejunum	27									
normal	2	0–2	0 (0–0.5)		0–5	1.5 (0.5–2.5) [A]		0 *	0 (0) *	
inflammation	16	0–12	0 (0–2)	0.837	0–38	6 (0–21.5) [A,B]	0.045	0–6 [$]	0 (0–1) [$]	0.749
lymphoma	9	0–6	0.5 (0–0.5)		0–27	6.5 (3–12) [B]		0–8 [†]	0 (0–2.5) [†]	
Ileum	14									
normal	2	0	0 (0)		2–13	6 (3–9)		0–1	0.5 (0–0.5)	
inflammation	8	0–1	0 (0–0.5)	0.271	0–24	3.5 (0.5–12)	0.901	0–4	0.5 (0–1)	0.719
lymphoma	4	0	0 (0)		0–16	5 (3.5–7.5)		0–1 [‡]	0.5 (0–0.5) [‡]	
Colon	17									
normal	7	0	0 (0)		0–19	6.5 (2.5–8.5)		0–11	1.5 (0.5–2.5) [A]	
inflammation	9	0–2	0 (0–0.5)	0.222	0–23	4 (1–10.5)	0.239	0–1	0 (0–0.5) [B,§]	0.004
lymphoma	1	0	0 (0)		0–21	10 (10)		0–2	0.5 (0.5) [A,B]	

Note: number of cats. * Available from $n = 1$ cat; [$] available from $n = 10$ cats; [†] available from $n = 5$ cats; [‡] available from $n = 3$ cats; [§] available from $n = 7$ cats. Parameters in bold font and green indicate significant differences at $p < 0.05$, where values stacked within a column that are not sharing a common superscript letter (A, B) are significantly different at $p < 0.05$.

Table 2. Gastrointestinal tissue S100A12 positivity in the cats included in the study ($n = 40$).

Histologic Lesion	n	Epithelial S100A12 Positivity (Cell Counts)			Lamina Propria S100A12 Positivity (Cell Counts)			Submucosal S100A12 Positivity (Cell Counts)		
		Range in All Tissue Biopsies	Median (Range)	p	Range in All Tissue Biopsies	Median (Range)	p	Range in All Tissue Biopsies	Median (Range)	p
Stomach	34									
normal	19	0–1	0 (0)		0–17	1 (0–6) [A]		—	—	
inflammation	13	0–10	0 (0–1)	0.102	0–6	0.5 (0–2) [B]	0.032	—	—	—
lymphoma	2	0	0 (0)		0–1	0.5 (0–1) [A,B]		—	—	
Duodenum/Jejunum	27									
normal	2	0	0 (0)		0–7	1.5 (0.5–3)		0	0 (0) *	
inflammation	16	0–13	0.5 (0–3.5)	0.153	0–23	4 (0.5–11)	0.342	0–2	0 (0–0.5) $	0.574
lymphoma	9	0–1	0 (0–0.5)		0–23	3.5 (1.5–8)		0–2	0 (0–0.5) [†]	
Ileum	14									
normal	2	0	0 (0)		0–10	4 (2–6)		0–1	0.5 (0–0.5)	
inflammation	8	0–1	0 (0–0.5)	0.544	0–17	2.5 (0.5–10)	0.852	0–1	0.5 (0–0.5)	0.635
lymphoma	4	0–1	0 (0–0.5)		0–30	3 (0.5–7)		0–1 [‡]	0.5 (0–0.5) [‡]	
Colon	17									
normal	7	0	0 (0)		0–30	6 (0.5–11)		0–22	0.5 (0–4) [A]	
inflammation	9	0–3	0 (0–1)	0.641	0–24	3 (0.5–9)	0.599	0–1 [§]	0 (0–0.5) [B,§]	0.041
lymphoma	1	0	0 (0)		0–12	5.5 (5.5)		0–2	0.5 (0.5) [A,B]	

Note: number of cats. * Available from $n = 2$ cats; $ available from $n = 10$ cats; [†] available from $n = 5$ cats; [‡] available from $n = 3$ cats; [§] available from $n = 7$ cats. Parameters in bold font and green indicate significant differences at $p < 0.05$, where values stacked within a column that are not sharing a common superscript letter (A, B) are significantly different at $p < 0.05$.

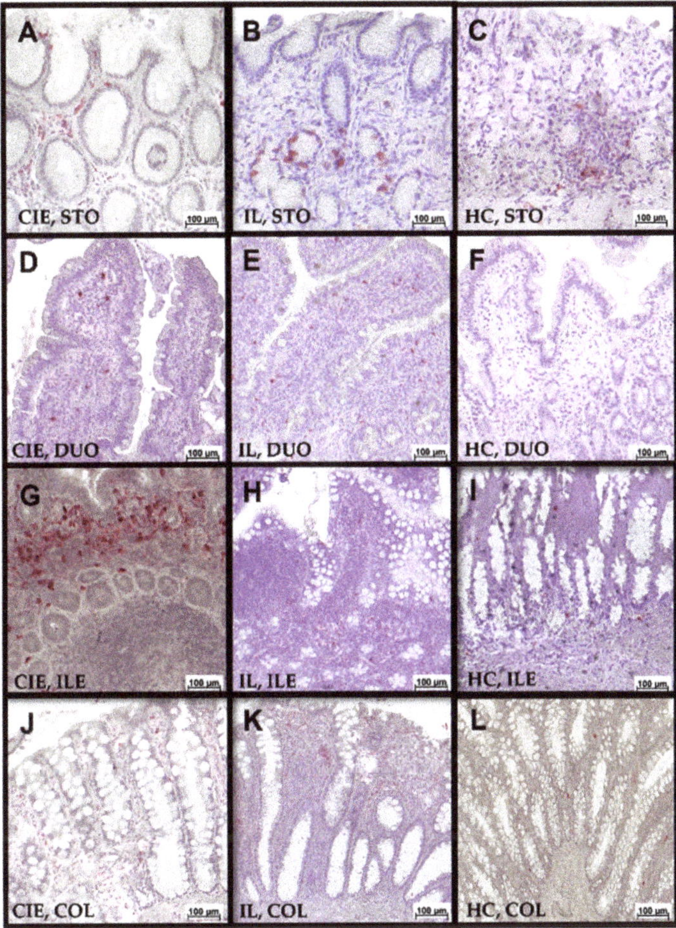

Figure 8. S100A12-IHC of gastrointestinal tissue biopsies with varying numbers of intraepithelial and lamina propria S100A12$^+$ cells. Left to right: chronic inflammatory enteropathy (CIE), intestinal lymphoma (IL), GI healthy control group (HC); stomach (STO, **A–C**), duodenum (DUO, **D–F**), ileum (ILE, **G–I**), colon (COL, **J–L**). Note positively stained epithelial cells in (**B**).

Table 3. Correlation of S100A12 and S100A8/A9 positivity with histologic findings in feline CIE. Shown are the correlations among epithelial and the lamina propria S100A8/A9$^+$ [and S100A12$^+$] cell counts and the severity of morphologic and inflammatory histologic lesions in the stomach, duodenum, ileum, and colon in cats with inflammatory lesions in these segments ($n = 20$).

Parameter	Correlated with	Spearman ρ Correlation Coefficient (p-Value, P_{corr})			
		Epithelial S100A8/A9$^+$ (Cell Counts)	Lamina Propria S100A8/A9$^+$ (Cell Counts)	Epithelial S100A12$^+$ (Cell Counts)	Lamina Propria S100A12$^+$ (Cell Counts)
		Stomach ($n = 13$)		Stomach ($n = 13$)	
Stomach (composite score) [#]		0.14 (0.655)	0.08 (0.801)	0.06 (0.844)	0.33 (0.279)
Morphologic criteria (sum)		0.49 (0.092)	−0.16 (0.609)	0.38 (0.199)	0.18 (0.559)
-	Surface epithelial injury	N/A	N/A	N/A	N/A
-	Gastric pit epithelial injury	N/A	N/A	N/A	N/A

Table 3. *Cont.*

Parameter	Correlated with	Epithelial S100A8/A9$^+$ (Cell Counts)	Lamina Propria S100A8/A9$^+$ (Cell Counts)	Epithelial S100A12$^+$ (Cell Counts)	Lamina Propria S100A12$^+$ (Cell Counts)
	- Fibrosis/glandular nesting/MA	0.33 (0.916)	−0.26 (0.394)	0.21 (0.494)	−0.09 (0.780)
Inflammatory criteria (sum)		−0.02 (0.948)	0.25 (0.403)	−0.08 (0.791)	0.38 (0.206)
	- Intraepithelial lymphocytes	−0.36 (0.221)	0.12 (0.689)	−0.54(0.059)	−0.25 (0.420)
	- Lamina propria LPC	−0.01 (0.965)	0.19 (0.528)	−0.18 (0.555)	0.42 (0.155)
	- Lamina propria eosinophils	N/A	N/A	N/A	N/A
	- Lamina propria neutrophils	0.27 (0.380)	0.12 (0.705)	0.47 (0.104)	0.31 (0.303)
	- Lamina propria MΦ	0.27 (0.380)	0.12 (0.705)	0.47 (0.104)	0.31 (0.303)
	- Lymphofollicular hyperplasia	0.59 (0.034; 0.204)	0.06 (0.852)	0.28 (0.357)	0.32 (0.295)
Duodenum/proximal jejunum (composite score)		Duodenum/proximal jejunum (*n* = 16) 0.09 (0.747)	0.18 (0.515)	Duodenum/proximal jejunum (*n* = 16) −0.08 (0.766)	0.13 (0.643)
Morphologic criteria (sum)		−0.02 (0.942)	0.11 (0.679)	−0.14 (0.614)	0.12 (0.671)
	- Villus stunting	−0.03 (0.927)	0.16 (0.546)	−0.06 (0.837)	0.14 (0.594)
	- Epithelial injury	−0.04 (0.880)	0.13 (0.632)	0.12 (0.667)	0.27 (0.320)
	- Crypt distension	0.07 (0.795)	−0.23 (0.401)	−0.06 (0.816)	−0.22 (0.423)
	- Lacteal dilation	−0.29 (0.289)	0.41 (0.115)	−0.42 (0.103)	0.21 (0.446)
	- Mucosal fibrosis	0.02 (0.931)	0.08 (0.763)	−0.19 (0.481)	0.25 (0.359)
Inflammatory criteria (sum)		0.34 (0.196)	0.44 (0.087)	0.19 (0.474)	0.29 (0.282)
	- Intraepithelial lymphocytes	0.12 (0.672)	0.06 (0.822)	−0.13 (0.620)	−0.13 (0.623)
	- Lamina propria LPC	0.14 (0.619)	0.23 (0.395)	0.01 (0.991)	0.14 (0.619)
	- Lamina propria eosinophils	0.10 (0.719)	0.26 (0.333)	0.22 (0.418)	0.17 (0.529)
	- Lamina propria neutrophils	0.42 (0.102)	0.41 (0.115)	0.25 (0.344)	0.45 (0.080)
	- Lamina propria MΦ	0.23 (0.402)	0.14 (0.605)	0.03 (0.916)	0.20 (0.467)
Ileum (composite score)		Ileum (*n* = 8) −0.70(0.053)	−0.34 (0.399)	Ileum (*n* = 8) −0.54 (0.168)	−0.35 (0.399)
Morphologic criteria (sum)		−0.58 (0.132)	−0.39 (0.346)	−0.61 (0.112)	−0.28 (0.506)
	- Villus stunting	−0.43 (0.284)	0.06 (0.882)	−0.44 (0.280)	−0.06 (0.882)
	- Epithelial injury	−0.43 (0.289)	−0.02 (0.971)	0.05 (0.899)	0.09 (0.826)
	- Crypt distension	−0.23 (0.585)	−0.55 (0.157)	−0.61 (0.111)	−0.35 (0.395)
	- Lacteal dilation	−0.28 (0.496)	0.25 (0.555)	−0.29 (0.493)	−0.25 (0.555)
	- Mucosal fibrosis	−0.02 (0.971)	−0.44 (0.275)	−0.07 (0.867)	−0.15 (0.721)
Inflammatory criteria (sum)		0.12 (0.783)	0.37 (0.367)	0.53 (0.176)	0.12 (0.786)
	- Intraepithelial lymphocytes	0.13 (0.768)	0.55 (0.162)	−0.19 (0.654)	0.22 (0.604)
	- Lamina propria LPC	−0.26 (0.528)	0.38 (0.359)	0.11 (0.805)	0.40 (0.326)
	- Lamina propria eosinophils	−0.23 (0.577)	0.03 (0.952)	0.31 (0.455)	−0.17 (0.694)
	- Lamina propria neutrophils	0.66(0.074)	−0.25 (0.555)	0.57 (0.139)	0.08 (0.846)
	- Lamina propria MΦ	0.66(0.074)	−0.25 (0.555)	0.57 (0.139)	0.08 (0.846)
Colon (composite score)		Colon (*n* = 9) 0.24 (0.527)	0.44 (0.241)	Colon (*n* = 9) 0.49 (0.179)	0.80 (0.009; 0.009)
Morphologic criteria (sum)		−0.32 (0.399)	0.12 (0.760)	−0.31 (0.430)	0.49 (0.184)
	- Epithelial injury	−0.37 (0.329)	0.00 (1.000)	−0.19 (0.626)	0.41 (0.268)
	- Goblet cell loss or hyperplasia	−0.24 (0.527)	−0.27 (0.476)	−0.13 (0.749)	0.14 (0.725)
	- Crypt dilation and distortion	−0.37 (0.333)	0.05 (0.907)	−0.19 (0.626)	0.30 (0.438)
	- Mucosal fibrosis and atrophy	0.06 (0.875)	0.21 (0.593)	−0.19 (0.626)	0.21 (0.593)
Inflammatory criteria (sum)		0.31 (0.427)	0.48 (0.197)	0.60(0.088)	0.68 (0.046; 0.092)
	- Intraepithelial lymphocytes	0.62(0.077)	−0.26 (0.500)	0.32 (0.407)	−0.35 (0.361)
	- Lamina propria LPC	0.07 (0.856)	0.51 (0.157)	0.53 (0.144)	0.84 (0.005; 0.025)
	- Lamina propria eosinophils	0.65(0.058)	0.55 (0.127)	1.00 (<0.001; <0.005)	0.55 (0.127)
	- Lamina propria neutrophils	−0.24 (0.527)	0.41 (0.272)	−0.13 (0.749)	0.27 (0.476)
	- Lamina propria MΦ	−0.24 (0.527)	0.41 (0.272)	−0.13 (0.749)	0.27 (0.476)

Note: LPC: lymphocytes/plasma cells; MA: mucosal atrophy; MΦ: macrophages; N/A: not applicable; [#] lesions evaluated for fundus and antrum (combined). Cells with numbers in red indicate a statistically significant correlation ($p < 0.05$), those cells with numbers in green indicate a trend for a significant correlation ($p < 0.1$). Correlation coefficients in bold font are >0.5 or <−0.5.

3.6. Association of Patient Characteristics with Gastrointestinal S100/Calgranulin Expression

S100/calgranulin-positive cell counts showed no association with the FCEAI scores in cats with CIE, but more severe diarrhea was associated with higher numbers of lamina propria S100A12$^+$ cells in the colon ($\rho = 0.78$, $p = 0.021$). In cats with GI lymphoma, higher duodenal/proximal jejunal lamina propria S100A12$^+$ cell counts correlated with higher

FCEAI scores ($\rho = 0.69$, $p = 0.042$) and more severe diarrhea was linked to higher numbers of lamina propria S100A8/A9$^+$ cells in the same GI segment ($\rho = 0.71$, $p = 0.032$). Given the small sample size, possible associations with colonic S100/calgranulin-positive cell counts could not be evaluated in cats with large intestinal lymphoma.

Serum cobalamin concentrations were positively correlated with the number of ileal epithelial S100A8/A9$^+$ cells ($\rho = 0.73$, $p = 0.008$). This relationship and the same trend for lamina propria S100A8/A9$^+$ and epithelial and lamina propria S100A12$^+$ cell counts remained in the cats with CIE, whereas these trends were throughout inverse in cats with ileal lymphoma. Serum folate concentrations were not correlated with S100/calgranulin-positive cell counts in any of the GI segments evaluated (all $p > 0.05$). Cats with leukocytosis had higher duodenal/proximal jejunal numbers of S100A8/A9$^+$ (median: 8.5 vs. 4; $p = 0.012$) and S100A12$^+$ cells (median: 6.5 vs. 3; $p = 0.014$) than cats with normal peripheral leukocyte counts.

Lamina propria S100A12$^+$ counts in the ileum were higher in hypoalbuminemic cats (median: 7.5) than in cats with normoalbuminemia (median: 1.5; $p = 0.027$). Compared to cats with normoglobulinemia, hyperglobulinemic cats had significantly higher duodenal/proximal jejunal S100A8/A9$^+$ (median: 12 vs. 5; $p = 0.006$) and S100A12$^+$ counts (median: 7 vs. 3; $p = 0.016$) and also higher ileal S100A8/A9$^+$ (median: 11 vs. 3.5; $p = 0.041$) and S100A12$^+$ counts (median: 8.5 vs. 1.5; $p = 0.041$). Endoscopic lesions in any GI segment were linked to higher duodenal/proximal jejunal lamina propria S100A8/A9$^+$ counts (median: 8 vs. 1.5; $p = 0.049$), and some endoscopic lesions were linked to higher numbers of segmental S100/calgranulin-positive cells (Supplementary Table S2). Linear hyperemic lesions on endoscopy were rarely seen and could not be assessed for a relationship with S100/calgranulin-positive cell counts. There were also no differences in any segmental S100A8/A9$^+$ or S100A12$^+$ counts between cats with FRE and those classified as IRE.

4. Discussion

In this study, mucosal expression of S100A8/A9 and S100A12 in cats with CIE, GI lymphoma, and controls without GI lesions were evaluated and compared among groups as well as to the severity of clinical signs and histologic lesions. After careful scrutiny of the bibliographic databases, the authors believe that this is the first investigation of S100/calgranulin expression in cats with CIE and IL.

The results support our hypothesis that S100/calgranulins play a role in the pathogenesis of FCE, but their expression in GI tissue biopsies did not differentiate between CIE and IL. The lack of a correlation between numbers of mucosal S100/calgranulin-positive cells and polymorphonuclear cells on routine histology agrees with studies in dogs and could be explained by the plasticity of these cells. This plasticity of morphology can present a challenge for their routine detection and lead to misclassifications of the different types of polymorphonuclear cells. The presence of different grades of cellular infiltration, different stages of maturity, and inactive and activated cells during inflammation with spatiotemporal differences in their S100/calgranulin expression could be an alternative explanation [35,41]. S100/calgranulins are expressed in neutrophils and activated MΦ infiltrating the intestinal mucosa of patients with CIE [56]. More specifically, MΦ plays a central role in maintaining intestinal homeostasis and intestinal protective immunity. In the healthy intestines, MΦ predominantly serves an anti-inflammatory role, whereas increased recruitment of blood monocytes or activated resident cells differentiating into MΦ that display a proinflammatory phenotype are linked to inflammation [42,57,58]. Expression of the S100A8/A9 protein complex (L1 antigen) is restricted to immature proinflammatory MΦ in humans, and L1 expression is down-regulated during MΦ maturation [59]. Thus, increased L1$^+$-MΦ counts reflect an influx of monocytes into the inflamed mucosa [35]. S100A8/A9 has been evaluated in dogs with CIE as a marker of early differentiated MΦ, whereby CD163 expression served as a marker for resident MΦ, showing decreased resident MΦ and increased early differentiated MΦ counts, and thus, more mucosal S100A8/A9$^+$ cells in IRE and FRE. With S100A8/A9 expression in dogs presumed to decrease during the maturation of canine MΦ, the protein complex would be useful to detect migrating monocytes and early-stage MΦ rather than the resident MΦ population [60]. Based on these former results in dogs, it is likely that the same holds true for felines. The functional implications of increased intestinal S100A8/A9 expression in canine and feline CIE (e.g., link to TLR4 expression and/or responsiveness) require further research.

Few significant associations were seen for S100A12 expression in our study, with slightly more S100A12$^+$ cells in the gastric lamina propria and colonic submucosa in the control group (Table 3). However, these results are based on overall low numbers of S100A12$^+$ cells in addition to small group sizes. The results are consistent with previous findings in dogs, showing an inverse relationship between duodenal lamina propria MΦ and S100A12$^+$ cell counts [35], with the lack of a correlation between fecal S100A12 abundance and the severity or presence of intestinal lamina propria phagocyte infiltration [29,30]. The hypothesis that the intestinal mucosa of dogs with CIE contains higher numbers of newly recruited (activated, proinflammatory) MΦ contributing to chronic inflammation [35] and expressing the S100A12 protein [58,61,62], whereas the MΦ population detected on routine histopathology is predominantly mature (anti-inflammatory, tissue-resident anergic) with reduced or absent S100A12 expression [62], is also reasonable in cats. Another theory is based on detecting higher Casp3 levels reflecting apoptotic activity in dogs with CIE compared to healthy controls [63], which may account for a shorter lifespan and lower S100A12 expression levels in infiltrating cells in CIE. Our findings are also in line with a study in human medicine, demonstrating higher S100-positive MΦ and neutrophil counts in non-inflamed than inflamed intestinal mucosa in patients with inflammatory bowel disease [64]. The association with inflammation is also supported by the higher numbers of S100/calgranulin-positive cells in cats with leukocytosis. Proinflammatory MΦ playing a role in CE is further supported by a study in dogs where a decrease in MHC II$^+$, Iba-1$^+$, CD204$^+$, and CD162$^+$ cells was seen in the duodenum and overall less MΦ in all GI segments [42].

However, this hypothesis requires further research in cats and dogs, and the exact population and activation status of S100A8/A9- and S100A12-expressing cells in feline CIE and IL remain to be studied. In addition, numbers of S100/calgranulin-positive cells were analyzed in this study but not the intensity of staining. A standardized, automated method determining the intensity of IHC staining could have quantified the mucosal S100/calgranulin expression but not distinguished intravascular cells from infiltrating or resident cells of the GI mucosa. Using a manual approach, only cells outside the intestinal blood vessels staining clearly positive for S100A12 or S100A8/A9 were considered and counted, but there was also some staining for both in the surrounding tissue outside defined cellular margins, which agrees with the findings of a previous study in dogs [35] and suggests secretion and localization of S100/calgranulins to the extracellular matrix [64]. Subjectively, different signal intensities for S100A8/A9 and S100A12 were noted in individual cats, which might result from higher expression levels in activated compared to non-activated cells.

The lack of correlation of the S100/calgranulin expression between cats with CIE and IL supports the recent proposal that CIE and alimentary lymphoma in cats might present different stages along the continuum or spectrum of the same disease process, in which CIE cases progress to SCL over time initiated by a chronic stimulation by a food or bacterial antigen or (unknown) viral pathogen [6,13,18,48,65]. This hypothesis of a continuum also exists in human celiac diseases that are suggested to progress from low-grade epitheliotropic lymphoma instead of inflammatory processes [66,67]. Alternatively, CIE and SCL might also be two distinct processes, and the inflammatory component in SCL could present a secondary process or an anti-tumor reaction.

It proved very difficult to identify cats that did not have any histologic lesions in any of the GI segments, especially in the duodenum and ileum. This agrees with a recent investigation of gastric and duodenal endoscopic biopsies from 20 clinically healthy cats [68]. With a median age of 9.5 years (range: 3–18 years), the demographic characteristics of that feline population were similar to the cats presenting with FCE in our study. Every cat in this previous study had abnormal histologic findings based on the WSAVA criteria. After a median follow-up time of 709 days, only 3 of the 20 cats (15%) developed GI signs, whereas the remaining 17 cats (85%) remained free of any clinical signs of GI disease during the follow-up period of the study (median: 709 days) [68].

The definition of a "normal intestinal microarchitecture" remains challenging in cats despite the WSAVA GI Standardization Group's attempt to standardize the histologic assessment of GI biopsies with a grading scheme, evaluating cellular infiltration (e.g., lymphocytes/plasma cells, eosinophils, neutrophils, MΦ) and morphologic lesions (e.g., epithelial injury, crypt distention, lacteal dilation, villus blunting, fibrosis) on a scale from 0 (normal) to 3 (severely abnormal) [3,39]. The results of different observers still vary, and some investigators suggest simplifying the WSAVA grading scheme [47]. Cats defined as "normal" and used to establish the WSAVA guidelines were approximately 1.5 years old, specific pathogen-free (SPF) colony cats, which are hardly representative of the typical cat presented with GI signs in clinical practice. Thus, the definition of "GI healthy" may need to be revised, and more studies, including clinically healthy cats with no signs of GI disease and long-term outcomes (ideally followed until the time of death or euthanasia), would be very useful to establish a more realistic "healthy control" reference for future comparison.

We could not include healthy cats biopsied without a diagnostic and/or therapeutic indication for ethical reasons. Thus, the tissue samples from all cats in the control group (except for one cat) were obtained post-mortem, making autolysis and other post-mortem artifacts more likely than in the disease groups of cats. All control cats died or were euthanized for reasons other than GI diseases. However, none of the cats were completely healthy, and the underlying (non-GI) condition (e.g., cardiac failure, resulting in central venous congestion and intestinal dysfunction [69], spinal trauma potentially causing intestinal injury) could have contributed to higher intestinal S100/calgranulin-positive cell counts. Furthermore, during the process of dying, an intestinal stress response due to local ischemia accompanied by an acute inflammatory response with the recruitment of neutrophils, macrophages, and other immune cells may occur [70].

In addition, post-mortem examination in the control cats offered the entire GI tract to be examined and generally larger (full-thickness) specimens than those in the FCE groups, where sampling was restricted to certain areas of the GI segments.

Cats with FRE generally had better outcomes than those with IRE and significantly longer survival times than cats with lymphoma. Despite the small number of animals with FRE in this study (cats that respond to diligent dietary trials are rarely biopsied), this result should alert clinicians to always perform an elimination diet trial in cats with CIE before initiating immunosuppressive treatment. No difference in S100A8/A9$^+$ and S100A12$^+$ cells was seen between both CIE groups, except for a significantly higher maximum S100A8/A9$^+$ cell count in the gastric lamina propria in cats with IRE.

Cats with CIE were younger than cats with GI lymphoma, although an overlap in the age ranges was noted, and these ranges agree with previous reports [5,13–15]. Similar to other studies, Domestic shorthair cats were overrepresented in both groups, likely due to being the most common feline breed seen at UL-CVM. Clinical signs have been reported to be indistinguishable between cats with GI lymphoma and those with CIE, with the most common clinical signs in both groups being weight loss (prevalence: 80–90%), followed by vomiting (70–80%), anorexia (60–70%), and diarrhea (50–65%) [16–18].

We hypothesized intestinal S100/calgranulin expression to correlate with the severity of histologic lesions and clinical disease. This was partially confirmed as S100A12$^+$ cell counts correlated positively with the severity of lymphoplasmacytic and eosinophilic infiltration and the resulting inflammatory and cumulative histologic lesion scores in the colon, with similar trends seen for S100A8/A9$^+$ cells. This finding agrees with the previously reported correlation of S100A12 levels in colonic mucosal tissue with the severity of overall histologic lesions and epithelial injury in the canine colon [71]. FCEAI scores and the severity of weight loss were significantly higher in cats with lymphoma than in those with

CIE. Together with more severe inflammatory lesions in the duodenum/proximal jejunum in cats with lymphoma, this may also support the theory of a spectrum of diseases in which clinical signs progressively worsen over time, and the transition from inflammation to lymphoma leads to emaciation of the affected cat.

FCEAI scores did not correlate with S100/calgranulin expression in the CIE group, but cats with marked diarrhea had more colonic $S100A12^+$ cells, similar to $S100A8/A9^+$ cell counts in cats with IL. FCEAI scores in cats with IL also correlated with $S100A12^+$ cell counts in the duodenum, suggesting a link between clinical disease severity and S100/calgranulin expression. The lack of an association between the FCEAI score and intestinal mucosal S100A12 positivity in cats with CIE is consistent with findings in dogs [35,71] and, again, the presence of select MΦ populations [42]. Correlation of segmental S100/calgranulin expression with clinical data (e.g., leukocytosis, endoscopic lesions) is also reported in human patients with IBD [72–74]. From a clinical perspective, it remains to be determined if intestinal S100/calgranulin expression reflects the fecal concentrations of these molecules. The presence of some staining for S100A8/A9 and/or S100A12 in the surrounding tissue outside defined cellular margins could indicate a release of inflammatory cells secreting S100/calgranulins to the extracellular matrix [64] and therefore leading to potentially higher S100/calgranulin levels in the feces [44] compared to intestinal tissue.

Histologic differentiation of severe LPE (CIE) from SCL can be challenging and often requires IHC and, in some cases, PCR-based methods (PARR) to confirm the diagnosis [6,75]. IHC employs specific stains (e.g., against CD3 for T-cells and CD20 for B-cells) to determine whether the infiltrating lymphocytes are of a single lineage, supporting a diagnosis of SCL. Infiltrates equally consisting of T and B cells support a diagnosis of CIE. However, SCL is often accompanied by inflammation, and even with IHC, establishing a diagnosis can be difficult [6]. Currently, PARR performed on formalin-fixed paraffin-embedded tissue samples is applied to determine the clonality of lymphocyte receptors [23]. Polyclonality (diverse receptors) on PARR suggests inflammatory lesions, whereas clonality (uniform receptors) indicates neoplasia. Because cell lineages cannot be determined by PARR, IHC and PARR analysis must be interpreted together and in the context of all diagnostics performed [6]. Histology and, in some cases, IHC for CD3/CD20 and PARR were performed and interpreted by several pathologists (CG, WvB, DB) blinded to each other and for CD-IHC and PARR also to the identity, clinical, clinicopathologic, and histologic results. Moderate to high inter-observer variance between different pathologists has been described [76,77], but very small discrepancies occurred in our study. Thus, all pathologic diagnoses were confirmed, except for two cases diagnosed by the pathologists' consensus. This included a cat in the CIE group and a healthy control cat. In addition, the results of the PARR analysis in our study supported previous investigations questioning the reclassification of CE cases in cats based on clonality testing alone [5,11,48,78].

Due to the retrospective nature of this study and to evaluate a realistic study population similar to that expected in clinical practice, we did not limit IL cases to include SCL but also two cases of LCL. No differences were seen between SCL and LCL, but the statistical power to detect any differences between both was low. One of the cats with IL was re-biopsied 5 months after the first endoscopy and was diagnosed with LCL after an initial diagnosis of SCL. The first endoscopy diagnosed SCL in the duodenum, whereas the gastric mucosa showed only inflammation. This had progressed to lymphoma in all examined segments at the time of reevaluation, suggesting the potential for (possibly rapid) worsening of this condition and potentially supporting the theory of a disease spectrum of FCE characterized by the transition from LPE (CIE) to GI lymphoma. However, this warrants further studies. In addition, one cat of the lymphoma group diagnosed with duodenal LPE showed lymphoma in the ileum, which is the most common location for SCL [9] and should always be examined if lymphoma is suspected. For various reasons, we did not have tissue samples from all segments of the GI tract in all cats. Thus, we cannot exclude the possibility of misclassification of cats as CIE with "hidden" lymphoma in the ileum (if ileal biopsies were unavailable) or segments of the GI tract outside the reach of the endoscope.

In addition, missing lymphoma in any GI segment during sampling cannot be completely ruled out. However, the study focused on the S100/calgranulin expression in the GI tract, where each GI segment was considered separately for analysis, leveraging this limitation. Still, it warrants reminding clinicians of the importance of obtaining tissue biopsies from all GI segments within reach of the endoscope (i.e., stomach, duodenum/proximal jejunum, ileum, and colon) for a reliable diagnosis. In addition, multiple tissue biopsy samples should be obtained from each GI segment, especially the ileum, to optimize the histologic assessment and reduce the risk of false results due to variation (e.g., patchy distribution of the mucosal disease process). It is generally recommended to obtain at least 6–8 adequate endoscopic biopsy samples per GI segment [79,80]. This number was not reached in all endoscopically obtained tissue samples, particularly due to using surplus material. Even fewer biopsies (max. 1–2 per segment, except for the colon) are obtained with a surgical approach, as was the case for 6 cats in this study. The lack of surgical colonic biopsies (usually obtained via colonoscopy) reflects the very high risk of suture dehiscence and colonic wall perforation after full-thickness incision of the large intestine in cats [81], which should always be weighed against the expected benefit for the diagnostic evaluation.

Due to the retrospective nature of this study, the patient medical data were incomplete in some cases, and the diagnostic evaluation, individual case management, and times of follow-up evaluations varied. In four cats, one FCEAI parameter (serum phosphorus concentration and serum ALP activity in 2 cats each) was not available and thus entered as "0" for FCEAI calculation. This was considered reasonable, given that, if at all, it could only have changed the cumulative FCEAI score by one point and would not have misrepresented the score for any of those four cats. As an advantage of the retrospective study design, a more reliable classification into the subgroups IRE, FRE, and lymphoma, which is usually challenging, was possible with the resulting long follow-up times and detailed information about the clinical course of the disease, response to treatment, and monitoring for any long-term consequences and/or other conditions. Still, some potential confounding factors cannot be excluded for all cats. In one cat, an intoxication (unknown toxic agent) could not be ruled out as a cause of worsening clinical signs. However, this cat had a chronic history of GI signs and laboratory, sonographic, endoscopic, and histopathologic findings, and the response to treatment was consistent with a diagnosis of IRE. Another cat (long-haired breed) was presented with acute ileus and a history of recurrent constipation and vomiting. After the endoscopic removal of a hairball, a dietary change to a commercial "anti-hairball" elimination diet, lactulose, and daily grooming led to clinical remission. Together with histopathology and CD3/CD20-IHC, this cat was classified as FRE. In this case and another cat with ileus as a primary cause for immediate endoscopy, the clinical signs just prior to the acute incidence were considered for calculating the FCEAI score. Interestingly, one cat in the CIE group responded well long-term to the combination of an antibiotic (metronidazole) and a probiotic for 4 weeks. While this might implicate a diagnosis of ARE, the cat also received an elimination diet, and the owners reduced several stress factors, resulting in the definitive classification as FRE because the dietary intervention was considered the main part of therapy in this cat.

Achieving remission in this study was based on clinical signs and non-invasive diagnostics alone because reinvestigation with endoscopy is often not performed even after unsuccessful treatment. This is a disadvantage of using the FCEAI score, which incorporates results of the cat's endoscopic evaluation, and calls for a new (i.e., simplified) scoring system for FCE that allows for reevaluating cats without needing repeated endoscopy (similar to the currently used canine inflammatory bowel disease activity index [82] or canine chronic enteropathy clinical activity index [83]). Finding higher S100/calgranulin-positive cell counts linked to higher serum cobalamin concentrations was an unexpected finding of this study. We speculate this to reflect dysregulations in cobalamin metabolism in FCE, which is currently subject to further investigation by our group.

We acknowledge that this study has some further limitations. Due to the clinical character of this study, randomized sampling and stercologic analysis could not be applied.

Additionally, specific staining was not performed to further characterize the cells expressing S100A12 and/or S100A8/A9. Thus, it cannot be determined if the expression of these proteins is localized to neutrophils, dendritic cells, and/or different types of MΦ in feline CE. Furthermore, the control group consisting of mostly autopsied cats was not ideal, and further evaluation of healthy cats would have been preferable but was not feasible for ethical reasons. Lastly, the sample size was small, and a type II error for finding no difference or association cannot be excluded.

Further studies are warranted to further characterize the population of cells expressing S100/calgranulins, for example, via double-staining for S100/calgranulins and polymorphonuclear cells and macrophages (particularly the proinflammatory phenotype), evaluate serial biopsies, and phenotypically characterize S100/calgranulin-positive stained cells. Stratification of the patient groups based on the level of inflammation and examination of fecal S100/calgranulin levels in cats are also needed.

5. Conclusions

In summary, the findings of this study support that S100A8/A9 and S100A12 do play a role in the pathogenesis of FCE and do not differ between the two main groups of FCE, which supports the theory of CIE and SCL presenting different stages along a spectrum of the same disease process. Additional studies are warranted to characterize further the population of cells expressing S100A8/A9 and/or S100A12 and to determine the functional implications of the differential expression of these proteins and their possible potential for diagnostic or monitoring purposes in FCE.

Supplementary Materials: The following supporting information can be downloaded at: https://www.mdpi.com/article/10.3390/ani12162044/s1, Figure S1: CD staining for T and B lymphocyte populations. Positive staining (brown) for CD3 (T cells; panels A&C) and CD20 (B cells; B&D) in the duodenal mucosa of a cat diagnosed with small-cell lymphoma (A&B) and a cat with CIE (C&D); Figure S2: Intestinal blood vessels containing polymorphonuclear cells. (A) S100A8/A9 immunohistochemistry (IHC), (B) S100A12+ cells within an intestinal blood vessel, (C) negative IHC control; Figure S3: Slight nonspecific background staining (S100A8/A9-IHC) of gastric parietal cells. Table S1: Patient characteristics, clinical findings, and clinicopathologic parameters in cats with chronic inflammatory enteropathy (CIE; *n* = 16), alimentary lymphoma (*n* = 8), and controls without histologic lesions (*n* = 16) included in the study; Table S2: Correlation of mucosal S100/calgranulin-positive cell counts with endoscopic lesions in cats with chronic enteropathy in this study.

Author Contributions: Conceptualization, D.S.R. and R.M.H.; methodology, D.S.R., C.G., M.P. and D.B.; software, R.M.H. and K.W.; validation, M.P., D.S.R. and R.M.H.; formal analysis, R.M.H. and D.S.R.; investigation, D.S.R., C.G., D.B. and W.v.B.; resources, J.M.S., G.A., D.B., W.v.B. and R.M.H.; data curation, D.S.R. and R.M.H.; writing—original draft preparation, D.S.R.; writing—review and editing, R.M.H., M.P., C.G., D.B., G.A., J.M.S., W.v.B. and K.W.; visualization, D.S.R., C.G., R.M.H. and K.W.; supervision, R.M.H., G.A., C.G. and M.P.; project administration, R.M.H. and D.S.R.; funding acquisition, R.M.H. All authors have read and agreed to the published version of the manuscript.

Funding: This project was supported by the EveryCat Health Foundation (W21-030). The contents of this publication are solely the responsibility of the authors and do not necessarily represent the views of EveryCat.

Institutional Review Board Statement: Ethical review and approval were waived for this study for the cases enrolled at the University of Leipzig College of Veterinary Medicine (UL-CVM) as the patient data and archived sample materials were retrospectively included.

Informed Consent Statement: Owners provided their written consent for using data and surplus biological specimens on the standard patient admission form used at the UL-CVM Department for Small Animals.

Data Availability Statement: Data (anonymized) are available from the first or last author upon reasonable request.

Acknowledgments: We acknowledge Barbara Rütgen and Nicole Hammer at the Department of Pathobiology, University of Veterinary Medicine Vienna, Austria, for performing the PARR analysis (fee-for-service) for this study. We thank Johannes Seeger for assisting with digitalizing the IHC slides and Michael Sieg and Anja Reinert for their assistance with the collection of tissue samples from control cats. Part of the data was presented at the 2022 Annual Meeting of the Working group Internal Medicine and Laboratory Diagnostics (InnLab) of the German Veterinary Society (DVG; Berlin, Germany, January 2022) and at the 2022 Annual Forum of the American College of Veterinary Internal Medicine (ACVIM; Austin, TX, USA, June 2022).

Conflicts of Interest: The authors declare no conflict of interest. The funders had no role in the design of the study, in the collection, analyses, or interpretation of data, in the writing of the manuscript, or in the decision to publish the results. RM.H. served as co-editor for this special issue but was not involved in the editing of this manuscript.

References

1. Jergens, A.E. Feline idiopathic inflammatory bowel disease: What we know and what remains to be unraveled. *J. Fel. Med. Surg.* **2012**, *14*, 445–458. [CrossRef] [PubMed]
2. Richter, K.P. Feline gastrointestinal lymphoma. *Vet. Clin. North. Am. Small. Anim. Pract.* **2003**, *33*, 1083–1098. [CrossRef]
3. Washabau, R.J.; Day, M.J.; Willard, M.D.; Hall, E.J.; Jergens, A.E.; Mansell, J.; Minami, T.; Bilzer, T.W. Endoscopic, biopsy, and histopathologic guidelines for the evaluation of gastrointestinal inflammation in companion animals. *J. Vet. Intern. Med.* **2010**, *24*, 10–26. [CrossRef]
4. Jergens, A.E.; Simpson, K.W. Inflammatory bowel disease in veterinary medicine. *Front. Biosci.* **2012**, *4*, 1404–1419. [CrossRef]
5. Jergens, A.E.; Crandell, J.M.; Evans, R.; Ackermann, M.; Miles, K.G.; Wang, C. A clinical index for disease activity in cats with chronic enteropathy. *J. Vet. Intern. Med.* **2010**, *24*, 1027–1033. [CrossRef]
6. Marsilio, S. Differentiating inflammatory bowel disease from alimentary lymphoma in cats: Does it matter? *Vet. Clin. North. Am. Small. Anim. Pract.* **2021**, *51*, 93–109. [CrossRef] [PubMed]
7. Albert, E.J. Inflammatory bowel disease: Current perspectives. *Vet. Clin. North. Am. Small Anim. Pract.* **1999**, *29*, 501–521. [CrossRef]
8. Ewald, N.; Rödler, F.; Heilmann, R.M. Chronische Enteropathien bei der Katze—Diagnostische und therapeutische Aspekte. *Tierarztl. Prax. Ausg. K. Kleintiere. Heimtiere.* **2021**, *49*, 363–376. [CrossRef] [PubMed]
9. Barrs, V.R.; Beatty, J.A. Feline alimentary lymphoma: 2. Further diagnostics, therapy and prognosis. *J. Fel. Med. Surg.* **2012**, *14*, 191–201. [CrossRef]
10. Vail, D.M.; Moore, A.S.; Ogilvie, G.K.; Volk, L.M. Feline lymphoma (145 cases): Proliferation indices, cluster of differentiation 3 immunoreactivity, and their association with prognosis in 90 cats. *J. Vet. Intern. Med.* **1998**, *12*, 349–354. [CrossRef] [PubMed]
11. Zwahlen, C.H.; Lucroy, M.D.; Kraegel, S.A.; Madewell, B.R. Results of chemotherapy for cats with alimentary malignant lymphoma: 21 cases (1993-1997). *J. Am. Vet. Med. Assoc.* **1998**, *213*, 1144–1149. [PubMed]
12. Heilmann, R.M.; Suchodolski, J.S. Is inflammatory bowel disease in dogs and cats associated with a Th1 or Th2 polarization? *Vet. Immunol. Immunopathol.* **2015**, *168*, 131–134. [CrossRef]
13. Lingard, A.E.; Briscoe, K.; Beatty, J.A.; Moore, A.S.; Crowley, A.M.; Krockenberger, M.; Churcher, R.K.; Canfield, P.J.; Barrs, V.R. Low-grade alimentary lymphoma: Clinicopathological findings and response to treatment in 17 cases. *J. Fel. Med. Surg.* **2009**, *11*, 692–700. [CrossRef] [PubMed]
14. Kiselow, M.A.; Rassnick, K.M.; McDonough, S.P.; Goldstein, R.E.; Simpson, K.W.; Weinkle, T.K.; Erb, H.N. Outcome of cats with low-grade lymphocytic lymphoma: 41 cases (1995–2005). *J. Am. Vet. Med. Assoc.* **2008**, *232*, 405–410. [CrossRef] [PubMed]
15. Marsilio, S.; Pilla, R.; Sarawichitr, B.; Chow, B.; Hill, S.L.; Ackermann, M.R.; Estep, J.S.; Lidbury, J.A.; Steiner, J.M.; Suchodolski, J.S. Characterization of the fecal microbiome in cats with inflammatory bowel disease or alimentary small cell lymphoma. *Sci. Rep.* **2019**, *9*, 19208. [CrossRef] [PubMed]
16. Norsworthy, G.D.; Estep, J.S.; Hollinger, C.; Steiner, J.M.; Lavallee, J.O.; Gassler, L.N.; Restine, L.M.; Kiupel, M. Prevalence and underlying causes of histologic abnormalities in cats suspected to have chronic small bowel disease: 300 cases (2008–2013). *J. Am. Vet. Med. Assoc.* **2015**, *247*, 629–635. [CrossRef] [PubMed]
17. Burke, K.F.; Broussard, J.D.; Ruaux, C.G.; Suchodolski, J.S.; Williams, D.A.; Steiner, J.M. Evaluation of fecal α_1-proteinase inhibitor concentrations in cats with idiopathic inflammatory bowel disease and cats with gastrointestinal neoplasia. *Vet. J.* **2013**, *196*, 189–196. [CrossRef] [PubMed]
18. Waly, N.E.; Gruffydd-Jones, T.J.; Stokes, C.R.; Day, M.J. Immunohistochemical diagnosis of alimentary lymphomas and severe intestinal inflammation in cats. *J. Comp. Pathol.* **2005**, *133*, 253–260. [CrossRef]
19. Reed, N.; Gunn-Moore, D.; Simpson, K. Cobalamin, folate and inorganic phosphate abnormalities in ill cats. *J. Fel. Med. Surg.* **2007**, *9*, 278–288. [CrossRef] [PubMed]
20. Dennis, J.S.; Kruger, J.M.; Mullaney, T.P. Lymphocytic/plasmacytic gastroenteritis in cats: 14 cases (1985–1990). *J. Am. Vet. Med. Assoc.* **1992**, *200*, 1712–1718. [PubMed]

21. Jergens, A.E.; Moore, F.M.; Haynes, J.S.; Miles, K.G. Idiopathic inflammatory bowel disease in dogs and cats: 84 cases (1987–1990). *J. Am. Vet. Med. Assoc.* **1992**, *201*, 1603–1608.

22. Norsworthy, G.D.; Scot Estep, J.; Kiupel, M.; Olson, J.C.; Gassler, L.N. Diagnosis of chronic small bowel disease in cats: 100 cases (2008–2012). *J. Am. Vet. Med. Assoc.* **2013**, *243*, 1455–1461. [CrossRef] [PubMed]

23. Moore, P.F.; Woo, J.C.; Vernau, W.; Kosten, S.; Graham, P.S. Characterization of feline T cell receptor gamma (TCRG) variable region genes for the molecular diagnosis of feline intestinal T cell lymphoma. *Vet. Immunol. Iimmunopathol.* **2005**, *106*, 167–178. [CrossRef]

24. Moore, P.F.; Rodriguez-Bertos, A.; Kass, P.H. Feline gastrointestinal lymphoma: Mucosal architecture, immunophenotype, and molecular clonality. *Vet. Pathol.* **2012**, *49*, 658–668. [CrossRef] [PubMed]

25. Kiupel, M.; Smedley, R.C.; Pfent, C.; Xie, Y.; Xue, Y.; Wise, A.G.; DeVaul, J.M.; Maes, R.K. Diagnostic algorithm to differentiate lymphoma from inflammation in feline small intestinal biopsy samples. *Vet. Pathol.* **2011**, *48*, 212–222. [CrossRef] [PubMed]

26. Marsilio, S.; Newman, S.J.; Estep, J.S.; Giaretta, P.R.; Lidbury, J.A.; Warry, E.; Flory, A.; Morley, P.S.; Smoot, K.; Seeley, E.H.; et al. Differentiation of lymphocytic-plasmacytic enteropathy and small cell lymphoma in cats using histology-guided mass spectrometry. *J. Vet. Intern. Med.* **2020**, *34*, 669–677. [CrossRef]

27. Heilmann, R.M.; Allenspach, K. Pattern-recognition receptors: Signaling pathways and dysregulation in canine chronic enteropathies-brief review. *J. Vet. Diagn. Investig.* **2017**, *29*, 781–787. [CrossRef] [PubMed]

28. Heilmann, R.M.; Volkmann, M.; Otoni, C.C.; Grützner, N.; Kohn, B.; Jergens, A.E.; Steiner, J.M. Fecal S100A12 concentration predicts a lack of response to treatment in dogs affected with chronic enteropathy. *Vet. J.* **2016**, *215*, 96–100. [CrossRef]

29. Heilmann, R.M.; Berghoff, N.; Mansell, J.; Grützner, N.; Parnell, N.K.; Gurtner, C.; Suchodolski, J.S.; Steiner, J.M. Association of fecal calprotectin concentrations with disease severity, response to treatment, and other biomarkers in dogs with chronic inflammatory enteropathies. *J. Vet. Intern. Med.* **2018**, *32*, 679–692. [CrossRef]

30. Heilmann, R.M.; Grellet, A.; Allenspach, K.; Lecoindre, P.; Day, M.J.; Priestnall, S.L.; Toresson, L.; Procoli, F.; Grützner, N.; Suchodolski, J.S.; et al. Association between fecal S100A12 concentration and histologic, endoscopic, and clinical disease severity in dogs with idiopathic inflammatory bowel disease. *Vet. Immunol. Immunopathol.* **2014**, *158*, 156–166. [CrossRef]

31. Otoni, C.C.; Heilmann, R.M.; García-Sancho, M.; Sainz, A.; Ackermann, M.R.; Suchodolski, J.S.; Steiner, J.M.; Jergens, A.E. Serologic and fecal markers to predict response to induction therapy in dogs with idiopathic inflammatory bowel disease. *J. Vet. Intern. Med.* **2018**, *32*, 999–1008. [CrossRef]

32. Heilmann, R.M.; Steiner, J.M. Clinical utility of currently available biomarkers in inflammatory enteropathies of dogs. *J. Vet. Intern. Med.* **2018**, *32*, 1495–1508. [CrossRef] [PubMed]

33. Goyette, J.; Geczy, C.L. Inflammation-associated S100 proteins: New mechanisms that regulate function. *Amino Acids* **2011**, *41*, 821–842. [CrossRef]

34. Foell, D.; Frosch, M.; Sorg, C.; Roth, J. Phagocyte-specific calcium-binding S100 proteins as clinical laboratory markers of inflammation. *Clin. Chim. Acta.* **2004**, *344*, 37–51. [CrossRef]

35. Heilmann, R.M.; Nestler, J.; Schwarz, J.; Grützner, N.; Ambrus, A.; Seeger, J.; Suchodolski, J.S.; Steiner, J.M.; Gurtner, C. Mucosal expression of S100A12 (calgranulin C) and S100A8/A9 (calprotectin) and correlation with serum and fecal concentrations in dogs with chronic inflammatory enteropathy. *Vet. Immunol. Immunopathol.* **2019**, *211*, 64–74. [CrossRef]

36. Vogl, T.; Tenbrock, K.; Ludwig, S.; Leukert, N.; Ehrhardt, C.; van Zoelen, M.A.D.; Nacken, W.; Foell, D.; van der Poll, T.; Sorg, C.; et al. Mrp8 and Mrp14 are endogenous activators of Toll-like receptor 4, promoting lethal, endotoxin-induced shock. *Nature. Med.* **2007**, *13*, 1042–1049. [CrossRef]

37. Hofmann, M.A.; Drury, S.; Fu, C.; Qu, W.; Taguchi, A.; Lu, Y.; Avila, C.; Kambham, N.; Bierhaus, A.; Nawroth, P.; et al. RAGE mediates a novel proinflammatory axis. *Cell* **1999**, *97*, 889–901. [CrossRef]

38. Cabrera-García, A.I.; Suchodolski, J.S.; Steiner, J.M.; Heilmann, R.M. Association between serum soluble receptor for advanced glycation end-products (RAGE) deficiency and severity of clinicopathologic evidence of canine chronic inflammatory enteropathy. *J. Vet. Diagn. Investig.* **2020**, *32*, 664–674. [CrossRef]

39. Day, M.J.; Bilzer, T.; Mansell, J.; Wilcock, B.; Hall, E.J.; Jergens, A.; Minami, T.; Willard, M.; Washabau, R. Histopathological standards for the diagnosis of gastrointestinal inflammation in endoscopic biopsy samples from the dog and cat: A report from the World Small Animal Veterinary Association Gastrointestinal Standardization Group. *J. Comp. Pathol.* **2008**, *138*, S1–S43. [CrossRef] [PubMed]

40. Tucker, S.; Penninck, D.G.; Keating, J.H.; Webster, C.R.L. Clinicopathological and ultrasonographic features of cats with eosinophilic enteritis. *J. Fel. Med. Surg.* **2014**, *16*, 950–956. [CrossRef] [PubMed]

41. German, A.J.; Hall, E.J.; Kelly, D.F.; Watson, A.D.; Day, M.J. An immunohistochemical study of histiocytic ulcerative colitis in boxer dogs. *J. Comp. Pathol.* **2000**, *122*, 163–175. [CrossRef]

42. Wagner, A.; Junginger, J.; Lemensieck, F.; Hewicker-Trautwein, M. Immunohistochemical characterization of gastrointestinal macrophages/phagocytes in dogs with inflammatory bowel disease (IBD) and non-IBD dogs. *Vet. Immunol. Immunopathol.* **2018**, *197*, 49–57. [CrossRef] [PubMed]

43. Truar, K.; Nestler, J.; Schwarz, J.; Grützner, N.; Gabris, C.; Kock, K.; Niederberger, C.; Heilmann, R.M. Feasibility of measuring fecal calprotectin concentrations in dogs and cats by the fCAL®turbo immunoassay. abstract. *J. Vet. Intern. Med.* **2018**, *32*, 580.

44. Zornow, K.A.; Slovak, J.E.; Lidbury, J.A.; Suchodolski, J.S.; Steiner, J.M. Fecal S100A12 (calgranulin C) concentrations in cats with chronic enteropathies. *J. Vet. Int. Med.* **2021**, *3151*.

45. Hsu, K.; Passey, R.J.; Endoh, Y.; Rahimi, F.; Youssef, P.; Yen, T.; Geczy, C.L. Regulation of S100A8 by glucocorticoids. *J. Immunol.* **2005**, *174*, 2318–2326. [CrossRef] [PubMed]
46. Guttin, T.; Walsh, A.; Durham, A.C.; Reetz, J.A.; Brown, D.C.; Rondeau, M.P. Ability of ultrasonography to predict the presence and location of histologic lesions in the small intestine of cats. *J. Vet. Intern. Med.* **2019**, *33*, 1278–1285. [CrossRef]
47. Freiche, V.; Fages, J.; Paulin, M.V.; Bruneau, J.; Couronné, L.; German, A.J.; Penninck, D.; Hermine, O. Clinical, laboratory and ultrasonographic findings differentiating low-grade intestinal T-cell lymphoma from lymphoplasmacytic enteritis in cats. *J. Vet. Intern. Med.* **2021**, *35*, 2685–2696. [CrossRef] [PubMed]
48. Freiche, V.; Paulin, M.V.; Cordonnier, N.; Huet, H.; Turba, M.-E.; Macintyre, E.; Molina, T.-J.; Hermine, O.; Couronné, L.; Bruneau, J. Histopathologic, phenotypic, and molecular criteria to discriminate low-grade intestinal T-cell lymphoma in cats from lymphoplasmacytic enteritis. *J. Vet. Intern. Med.* **2021**, *35*, 2673–2684. [CrossRef]
49. Heilmann, R.M.; Suchodolski, J.S.; Steiner, J.M. Development and analytic validation of a radioimmunoassay for the quantification of canine calprotectin in serum and feces from dogs. *Am. J. Vet. Res.* **2008**, *69*, 845–853. [CrossRef] [PubMed]
50. Heilmann, R.M.; Suchodolski, J.S.; Steiner, J.M. Purification and partial characterization of canine calprotectin. *Biochimie* **2008**, *90*, 1306–1315. [CrossRef]
51. Heilmann, R.M.; Suchodolski, J.S.; Steiner, J.M. Purification and partial characterization of canine S100A12. *Biochimie* **2010**, *92*, 1914–1922. [CrossRef] [PubMed]
52. Heilmann, R.M.; Cranford, S.M.; Ambrus, A.; Grützner, N.; Schellenberg, S.; Ruaux, C.G.; Suchodolski, J.S.; Steiner, J.M. Validation of an enzyme-linked immunosorbent assay (ELISA) for the measurement of canine S100A12. *Vet. Clin. Pathol.* **2016**, *45*, 135–147. [CrossRef] [PubMed]
53. Heilmann, R.M.; Lanerie, D.J.; Ruaux, C.G.; Grützner, N.; Suchodolski, J.S.; Steiner, J.M. Development and analytic validation of an immunoassay for the quantification of canine S100A12 in serum and fecal samples and its biological variability in serum from healthy dogs. *Vet. Immunol. Immunopathol.* **2011**, *144*, 200–209. [CrossRef]
54. Heilmann, R.M.; Grützner, N.; Handl, S.; Suchodolski, J.S.; Steiner, J.M. Preanalytical validation of an in-house radioimmunoassay for measuring calprotectin in feline specimens. *Vet. Clin. Pathol.* **2018**, *47*, 100–107. [CrossRef]
55. Bridges, C.S.; Grützner, N.; Suchodolski, J.S.; Steiner, J.M.; Heilmann, R.M. Analytical validation of an enzyme-linked immunosorbent assay for the quantification of S100A12 in the serum and feces of cats. *Vet. Clin. Pathol.* **2019**, *48*, 754–761. [CrossRef]
56. Fukunaga, S.; Kuwaki, K.; Mitsuyama, K.; Takedatsu, H.; Yoshioka, S.; Yamasaki, H.; Yamauchi, R.; Mori, A.; Kakuma, T.; Tsuruta, O.; et al. Detection of calprotectin in inflammatory bowel disease: Fecal and serum levels and immunohistochemical localization. *Int. J. Mol. Med.* **2018**, *41*, 107–118. [CrossRef]
57. Tamoutounour, S.; Henri, S.; Lelouard, H.; de Bovis, B.; de Haar, C.; van der Woude, C.J.; Woltman, A.M.; Reyal, Y.; Bonnet, D.; Sichien, D.; et al. CD64 distinguishes macrophages from dendritic cells in the gut and reveals the Th1 inducing role of mesenteric lymph node macrophages during colitis. *Eur. J. Immunol.* **2012**, *42*, 3150–3166. [CrossRef]
58. Nolte, A.; Junginger, J.; Baum, B.; Hewicker-Trautwein, M. Heterogeneity of macrophages in canine histiocytic ulcerative colitis. *Innate. Immun.* **2017**, *23*, 228–239. [CrossRef]
59. Zwadlo, G.; Brüggen, J.; Gerhards, G.; Schlegel, R.; Sorg, C. Two calcium-binding proteins associated with specific stages of myeloid cell differentiation are expressed by subsets of macrophages in inflammatory tissues. *Clin. Exp. Immunol.* **1988**, *72*, 510–515.
60. Dandrieux, J.R.; Martinez Lopez, L.M.; Stent, A.; Jergens, A.; Allenspach, K.; Nowell, C.J.; Firestone, S.M.; Kimpton, W.; Mansfield, C.S. Changes in duodenal CD163-positive cells in dogs with chronic enteropathy after successful treatment. *Innate Immun.* **2018**, *24*, 400–410. [CrossRef] [PubMed]
61. Bain, C.C.; Mowat, A.M. Intestinal macrophages—Specialised adaptation to a unique environment. *Eur. J. Immunol.* **2011**, *41*, 2494–2498. [CrossRef] [PubMed]
62. Bujko, A.; Atlasy, N.; Landsverk, O.J.B.; Richter, L.; Yaqub, S.; Horneland, R.; Øyen, O.; Aandahl, E.M.; Aabakken, L.; Stunnenberg, H.G.; et al. Transcriptional and functional profiling defines human small intestinal macrophage subsets. *J. Exp. Med.* **2018**, *215*, 441–458. [CrossRef] [PubMed]
63. Dandrieux, J.R.; Bornand, V.F.; Doherr, M.G.; Kano, R.; Zurbriggen, A.; Burgener, I.A. Evaluation of lymphocyte apoptosis in dogs with inflammatory bowel disease. *Am. J. Vet. Res.* **2008**, *69*, 1279–1285. [CrossRef] [PubMed]
64. Leach, S.T.; Yang, Z.; Messina, I.; Song, C.; Geczy, C.L.; Cunningham, A.M.; Day, A.S. Serum and mucosal S100 proteins, calprotectin (S100A8/S100A9) and S100A12, are elevated at diagnosis in children with inflammatory bowel disease. *Scand. J. Gastroenterol.* **2007**, *42*, 1321–1331. [CrossRef] [PubMed]
65. Marsilio, S.; Ackermann, M.R.; Lidbury, J.A.; Suchodolski, J.S.; Steiner, J.M. Results of histopathology, immunohistochemistry, and molecular clonality testing of small intestinal biopsy specimens from clinically healthy client-owned cats. *J. Vet. Intern. Med.* **2019**, *33*, 551–558. [CrossRef]
66. Jergens, A.E.; Evans, R.B.; Ackermann, M.; Hostetter, J.; Willard, M.; Mansell, J.; Bilzer, T.; Wilcock, B.; Washabau, R.; Hall, E.J.; et al. Design of a simplified histopathologic model for gastrointestinal inflammation in dogs. *Vet. Pathol.* **2014**, *51*, 946–950. [CrossRef]

67. Harjola, V.-P.; Mullens, W.; Banaszewski, M.; Bauersachs, J.; Brunner-La Rocca, H.-P.; Chioncel, O.; Collins, S.P.; Doehner, W.; Filippatos, G.S.; Flammer, A.J.; et al. Organ dysfunction, injury and failure in acute heart failure: From pathophysiology to diagnosis and management. A review on behalf of the Acute Heart Failure Committee of the Heart Failure Association (HFA) of the European Society of Cardiology (ESC). *Eur. J. Heart. Fail.* **2017**, *19*, 821–836. [CrossRef] [PubMed]

68. Tsukamoto, T.; Chanthaphavong, R.S.; Pape, H.-C. Current theories on the pathophysiology of multiple organ failure after trauma. *Injury* **2010**, *41*, 21–26. [CrossRef] [PubMed]

69. Chott, A.; Dragosics, B.; Radaszkiewicz, T. Peripheral T-cell lymphomas of the intestine. *Am. J. Pathol.* **1992**, *141*, 1361–1371.

70. Dieter, R.S.; Duque, K. Enterotherapy associated T-cell lymphoma: A case report and literature review. *WMJ* **2000**, *99*, 28–31. [PubMed]

71. Hanifeh, M.; Sankari, S.; Rajamäki, M.M.; Syrjä, P.; Kilpinen, S.; Suchodolski, J.S.; Heilmann, R.M.; Guadiano, P.; Lidbury, J.; Steiner, J.M.; et al. S100A12 concentrations and myeloperoxidase activities are increased in the intestinal mucosa of dogs with chronic enteropathies. *BMC. Vet. Res.* **2018**, *14*, 125. [CrossRef] [PubMed]

72. Schoepfer, A.M.; Beglinger, C.; Straumann, A.; Trummler, M.; Vavricka, S.R.; Bruegger, L.E.; Seibold, F. Fecal calprotectin correlates more closely with the Simple Endoscopic Score for Crohn's disease (SES-CD) than CRP, blood leukocytes, and the CDAI. *Am. J. Gastroenterol.* **2010**, *105*, 162–169. [CrossRef] [PubMed]

73. Schoepfer, A.M.; Beglinger, C.; Straumann, A.; Trummler, M.; Renzulli, P.; Seibold, F. Ulcerative colitis: Correlation of the Rachmilewitz endoscopic activity index with fecal calprotectin, clinical activity, C-reactive protein, and blood leukocytes. *Inflamm. Bowel. Dis.* **2009**, *15*, 1851–1858. [CrossRef] [PubMed]

74. Önal, İ.K.; Beyazit, Y.; Şener, B.; Savuk, B.; Özer Etık, D.; Sayilir, A.; Öztaş, E.; Torun, S.; Özderın Özın, Y.; Tunç Demırel, B.; et al. The value of fecal calprotectin as a marker of intestinal inflammation in patients with ulcerative colitis. *Turk. J. Gastroenterol.* **2012**, *23*, 509–514. [CrossRef] [PubMed]

75. Welter, J.; Duckova, T.; Groiss, S.; Wolfesberger, B.; Fuchs-Baumgartinger, A.; Rütgen, B.C.; Hammer, S.E. Revisiting lymphocyte clonality testing in feline B-cell lymphoma. *Vet. Immunol. Immunopathol.* **2021**, *242*, 110350. [CrossRef] [PubMed]

76. Eaden, J.; Abrams, K.; McKay, H.; Denley, H.; Mayberry, J. Inter-observer variation between general and specialist gastrointestinal pathologists when grading dysplasia in ulcerative colitis. *J. Pathol.* **2001**, *194*, 152–157. [CrossRef]

77. Bonsembiante, F.; Martini, V.; Bonfanti, U.; Casarin, G.; Trez, D.; Gelain, M.E. Cytomorphological description and intra-observer agreement in whole slide imaging for canine lymphoma. *Vet. J.* **2018**, *236*, 96–101. [CrossRef]

78. Wilson, H.M. Feline alimentary lymphoma: Demystifying the enigma. *Top. Companion. Anim. Med.* **2008**, *23*, 177–184. [CrossRef]

79. Willard, M.D.; Mansell, J.; Fosgate, G.T.; Gualtieri, M.; Olivero, D.; Lecoindre, P.; Twedt, D.C.; Collett, M.G.; Day, M.J.; Hall, E.J.; et al. Effect of sample quality on the sensitivity of endoscopic biopsy for detecting gastric and duodenal lesions in dogs and cats. *J. Vet. Intern. Med.* **2008**, *22*, 1084–1089. [CrossRef]

80. Jergens, A.E.; Willard, M.D.; Allenspach, K. Maximizing the diagnostic utility of endoscopic biopsy in dogs and cats with gastrointestinal disease. *Vet. J.* **2016**, *214*, 50–60. [CrossRef]

81. Lux, C.N.; Roberts, S.; Grimes, J.A.; Benitez, M.E.; Culp, W.T.N.; Ben-Aderet, D.; Brown, D.C. Evaluation of short-term risk factors associated with dehiscence and death following full-thickness incisions of the large intestine in cats: 84 cases (1993-2015). *J. Am. Vet. Med. Assoc.* **2021**, *259*, 162–171. [CrossRef] [PubMed]

82. Jergens, A.E.; Schreiner, C.A.; Frank, D.E.; Niyo, Y.; Ahrens, F.E.; Eckersall, P.D.; Benson, T.J.; Evans, R. A scoring index for disease activity in canine inflammatory bowel disease. *J. Vet. Intern. Med.* **2003**, *17*, 291–297. [CrossRef] [PubMed]

83. Allenspach, K.; Wieland, B.; Gröne, A.; Gaschen, F. Chronic enteropathies in dogs: Evaluation of risk factors for negative outcome. *J. Vet. Intern. Med.* **2007**, *21*, 700. [CrossRef] [PubMed]

Review

Elucidating the Role of Innate and Adaptive Immune Responses in the Pathogenesis of Canine Chronic Inflammatory Enteropathy—A Search for Potential Biomarkers

Daniela Siel [1,2,*], Caroll J. Beltrán [3], Eduard Martínez [1,4,5], Macarena Pino [1,4,5], Nazla Vargas [1,4], Alexandra Salinas [1], Oliver Pérez [6], Ismael Pereira [7] and Galia Ramírez-Toloza [1,4,*]

1   Central Veterinary Research Laboratory (LaCIV), Facultad de Ciencias Veterinarias y Pecuarias, Universidad de Chile, Santiago 8820808, Chile; eduard.martinez@ug.uchile.cl (E.M.); macarena.pino@ug.uchile.cl (M.P.); nazla.vargas@ug.uchile.cl (N.V.); ale.ssoza@gmail.com (A.S.)
2   Escuela de Medicina Veterinaria, Facultad de Ciencias de la Vida, Universidad Andrés Bello, Santiago 8370251, Chile
3   Laboratory of Immunogastroenterology, Gastroenterology Unit, Medicine Department, Hospital Clínico Universidad de Chile, Santiago 8380420, Chile; carollbeltranm@uchile.cl
4   Department of Animal Preventive Medicine, Facultad de Ciencias Veterinarias y Pecuarias, Universidad de Chile, Santiago 8820808, Chile
5   Programa de Magister en Ciencias Animales y Veterinarias, Facultad de Ciencias Veterinarias y Pecuarias, Universidad de Chile, Santiago 8820808, Chile
6   Instituto de Ciencias Básicas y Preclínicas "Victoria de Girón", Universidad de Ciencias Médicas de la Habana, La Habana 11300, Cuba; oliverperezmartin@gmail.com
7   Hospital Veterinario MEDIVET, Centro de Diagnóstico Veterinario VETPOINT, Santiago 8320000, Chile; ismaelpereiraveterinario@gmail.com
*   Correspondence: daniela.siel@unab.cl (D.S.); galiaram@uchile.cl (G.R.-T.); Tel.: +56-2-2978-5532 (G.R.-T.)

Citation: Siel, D.; Beltrán, C.J.; Martínez, E.; Pino, M.; Vargas, N.; Salinas, A.; Pérez, O.; Pereira, I.; Ramírez-Toloza, G. Elucidating the Role of Innate and Adaptive Immune Responses in the Pathogenesis of Canine Chronic Inflammatory Enteropathy—A Search for Potential Biomarkers. *Animals* **2022**, *12*, 1645. https://doi.org/10.3390/ani12131645

Academic Editors: Aarti Kathrani, Romy M. Heilmann and Edward J. Hall

Received: 10 March 2022
Accepted: 14 June 2022
Published: 27 June 2022

Publisher's Note: MDPI stays neutral with regard to jurisdictional claims in published maps and institutional affiliations.

Simple Summary: Canine chronic inflammatory enteropathy (CIE) is a chronic disease affecting the small or large intestine and, in some cases, the stomach of dogs. This gastrointestinal disorder is common and is characterized by recurrent vomiting, diarrhea, and weight loss in affected dogs. The pathogenesis of IBD is not completely understood. Similar to human IBD, potential disease factors include genetics, environmental exposures, and dysregulation of the microbiota and the immune response. Some important components of the innate and adaptive immune response involved in CIE pathogenesis have been described. However, the immunopathogenesis of the disease has not been fully elucidated. In this review, we summarized the literature associated with the different cell types and molecules involved in the immunopathogenesis of CIE, with the aim of advancing the search for biomarkers with possible diagnostic, prognostic, or therapeutic utility.

Abstract: Canine chronic inflammatory enteropathy (CIE) is one of the most common chronic gastrointestinal diseases affecting dogs worldwide. Genetic and environmental factors, as well as intestinal microbiota and dysregulated host immune responses, participate in this multifactorial disease. Despite advances explaining the immunological and molecular mechanisms involved in CIE development, the exact pathogenesis is still unknown. This review compiles the latest reports and advances that describe the main molecular and cellular mechanisms of both the innate and adaptive immune responses involved in canine CIE pathogenesis. Future studies should focus research on the characterization of the immunopathogenesis of canine CIE in order to advance the establishment of biomarkers and molecular targets of diagnostic, prognostic, or therapeutic utility.

Keywords: canine chronic inflammatory enteropathy; inflammatory bowel disease; microbiota; innate immune response; adaptive immune response

1. Introduction

Canine chronic inflammatory enteropathy (CIE) is a term used for gastrointestinal diseases present for 3 weeks or longer, after extraintestinal diseases and intestinal diseases of infectious origin or neoplastic conditions have been ruled out [1]. As in humans, a few years ago, the term IBD was used for canine CIE. However, nowadays there is consensus among experts that CIE is the most appropriate term [1–3], since, clinically, there are fundamental differences between dog and human diseases.

Human IBD includes at least two different chronic disorders characterized by different patterns of inflammation of the intestinal wall: Crohn's disease (CD) and ulcerative colitis (UC) [4]. Instead, CIE in dogs is defined by the response to medical treatment: food-responsive enteropathy (FRE); antibiotic-responsive enteropathy (ARE); immunosuppressant-responsive enteropathy (IRE/SRE); and non-responsive enteropathy (NRE) [1]. Specifically for ARE, Cerquetella et al. recently provided evidence establishing that the empirical use of antibiotics in dogs with IBD can have detrimental effects. Thus, the authors suggested the use of antibacterials only after histopathological evaluation of gastrointestinal biopsies, when endoscopy is not possible, after other therapeutic trials have been unsuccessful, or when there is evidence of adherent-invasive bacteria [5].

Additionally, canine CIE is histopathologically classified according to the affected intestinal segment (stomach, small intestine, or large intestine) and the predominant type of cellular infiltrate (lymphocytic plasmacytic, eosinophilic, neutrophilic, or granulomatous). For example, when the small intestine is affected, lymphoplasmacytic or eosinophilic enteritis is dominant, while when the large intestine is affected, lymphoplasmacytic, eosinophilic, histiocytic, ulcerative, and regional granulomatous colitis has been identified [6]. In this regard, a simplified histopathologic scoring system has recently been proposed [7]. This score provides objective and descriptive information on the extent of mucosal inflammation in the gastrointestinal tract of dogs with CIE, demonstrating great diagnostic utility and important correlation with clinical findings in canine CIE [7].

Therefore, the classification of the disease is very different in dogs and humans. There is no Crohn's-like disease in dogs, and canine histiocytic ulcerative colitis (HUC), an enteropathy previously compared with human enteropathies affecting the boxer breed, is different from human UC. HUC is considered to be caused by enteroinvasive *E. coli* and, therefore, does not belong to the canine idiopathic CIE [8].

The therapeutic approach also differs between humans and dogs. In human IBD, the treatment goals are defined by STRIDE-II, the latest update of Selecting Therapeutic Targets in Inflammatory Bowel Disease (STRIDE), initiative of the International Organization for the Study of Inflammatory Bowel Diseases (IOIBD) [9]. Medical therapy based on different types of immunosuppressant drugs is central to the management of the human condition and is aimed at controlling the inflammation and achieving mucosal remission [4]. In contrast, treatment in dogs is currently mainly aimed at clinical remission and many dogs do not require any treatment other than dietary modification (FRE). A further subset improves with antibiotic therapy, and a small proportion requires immunosuppressant treatment [10]. Several human patients with IBD have undergone surgery [11]. In contrast, surgery is not an indicated treatment for canine CIE [1].

The exact pathogenesis of canine CIE is still unknown and likely multifactorial, involving genetic and environmental factors, intestinal microbiota disarrangement (dysbiosis), and a dysregulated host immune response [4]. There are still significant knowledge gaps as to the role of innate and acquired immune response and microbiota in canine CIE pathogenesis. In order to identify potential novel biomarkers to determine the diagnosis, prognosis, or therapeutic approach, this review highlights some of the most relevant immunological findings of innate and acquired immune response concerning canine CIE.

2. Innate Immune Response

Local innate immune response is the first line of defense against commensal and opportunistic pathogens. However, its role in the pathogenesis of CIE has not been completely elucidated.

2.1. Intestinal Microbiota

Several studies have described differences in the gastrointestinal microbiota among dogs with various gastrointestinal diseases [12,13]. Thus, intestinal dysbiosis has been proposed as an important factor involved in CIE pathogenesis [14]. In addition, the use of a dysbiosis index (DI) to assess changes in the intestinal microbiota has been shown to be useful in evaluating microbial changes in fecal samples from dogs with CIE. [15].

Each intestinal segment has a specific microbiota; the colon and rectum contain the most diverse populations [14]. *Bacteroides, Clostridium, Lactobacillus, Bifidobacterium* spp., and *Enterobacteriaceae* are predominant genera. By 16S rRNA sequencing, the phyla *Firmicutes, Bacteroidetes*, and *Fusobacteria* have been identified as 95% of the total bacterial population, followed by *Proteobacteria* and *Actinobacteria* (1–5%) [16].

Phylogenetic studies have identified a decrease in the proportion of Clostridia and an increase in Proteobacteria in the duodenum of dogs with CIE [12]. Data concerning the genera present in the large intestine are limited. Suchodolski et al. (2012) concluded that dogs with active CIE present with a decrease in *Faecalibacterium* spp., which produces anti-inflammatory peptides in vitro, and *Fusobacterium* phyla. There are no differences in the *Proteobacteria* members [8].

2.2. Mucosal Epithelial Barrier

Gastrointestinal mucus of the intestinal epithelia is the first physical barrier to reduce the exposure to aggressors [17]. In the small intestine, mucus forms a single removable layer and, in the colon, a double layer. Here, mucins are the major barrier with the transmembrane and gel-forming mucins [18–20], which have direct immunological effects by binding to the numerous lectin like proteins found in immune cells [21].

Homeostatic maintenance of the barrier is central to preventing the entry of bacteria and toxins from the lumen [22–26]. In dogs with CIE, it is possible that the breakdown of barrier integrity and the immunological tolerance against intestinal symbionts lead to deregulated inflammation and disease. In turn, this breakdown may also be amplified by CIE. Additionally, pathophysiological or environmental factors may induce loss of mucus barrier integrity [27,28].

Goblet cells, crucial for epithelial restitution, produce small peptides called trefoil factors (TFFs) which protect and repair the epithelial surfaces. The expression of TFFs is upregulated in human IBD [29,30]. In dogs with CIE, TFF1 expression is elevated in the duodenum, where TFF3 expression is down-regulated in the colon, suggesting they may contribute to the deterioration of the epithelial barrier [31].

Another epithelial barrier component is P-glycoprotein (P-gp), a membrane-bound efflux pump involved in the transport of a wide range of small molecules, whose abnormal expression is observed in dogs with lymphoplasmacytic enteritis (LPE). Some LPE patients have increased P-gp expression in the apical surface membrane of villus epithelial cells in the duodenum, jejunum, and/or ileum. In other patients, P-gp expression is decreased [32]. An upregulation in P-gp expression has been identified in lymphocytes from lamina propria after prednisolone treatment in dogs with CIE, which may be considered a predictor of response to therapy [33].

2.3. Innate Immune Cells and Their Derived Molecules

The barrier between blood and endothelial cells is tightly controlled physiologically. The barrier establishes the type and numbers of inflammatory cells that migrate to the interstitial space where dendritic cells (DCs), macrophages and mast cells in the lamina propria, and intraepithelial lymphocytes (IELs) monitor tissues, contributing to intestinal homeostasis [34,35].

2.3.1. Integrins

In human IBD, when the barrier is disrupted, an uncontrolled transfer of inflammatory cells from the blood to the intestinal tissue occurs [36]. The extravasation is mediated mainly by integrins that bind their counterpart receptors on the endothelial cells. The molecules mediating normal endothelial–leukocyte interaction are the same as the molecules engaged in human IBD ($\alpha4\beta1$ (VLA-4), $\alpha4\beta7$, $\alpha D\beta2$, JAM-A, E-selectin, P-selectin, CD31, and CD99), although their expression levels are upregulated by inflammation [37–39]. There is not much information about specific integrins overexpressed in dogs with CIE. However, a reduced expression of the β-integrin CD11c has been described. This finding suggests that canine CIE may have an imbalance in the intestinal CD11c+ DCs. However, further studies are needed to determine whether CD11c could be a useful diagnostic biomarker for canine IBD [40].

2.3.2. Cytokines

In canine CIE, IL-8 may stimulate transmigration of neutrophils to the mucosa and luminal contents to eliminate microbes during intestinal inflammation [28,37]. In a study on German shepherd dogs with CIE, the mRNA expression of many cytokines such as IL-2, IL-5, IL-12p40, interferon-γ (IFN-γ), tumor necrosis factor-α (TNF-α), and transforming growth factor-$\beta1$ (TGF-$\beta1$) was higher in diseased animals compared to controls [41]. However, Jergens et al. previously described through a meta-analysis that healthy dogs showed mRNA expression for most cytokines including IL-2, IL-4, IL-5, IL-10, IL-12, IFN-γ, TNF-α, and TGF-β. They determined that only IL-12 mRNA expression was increased consistently in small-intestinal CIE, whereas CIE lacked consistent patterns of expression [42]. Additionally, it remains unclear whether epithelial cell-derived cytokines such as IL-25, IL-33, and thymic stromal lymphopoietin (TSLP) contribute to the development of canine CIE [43].

2.3.3. Metalloproteinases

Matrix metalloproteinases (MMPs) 2 and 9 are endopeptidases that play an important role in the turnover of extracellular matrix and cell migration and activate and degrade chemokines, cytokines, growth factors, and junction proteins [44]. MMP-2 is produced by stromal cells [45,46] and MMP-9 mainly by neutrophils, followed by eosinophils, monocytes/macrophages, lymphocytes, and epithelial cells [45,47–50]. Both MMPs could be involved in the pathogenesis of canine CIE. Although their role in this pathology has not been completely elucidated, they are upregulated in dogs with CIE [51].

2.3.4. Neutrophils

In the duodenal mucosa of dogs with CIE, an increase in neutrophils is associated with disease severity [48]. Serum perinuclear anti-neutrophilic cytoplasmatic autoantibodies (pANCA) [52] and blood neutrophil-to-lymphocyte ratio (NLR) [53] have been proposed as biomarkers of canine CIE severity. NLR has also been proposed as a useful marker to differentiate FRE from IRE, with clinical utility to subclassify the canine CIE. However, it is important to note that NLR may not be useful for NRE subclassification [54].

In addition, calgranulin-C, a protein secreted by activated neutrophils and monocytes/macrophages, and myeloperoxidase (MPO) activities increase in the mucosa of the duodenum and colon of dogs with CIE, and MPO also increases in the ileum and cecum. However, none have been related to the clinical outcome of patients [55].

Calprotectin, another protein released by activated mononuclear cells, has increased expression in canine intestinal mucosa [56] and is used as a diagnostic and prognostic factor in human IBD [57,58]. Determination of fecal calprotectin concentration is a useful screening test for human IBD diagnosis, reducing the need for colonoscopy by 66.7% [59]. Serum calprotectin concentrations may also be a useful biomarker for the detection of inflammation in dogs, but the use of certain drugs such as glucocorticoids could limit clinical usefulness [60]. The authors also showed that fecal calprotectin could be used as a possible marker for assessing the severity of gastrointestinal inflammation in dogs with CIE [61]. Additionally, a recent meta-analysis concluded that fecal calprotectin concentration is one of the most promising biomarkers of gastrointestinal functionality in dogs [62].

2.3.5. Macrophages

Macrophages participate in the host defense against infections and also remove apoptotic cells and remodel the extracellular matrix [63]. These cells are differentiated into two subtypes, termed M1 and M2. M1 macrophages initiate and maintain inflammatory processes, whereas M2 are associated with the resolution of chronic inflammation and the promotion of tissue repair [64].

Macrophages are present in large amounts in the intestine, primarily the colon, which has a high bacterial load. Intriguingly, these macrophages release mediators that promote homeostasis and thereby do not contribute to a proinflammatory environment [65–68]. This selective inertia is important to maintain the homeostasis and epithelial integrity. Disturbances in this condition may be involved in the pathogenesis of UC and CD in humans [69–71].

An IBD murine model determined that under healthy conditions, macrophages display an anti-inflammatory phenotype (M2) with expression of MHC II, CD163, and IL-10 production. Under pathological conditions, monocytes differentiate into pro-inflammatory macrophages (M1), characterized by the expression of inducible NO synthase (iNOs), $CD64^+HLA-DR^{hi}$ $CD14^{lo}$, producing pro-inflammatory cytokines and chemokines such as TNFα, IL-1β, IL-6, IL-12, IL-23, and CCL11 [69,72].

Increased numbers of macrophages have also been identified in the duodenal mucosa of dogs with CIE [73]. Similar to humans with UC, boxer breed dogs with histiocytic ulcerative colitis (HUC) have higher infiltration of periodic acid-Schiff (PAS)-positive macrophages in the lamina propria in colonic and non-colonic affected regions, with a decrease in Goblet cells and an increase in MHC class II expression in enterocytes [74]. However, a later study revealed that HUC in boxer dogs is caused by enteroinvasive *E. coli* and can be successfully treated with fluoroquinolones such as enrofloxacin. Therefore, canine HUC is considered rather an infectious disease than belonging to the canine idiopathic CIE complex [8,75]. Similarly, granulomatous colitis in young French bulldogs has been also associated with the presence of invasive *E. coli* [75].

A recent study on dogs of different breeds and with or without CIE determined a reduced number of total macrophages but a slightly increased number of $CD64^+$ macrophages, contributing to CIE pathogenesis [76]. Another study characterizing macrophages in the duodenum by immunohistochemistry, evaluated calprotectin (CAL) as a marker of early differentiated macrophages (M1) and CD163 expression as a marker of duodenal resident macrophages (M2), before and after treatment. This study demonstrated that macrophages play an important role in dogs with CIE. In particular, in dogs with FRE and IRE, the $CD163^+$/CAL ratio is lower than in healthy dogs at diagnosis, and it is normalized after treatment in dogs with FRE. No significant differences were observed in dogs with ARE [77].

A study analyzing transcription nuclear factor (NF-κB) activation during mucosal inflammation in situ in dogs with CIE, identified significantly more macrophages/mm^2 with increased activity of the NF-κB pathway in the lamina propria [78], suggesting a role of NF-κB and derived pro-inflammatory cytokines in CIE.

2.3.6. Eosinophils

Eosinophils are granulated cells that contribute to the host defense against parasites and play an important role in local immune regulation. In humans with IBD, eosinophils are increased in number along with IL-5 production, prompting circulation and activation [79,80]. Eosinophils infiltrating the intestinal mucosa release granules composed of crystalloid-containing core encapsulating cationic proteins and leukotriene C4 [80–82].

Canine CIE classification is based on the predominant type of inflammatory cells. The second most diagnosed form of CIE is in the small intestine where lymphoplasmacytic and eosinophilic enteritis are observed [83–85]. A study demonstrated that dogs with CIE have a significantly higher number of degranulated eosinophils in the lower region of the lamina propria, while the upper region has a significantly higher number of degranulated and intact eosinophils [6]. Comparing human and canine eosinophilic gastroenteritis showed that both pathologies have clinical and histopathological similarities [85], and canine eosinophilic gastroenteritis could be a good model for its human counterpart.

Various noninvasive tools and markers have been studied to identify eosinophil activation in the GI tract, including peripheral eosinophil counts and serum 3-bromotyrosine concentrations (3-BrY) [86]. Serum 3-BrY concentrations were also higher in dogs with SRE/IRE than in those with FRE or healthy control dogs. Thus, 3-BrY may serve as a noninvasive biomarker for CIE diagnosis and prognosis [87].

Another potential marker of eosinophil activity is the soluble epoxide hydrolase (sEH), a molecule with a proinflammatory role by metabolizing anti-inflammatory epoxyeicosatrienoic acid to proinflammatory diols. In a murine model, the use of a specific inhibitor of sEH significantly inhibited eosinophil migration, suggesting that sEH plays an important role in the migration of eosinophils to the gastrointestinal system [88].

Interestingly, a study of 30 dogs with CIE found that a significant number of dogs with CIE showed severe (n = 8) or moderate and mixed eosinophilic inflammation (n = 12). Future studies should be performed to further characterize the role of these eosinophils in canine CIE [89].

2.3.7. Mast Cells

Mast cells, involved in the immediate and delayed defense against foreign antigens, also release mediators that affect the mucosal barrier [90]. More recently, a possible relationship between mast cells and host microbiota in human IBD pathogenesis has been proposed. This interaction is crucial to prevent mast cell hyper-reactivity. However, when microbiota genera are expanded, the interaction increases, favoring permeability and release of immunomodulatory molecules that promote inflammation [91].

In dogs with CIE, an increase in mast cells in the area of the eosinophilic gastroenterocolitis has been described, suggesting a role of type I hypersensitivity [85]. Thus, dogs with CIE have significantly more cells positive for IgE protein and mast cells in the mucosa, but their main location is mesenteric lymph nodes [92]. Moreover, a study defining the distribution and types of mast cells in the normal gastrointestinal tract of canines detected fewer mast cells in the villus area compared to the crypt areas; tryptase-positive mast cells (MC_T) were the most abundant cell type, followed by chymase- and a few tryptase- and chymase-positive mast cells (MC_C, MC_{TC}) [93]. However, in dogs with lymphocytic-plasmacytic or eosinophilic gastroenterocolitis, there was a decrease in the number of metachromatically stained granule-containing mast cells and a decrease in the number of the three types of mast cells identified ($MC_{T, -C, -TC}$), suggesting a mast cell degranulation or a Th1 predominant pattern [94].

N-Methylhistamine (NMH) is a stable metabolite of histamine and may be used as a marker of mast cell degranulation and gastrointestinal inflammation [95]. A study by Berghoff et al. showed that some dogs with CIE have increased fecal and/or urinary NMH concentrations, which could indicate increased mast cell activity. However, they were unable to definitively demonstrate such an association. In the same study, the authors suggest that urinary NMH concentrations could have clinical utility as a biomarker of chronic gastrointestinal inflammation, but this area remains to be explored further [96].

2.3.8. Natural Killer Lymphocytes and Natural Killer Cells

Natural killer lymphocyte (NKT) and natural killer (NK) cells also have a role in human IBD. Th2 cytokines such as IL-13, IL-5, and IL-4, involved in UC, are partly produced by NKTs [97–100]. However, NKTs have a dual role as they play a protective role in a dextran sulfate sodium-induced colitis model [101] and a detrimental role in an oxazolone-induced model [102]. An increase in the cytotoxic $CD56^+CD16^+$ NK cell subset in the lamina propria in human IBD patients [103] and a decrease in the $NKp44^+/NKp46^+$ ratio in biopsies of CD patients have been demonstrated [104]. Recently, a meta-analysis evaluating the role of killer-cell immunoglobulin-like receptor (*KIR*) genes of IBD susceptibility in humans found that *2DL5* and *2DLS1* genes are associated with an increased risk of UC, while the *2DS3* gene is associated with a decreased risk of CD development [105]. Additionally, experimental treatment with monoclonal antibodies against NK Group 2D (NKG2D), a constitutively expressed receptor whose ligand is highly expressed in human IBD, has resulted in remission of CD in some patients [106].

The role of NK and NKT cells in CIE development in dogs has not been studied. Due to the relevance in human IBD, further investigation in dogs could provide relevant information.

2.3.9. Natural Antibodies

Natural antibodies (Nabs), IgM and other pre-existent classes of immunoglobulins circling in plasma, are essential components of innate immunity reacting against foreign antigens and microbe-derived substances and activating the classical pathway of complement activation [107,108]. Its role in canine CIE has not been established but should be considered because, in a murine model, homeostatic intestinal IgAs are natural polyreactive antibodies with innate specificity to microbiota [109].

2.3.10. S100/Calgranulins and RAGE Receptors

S100/calgranulins are a group of three phagocyte-specific damage-associated molecular pattern molecules (DAMPs) [110,111] that include S100A12 (calgranulin C) and the S100A8/A9 (calprotectin or calgranulin A/B) complex. The proteins are produced by activated macrophages and neutrophils and accumulate at sites of inflammation [3]. Recently, it has been suggested that fecal S100A12 and fecal calprotectin concentrations are clinically useful markers of gastrointestinal inflammation in dogs [3]. Fecal canine S100A12 concentrations are increased in dogs with CIE, associated with clinical disease activity, the severity of endoscopic lesions, and the severity of colonic inflammation in dogs with CIE [112,113].

Additionally, a recent study evaluated the expression of gastrointestinal mucosal receptor for advanced glycation end products (RAGE), which are considered molecular pattern receptors with relevance to inflammation in dogs with CIE, and its binding to canine S100/calgranulin ligands. CIE in dogs is associated with decreased serum sRAGE concentrations [114] and an increase in epithelial RAGE expression in the duodenum and colon [113] suggesting a dysregulated sRAGE/RAGE axis [114]. The epithelial RAGE expression in the duodenum and colon was significantly higher in dogs with CIE than in healthy controls, with a pattern of overexpression in the ileum and underexpression in the stomach. Thus, although the role of this axis in canine CIE is not completely understood, this axis might be a possible therapeutic target for dogs with CIE, with utility as a therapeutic model for humans [115].

2.3.11. Pattern Recognition Receptors (PRRs): Toll-Like Receptor and NOD-Like Receptors

Both commensals and pathogenic bacteria express pathogen-associated molecular patterns (PAMPs) on their surface, which are recognized by host pattern recognition receptors PRRs [116]. Among the best characterized PRRs are Toll-like receptors (TLRs) and NOD-like receptors (NOD) [117,118]. Under normal conditions, PRRs recognize antigens from food and commensal bacteria inducing tolerogenic responses. In canine CIE, these antigens, which normally induce immune tolerance, trigger an inflammatory response, with proliferation of T lymphocytes and production of several pro-inflammatory cytokines [116]. Thus, an increased TLR2, TLR4, and TLR9 mRNA expression in dogs with CIE has been described [119,120].

In human IBD, some mutations in these PRRs have been associated with its development [121]. Similarly, several mutations in PRRs have been also associated with the development of canine CIE. Single-nucleotide polymorphisms (SNPs) associated with TLR4 and TLR5 [122] and NOD2 [123] have been identified in German shepherd dogs with CIE. Additionally, a genetic component has been established in canine CIE, with a predisposition of certain breeds. Among those predisposed breeds are German shepherd dogs, Weimaraners, Rottweilers, border collies, and boxers [124].

Subsequently, a TLR5 haplotype has been identified, which is associated with a hypersensitivity to flagellin, exacerbating inflammatory pathways in dogs carrying this haplotype, increasing the risk of developing CIE [125].

While all these results provide valuable information for the development of possible genetic markers of CIE, it should be considered that CIE is a polygenetic disorder. Furthermore, these potential genetic markers would in many cases be expected to be breed-dependent [116].

3. Adaptive Immunity

The intestinal adaptive immune response is composed of CD4$^+$ T cells, IgA-producing B cells in Peyer's patches (PPs) and lamina propria, and intestinal epithelial lymphocytes (IELs), which play a critical role in maintaining immune tolerance [126,127].

3.1. T Helper CD4$^+$ Lymphocytes

In homeostasis, intestinal professional antigen-presenting cells (APCs) migrate to mesenteric lymph nodes where they present antigens and activate CD4$^+$ T lymphocytes mediating pathogenic immunity or mucosal tolerance and barrier integrity contributing to the expansion of T helper cells or regulatory T-cells (Treg), respectively. Alteration of T cells leads to an imbalance between regulatory and effector cells, resulting in tissue inflammation, where pro-inflammatory cytokines such as TNF-α, IL-1, IL-6, IL-8, IL-12, IL-18, IL-23, and chemokines are secreted [28,128].

T helper cells (CD4$^+$) are a subpopulation of T lymphocytes. The main T helper subpopulations are Th1, Th2, Th17, and Treg. The differentiation of T helper lymphocytes varies according to the type of antigen that the APC faces and the length of exposure. These cells are essential components of adaptive immunity since they secrete specific cytokines in response to MHC-II-dependent peptide recognition and co-stimulatory signals from APCs [28,129].

Naive helper T-lymphocyte (Th0) differentiation into Th1 occurs when DCs or professional phagocytes primarily release IL-12 [128]. These Th1 cells induce cell-mediated effector responses, such as cytotoxicity and immunity to intracellular organisms, secreting IL-2, IL-12, INF-γ, and TNF-α [130]. The transcription factors associated with their differentiation are STAT4 and T-bet [129,130].

When activated DCs secrete IL-4, Th0 cells differentiate into Th2, with the participation of transcription factors GATA-3 and STAT6 [129]. Th2 cells produce IL-6, IL-4, IL-5, and IL-13, activating B lymphocytes to produce IgE and recruit eosinophils. This Th2 response is mainly associated with helminth infections and allergies [129,130].

If activated DCs mainly secrete IL-6 and TGF-β, Th0 differentiate to Th17. Th17 cells are characterized by the transcription factor RORγt and production of IL-17A, IL-17F, and IL-22 cytokines, which subsequently trigger inflammatory signaling cascades and lead to the recruitment of innate immune cells [129,131].

Finally, Th0 differentiate to Treg cells in the presence of transcription factor Foxp3. Treg cells secrete cytokines that have anti-inflammatory effects such as IL-10 and TGF-β [132].

In humans, an association between the predominant immune profile and the type of IBD induced has been established. CD is associated with a Th1/Th17 response, whereas UC has a predominant Th2/Th9 response [133–135].

In dogs, a predominant profile for CIE has not been determined [130]. A meta-analysis of intestinal cytokine mRNA expression showed a balance in the expression of pro-inflammatory and anti-inflammatory cytokines in German shepherd dogs with CIE [42]. Previously, an increase in the expression of IL-2 and TNF-α in dogs with colitis was observed [136]. A more recent study evaluated IL-25, IL-33, and TSLP mRNA expression in the intestinal epithelial cells (IECs) of the duodenal and colonic mucosa of dogs with FRE [43]. These three cytokines enhance Th2-dominated immunity by stimulating DCs, innate lymphoid cells, basophils, and mast cells [137]. This study showed that IL-33 mRNA expression was significantly lower in the duodenum of dogs with FRE than in healthy dogs. These results suggest that a Th2-response is not induced in canine CIE. However, further studies are needed to understand the role of IL-33 in canine CIE [43].

A Th17 response is involved in the pathogenesis of IBD in humans [138]. Although the role of these cells has not been completely described in canine CIE [139,140], an increase in the expression of IL-17A, IL-23p19, and Il-12p35 has been identified [141,142].

In addition, a low number of Treg cells and mRNA expression of IL-10 and TGF-β has been described in dogs with lymphoplasmacytic enteritis. Since TGF-β is essential for Th0 differentiation into Treg, a decrease in TGF-β expression may contribute to a decrease in Treg cells in the duodenum in dogs with CIE [143,144].

There are discrepancies associated with predominant Th profiles in canine CIE that could be explained by differences in the inclusion and exclusion criteria and the clinical heterogenicity of dogs in different studies [28,145].

Signal transducer and activator of transcription 3 (STAT3), an essential transcription factor for the differentiation of Th17 lymphocytes, plays an important role in the pathogenesis of human IBD [146]. For its activation, STAT3 is phosphorylated (pSTAT3), contributing to the intestinal homeostasis and intestinal wound healing, and stimulating the release of several anti-inflammatory cytokines [147,148].

A recent study found significant activation of STAT3 in the duodenal mucosa of dogs with different subtypes of CIE. Higher expression was found in the epithelium and lamina propria of the crypt area in the FRE group than in the PLE group, where pSTAT3 upregulation was more dominant in the epithelium of the crypt and villus area. Only the SRE group featured pSTAT3 upregulation in both areas compared to the control group. Thus, pSTAT3 upregulation has been proposed as a characteristic of SRE and an important clinical marker for active mucosal inflammation in CIE [149].

3.2. Intestinal Intraepithelial Lymphocytes

Intestinal intraepithelial lymphocytes (IELs) are an important cell population at mucosal sites. Two major subtypes of IELs have been described, involving both pro- and anti-inflammatory functions [126]. These subtypes correspond to the conventional IELs, characterized by the expression of the T-cell receptor (TCR) $\alpha\beta^+$ with co-receptor clusters CD4$^+$ and CD8$^+$ and a non-conventional IEL expressing TCR$\alpha\beta^+$ or TCR$\gamma\delta^+$ combined with co-receptor CD8$\alpha\alpha^+$ [150,151]. Human patients with CD show higher levels of TCR$\gamma\delta^+$ T cells in the inflamed colonic mucosa [152]. In dogs, IELs are diffusely scattered throughout the small-intestinal villus epithelium [93], but an increase in the number of TCR$\gamma\delta^+$ T cells has been observed in dogs with CIE [153].

3.3. B-Lymphocytes

B cells are an essential component of mucosal immunity with an important role as APCs, mainly in the secondary immune response, modulating the microbiota diversity, maintaining the integrity of the intestinal barrier [154,155], and secreting a variety of antibodies. IgA in the lamina propria neutralizes luminal microbes transporting them from the mucosal epithelium to the lumen, among other functions [155]. In humans, an increase in plasma and serum cell infiltration and local IgA, IgM, and IgG in the intestinal mucosa has been described in patients with CD [156,157] and UC [158–160]. Similarly, dogs with CIE show an increase in the number of B lymphocytes in the bloodstream [161] and intestinal mucosa [162] and an increase in IgG^+, $IgG3^+$, and $IgG4^+$ in plasma cells [73].

In the intestinal mucosa, dimeric IgA is secreted and transported through the epithelium into the intestinal lumen. IgA is important in mucosal defense through (1) preventing the passage of pathogens, (2) influencing mucosal defense mechanisms by preventing the spread of pathogens into intestinal tissue, and (3) preventing infections and dysbiosis [163,164].

Canine CIE has been associated with a reduction in IgA levels in intestinal mucosa and feces [144] and is associated with a decreased pattern of expression of the transmembrane activator and calcium-modulating cyclophilin–ligand interactor (TACI) and a decreased B cell-activating factor of the TNF family (BAFF-R). In addition, a hypermethylation of TNFRSF13B and TNFRSF13C loci has been observed, possibly associated with a defect in IgA class switching [165,166]. All this suggests a role of mucosal IgA deficit in CIE [28].

A recent study showed that dogs with CIE have higher levels of IgA specific to serological markers such as polynuclear leukocytes, bacterial OmpC, calprotectin, gliadins, and bacterial flagellins. However, further studies demonstrating its clinical relevance are required [167].

Interestingly, Soontararak et al. [168] showed that bacteria present in the gut of dogs with CIE had significantly higher levels of IgG fixation. Furthermore, these IgG levels appeared to be directed against certain dysbiotic bacteria, mainly Actinobacteria. These IgG-coated bacteria induce higher production of TNF-α by macrophages, indicating a pro-inflammatory effect in dogs with CIE, similar to that observed in humans [168].

4. Cross-Talk in the Immune Responses and Their Possible Role in the Pathogenesis of Chronic Inflammatory Enteropathy (CIE) in Dogs

The immune system is traditionally classified into the innate and adaptive immune response. However, both responses are part of a dynamic process, where molecules and cells from both innate and acquired responses are strongly integrated. Thus, cells of the innate immune response recognize pathogens and tissue damage triggering an inflammatory response, and DCs, T lymphocytes, and B lymphocytes drive an acquired immune response, which is concomitantly induced [169].

In CIE, barrier integrity breakdown leads to an exaggerated immune response and loss of immunological tolerance, facilitating microbiota influx and phagocytosis [27,28] initiating the innate immune response. This response modulates the expression of different molecules and cytokines contributing to recruiting and activating different types of inflammatory cells [33,40–43,72,73,76,89,92–94,170,171]. As a consequence, antigen-presenting cells (APCs), such as local DCs or macrophages, and phagocyte pathogens migrate to mesenteric lymph nodes to present antigens and activate different subpopulations of T lymphocytes. Although in dogs, a predominant profile of lymphocytes for CIE has not been determined [130], both pro-inflammatory and anti-inflammatory cytokines have been identified [42,43,136,137]. This difference from human IBD could be explained by the complex classification of CIE and other factors involved in its pathogenesis which may interfere with and modify the immune response.

The main findings related to the role of the innate and adaptive immune response in the pathogenesis of canine CIE and other enteropathies are summarized in Figure 1 and Table 1.

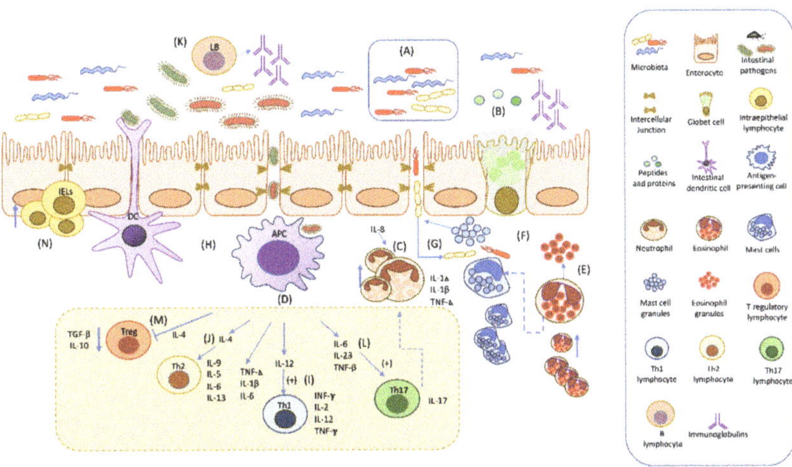

Figure 1. Cross-talk in the immune responses and their possible role in the pathogenesis of chronic inflammatory enteropathy (CIE) in dogs. (**A**) In CIE, barrier integrity breakdown leads to pathological inflammation and loss of immunological tolerance. A decrease in bacteria forming part of the physiologic microbiota such as Faecalibacterium spp. and Fusobacterium phyla has been identified. (**B**) Goblet-cell-derived peptide and glycoprotein expression from enterocytes is up- or down-regulated. (**C**) During intestinal inflammation, IL-8 contributes to neutrophil recruitment and secretion of IL-1α, IL-1β, and TNF-α. (**D**) Phagocytic cells, including antigen-presenting cells (APCs) such as dendritic cells (DCs) and macrophages phagocyte microorganisms and differentiate to a pro-inflammatory phenotype, secreting IL-12, TNF-α, IL-1β, and IL-6. Additionally, (**E**) the number and degranulation of eosinophils and (**F**) the number and granules of mast cells are increased. (**G**) When the barrier is disrupted, mucosal permeability facilitates microbiota influx and phagocytosis. The microbiota influx also promotes microbiota–mast-cell interaction and releasing of immunomodulatory molecules. (**H**) APCs phagocyte pathogens and migrate to mesenteric lymph nodes to present antigens and activate different subpopulations of CD4+ T lymphocytes. (**I**) Th1 is stimulated by IL-12 released by APC and secretes mainly IL-2, IL-12, INF-γ, and TNF-α, while (**J**) Th2 is stimulated by IL-4 released by APC to produce IL-6, IL-4, IL-5, IL-9, and IL-13. (**K**) There is an increase in B lymphocytes and in IgG, IgG3, and IgG4 levels but a decrease in IgA in the intestinal mucosa. (**L**) When activated, APCs secrete IL-6 and TGF-β, and a Th17 subpopulation is activated. This population secretes IL-17A, IL-17F, and IL-22, which participate in neutrophil recruitment. Finally, (**M**) Treg subpopulations secrete IL-10 and TGF-β, contributing to control of the immune response. However, in dogs, a Th predominant immune profile for CIE has not been completely determined. In addition, (**N**) an increase in intestinal intraepithelial lymphocytes has been identified in dogs with CIE.

Table 1. The innate and adaptive immune response in dogs with CIE.

Innate Immune Response	References
Microbiota	
A decrease in proportion of Clostridia and increase in proportion of Proteobacteria in the duodenum	[12]
A decrease in Faecalibacterium spp and Fusobacteria	[12]

Table 1. *Cont.*

Innate Immune Response	References
Mucosal epithelial barrier	
Pathophysiological or environmental factors could induce loss of the mucosal barrier integrity and immune tolerance against intestinal symbionts	[27,28]
Trefoil factor (TFF) 1 expression is elevated in the duodenum, whereas TFF3 expression is down-regulated in the colon, suggesting that it contributes to impaired epithelial barrier function	[31]
Abnormal P-glycoprotein (P-gp) expression is observed in dogs with lymphoplasmacytic enteritis (LPE)	[32]
Upregulation of P-gp expression in lamina propria lymphocytes after prednisolone treatment	[33]
Innate immune cells and derived molecules	
A reduced expression of the β-integrin CD11c	[40]
An increase in neutrophils as a factor associated with severity	[73]
Perinuclear anti-neutrophil cytoplasmic autoantibodies (pANCA) and neutrophil-to-lymphocyte ratio (NLR) as biomarkers of severity	[52,53]
Calgranulin-C and myeloperoxidase (MPO) activities are increased in the duodenum and colon of dogs with chronic enteropathies, and myeloperoxidase (MPO) is also increased in the ileum and cecum. Calprotectin is overexpressed and released by activated mononuclear cells in canine CIE	[55,56]
Matrix metalloproteinases (MMPs)-2 and -9 are upregulated in dogs with CIE	[51]
Increased numbers of macrophages in the duodenal mucosa	[73]
An increase in macrophage infiltration in the lamina propria in colonic and noncolonic affected regions, a decrease in Goblet cells, and an increase in MHC class II expression in enterocytes of boxer breed dogs with CIE	[74]
An increase in macrophages/mm^2 with increased NF-κB pathway activity in the lamina propria	[78]
Degranulated eosinophils in the lower region of the lamina propria and degranulated and intact eosinophils in the upper	[6]
Increased concentration of Serum 3-BrY (associated with eosinophil activation) in dogs with SRE/IRE compared to those with FRE or healthy control dogs	[87]
Increased mast cells in the area of eosinophilic gastroenterocolitis	[85]
More IgE-positive cells and mast cells in the mucosa and mesenteric lymph nodes	[92]
A decrease in metachromatically stained granules and mast cells in dogs with lymphocytic-plasmacytic or eosinophilic gastroenterocolitis	[94]
Increased fecal and/or urinary NMH concentrations in some dogs with CIE	[96]
Increased fecal S100A12 concentrations associated with clinical disease activity, the severity of endoscopic lesions, and the severity of colonic inflammation	[112]
Decreased serum sRAGE concentrations in canine CIE	[114]
Overexpression of epithelial RAGE along the gastrointestinal tract in dogs with CIE	[115]
Adaptive Immune Response	
T helper lymphocytes (CD4+)	
A balance in the expression of proinflammatory and anti-inflammatory cytokines in German shepherd dogs	[42]
An increase in IL-2 and TNF-α expression in dogs with colitis	[136]
An increase in IL12p40-associated mRNA in dogs with lymphocytic-plasmocytic enteritis and lymphocytic-plasmocytic colitis, when the duodenum is affected. An increase in IL-4 mRNA expression when the colon is affected	[42]
An increased expression of IL-17A, IL-23p19, and Il-12p35	[139–142]
Low number of Treg cell and IL-10 and TGF-β mRNA expression in dogs with lymphocytic-plasmocytic enteritis	[143,144]

Table 1. *Cont.*

Innate Immune Response	References
Intestinal intraepithelial T lymphocytes	
Increased numbers of TCRγδ+ cells	[93,153]
B lymphocytes	
Increased numbers of B lymphocytes in the bloodstream and intestinal mucosa. IgG+, IgG3+, and IgG4+ also increase in plasma cells	[73,161,162]
Reduced IgA levels in intestinal mucosa, feces, and peripheral blood	[28,144]
High levels of specific IgA against serological markers such as polynuclear leukocytes, bacterial OmpC, calprotectin, gliadins, and bacterial flagellins	[167]
An increase in IgG-coated gut bacteria, which induce increased production of TNF-α by macrophages	[168]

5. Concluding Remarks

Despite advances in the complex molecular mechanisms explaining CIE in dogs, the contribution of the immune response in the pathogenesis of CIE is not completely understood. To determine the role of barrier integrity breakdown and the loss of immunological tolerance against intestinal symbionts, the microbiota–immune-system interaction is essential to completely understand canine CIE pathogenesis and modulate the clinical consequences. There are important gaps in the knowledge about the role of some molecular and cellular components forming part of the innate immune response which has been identified as an experimental therapeutic target in human IBD. The adaptive immune response also plays a critical role in maintaining immune tolerance toward symbiotic bacteria, integrity of the intestine barrier, and gut homeostasis. In human IBD, CDs-Th1/Th17 and UC-Th2/Th9 associations have been described. However, in canine CIE, a predominant immune profile has not been clearly established due to the heterogenicity in the inclusion/exclusion criteria and clinical aspects of patients enrolled in different studies.

In this review, we characterized different cells and molecules which have been identified as playing a role in the immunopathogenesis of canine CIE. Future research should advance the characterization of canine CIE immunopathogenesis in order to identify future biomarkers and molecular targets of diagnostic, prognostic, and potential therapeutic utility.

Author Contributions: Conceptualization, D.S. and G.R.-T.; writing—original draft preparation, E.M., N.V., M.P. and A.S.; writing—review and editing, D.S., G.R.-T., O.P., I.P. and C.J.B.; illustrations and figures, G.R.-T.; supervision, G.R.-T.; project administration, D.S. and G.R.-T.; funding acquisition, D.S. and G.R.-T. All authors have read and agreed to the published version of the manuscript.

Funding: This research was funded by FONDECYT-ANID (Projects 3190649 and 1181699) (D.S., G.R.-T. and CJ.B.).

Institutional Review Board Statement: Not applicable.

Informed Consent Statement: Not applicable.

Data Availability Statement: Not applicable.

Conflicts of Interest: The authors declare no conflict of interest.

References

1. Dandrieux, J.R. Inflammatory Bowel Disease Versus Chronic Enteropathy in Dogs: Are They One and the Same? *J. Small Anim. Pract.* **2016**, *57*, 589–599. [CrossRef] [PubMed]
2. Dandrieux, J.R.S.; Mansfield, C.S. Chronic Enteropathy in Canines: Prevalence, Impact and Management Strategies. *Vet. Med.* **2019**, *10*, 203–214. [CrossRef] [PubMed]
3. Heilmann, R.M.; Steiner, J.M. Clinical Utility of Currently Available Biomarkers in Inflammatory Enteropathies of Dogs. *J. Vet. Intern. Med.* **2018**, *32*, 1495–1508. [CrossRef] [PubMed]
4. Grevenitis, P.; Thomas, A.; Lodhia, N. Medical Therapy for Inflammatory Bowel Disease. *Surg. Clin. N. Am.* **2015**, *95*, 1159–1182. [CrossRef] [PubMed]

5. Cerquetella, M.; Rossi, G.; Suchodolski, J.S.; Schmitz, S.S.; Allenspach, K.; Rodríguez-Franco, F.; Furlanello, T.; Gavazza, A.; Marchegiani, A.; Unterer, S.; et al. Proposal for Rational Antibacterial Use in the Diagnosis and Treatment of Dogs with Chronic Diarrhoea. *J. Small Anim. Pract.* **2020**, *61*, 211–215. [CrossRef]

6. Bastan, I.; Robinson, N.A.; Ge, X.N.; Rendahl, A.K.; Rao, S.P.; Washabau, R.J.; Sriramarao, P. Assessment of Eosinophil Peroxidase as a Potential Diagnostic and Prognostic Marker in Dogs with Inflammatory Bowel Disease. *Am. J. Vet. Res.* **2017**, *78*, 36–41. [CrossRef]

7. Allenspach, K.A.; Mochel, J.P.; Du, Y.; Priestnall, S.L.; Moore, F.; Slayter, M.; Rodrigues, A.; Ackermann, M.; Krockenberger, M.; Mansell, J.; et al. Correlating Gastrointestinal Histopathologic Changes to Clinical Disease Activity in Dogs with Idiopathic Inflammatory Bowel Disease. *Vet. Pathol.* **2019**, *56*, 435–443. [CrossRef]

8. Mansfield, C.S.; James, F.E.; Craven, M.; Davies, D.R.; O'Hara, A.J.; Nicholls, P.K.; Dogan, B.; MacDonough, S.P.; Simpson, K.W. Remission of Histiocytic Ulcerative Colitis in Boxer Dogs Correlates with Eradication of Invasive Intramucosal Escherichia Coli. *J. Vet. Intern. Med.* **2009**, *23*, 964–969. [CrossRef]

9. Turner, D.; Ricciuto, A.; Lewis, A.; D'Amico, F.; Dhaliwal, J.; Griffiths, A.M.; Bettenworth, D.; Sandborn, W.J.; Sands, B.E.; Reinisch, W.; et al. Stride-Ii: An Update on the Selecting Therapeutic Targets in Inflammatory Bowel Disease (Stride) Initiative of the International Organization for the Study of Ibd (Ioibd): Determining Therapeutic Goals for Treat-to-Target Strategies in Ibd. *Gastroenterology* **2021**, *160*, 1570–1583. [CrossRef]

10. Jergens, A.E.; Simpson, K.W. Inflammatory Bowel Disease in Veterinary Medicine. *Front. Biosci. Elite Ed.* **2012**, *4*, 1404–1419. [CrossRef]

11. Annese, V.; Duricova, D.; Gower-Rousseau, C.; Jess, T.; Langholz, E. Impact of New Treatments on Hospitalisation, Surgery, Infection, and Mortality in Ibd: A Focus Paper by the Epidemiology Committee of Ecco. *J. Crohn's Colitis* **2015**, *10*, 216–225. [CrossRef] [PubMed]

12. Suchodolski, J.S.; Markel, M.E.; Garcia-Mazcorro, J.F.; Unterer, S.; Heilmann, R.M.; Dowd, S.E.; Kachroo, P.; Ivanov, I.; Minamoto, Y.; Dillman, E.M.; et al. The Fecal Microbiome in Dogs with Acute Diarrhea and Idiopathic Inflammatory Bowel Disease. *PLoS ONE* **2012**, *7*, e51907. [CrossRef] [PubMed]

13. Wang, S.; Martins, R.; Sullivan, M.C.; Friedman, E.S.; Misic, A.M.; El-Fahmawi, A.; De Martinis, E.C.P.; O'Brien, K.; Chen, Y.; Bradley, C.; et al. Diet-Induced Remission in Chronic Enteropathy Is Associated with Altered Microbial Community Structure and Synthesis of Secondary Bile Acids. *Microbiome* **2019**, *7*, 126. [CrossRef] [PubMed]

14. Huang, Z.; Pan, Z.; Yang, R.; Bi, Y.; Xiong, X. The Canine Gastrointestinal Microbiota: Early Studies and Research Frontiers. *Gut Microbes* **2020**, *11*, 635–654. [CrossRef] [PubMed]

15. AlShawaqfeh, M.; Wajid, B.; Minamoto, Y.; Markel, M.; Lidbury, J.; Steiner, J.; Serpedin, E.; Suchodolski, J. A Dysbiosis Index to Assess Microbial Changes in Fecal Samples of Dogs with Chronic Inflammatory Enteropathy. *FEMS Microbiol. Ecol.* **2017**, *93*, 431–457. [CrossRef]

16. Suchodolski, J.S. Intestinal Microbiota of Dogs and Cats: A Bigger World Than We Thought. *Vet. Clin. Small Anim. Pract.* **2011**, *41*, 261–272. [CrossRef]

17. Pelaseyed, T.; Bergström, J.H.; Gustafsson, J.K.; Ermund, A.; Birchenough, G.M.; Schütte, A.; van der Post, S.; Svensson, F.; Rodríguez-Piñeiro, A.M.; Nyström, E.E.; et al. The Mucus and Mucins of the Goblet Cells and Enterocytes Provide the First Defense Line of the Gastrointestinal Tract and Interact with the Immune System. *Immunol. Rev.* **2014**, *260*, 8–20. [CrossRef]

18. Corfield, A.P. Mucins: A Biologically Relevant Glycan Barrier in Mucosal Protection. *Biochim. Biophys. Acta BBA Gen. Subj.* **2015**, *1850*, 236–252. [CrossRef]

19. Hattrup, C.L.; Gendler, S.J. Structure and Function of the Cell Surface (Tethered) Mucins. *Annu. Rev. Physiol.* **2008**, *70*, 431–457. [CrossRef]

20. Johansson, M.E.; Hansson, G.C. Immunological Aspects of Intestinal Mucus and Mucins. *Nat. Rev. Immunol.* **2016**, *16*, 639–649. [CrossRef]

21. Rosen, S.D. Ligands for L-Selectin: Homing, Inflammation, and Beyond. *Annu. Rev. Immunol.* **2004**, *22*, 129–156. [CrossRef]

22. Balimane, P.V.; Chong, S.; Morrison, R.A. Current Methodologies Used for Evaluation of Intestinal Permeability and Absorption. *J. Pharmacol. Toxicol. Methods* **2000**, *44*, 301–312. [CrossRef]

23. Farquhar, M.J.; McCluskey, E.; Staunton, R.; Hughes, K.R.; Coltherd, J.C. Characterisation of a Canine Epithelial Cell Line for Modelling the Intestinal Barrier. *Altern. Lab. Anim.* **2018**, *46*, 115–132. [CrossRef] [PubMed]

24. Halpern, M.D.; Denning, P.W. The Role of Intestinal Epithelial Barrier Function in the Development of Nec. *Tissue Barriers* **2015**, *3*, e1000707. [CrossRef] [PubMed]

25. Hollander, D. Intestinal Permeability, Leaky Gut, and Intestinal Disorders. *Curr. Gastroenterol. Rep.* **1999**, *1*, 410–416. [CrossRef] [PubMed]

26. Watson, A.J.; Hughes, K.R. Tnf-A-Induced Intestinal Epithelial Cell Shedding: Implications for Intestinal Barrier Function. *Ann. N. Y. Acad. Sci.* **2012**, *1258*, 1–8. [CrossRef] [PubMed]

27. Chelakkot, C.; Ghim, J.; Ryu, S.H. Mechanisms Regulating Intestinal Barrier Integrity and Its Pathological Implications. *Exp. Mol. Med.* **2018**, *50*, 103. [CrossRef] [PubMed]

28. Eissa, N.; Kittana, H.; Gomes-Neto, J.C.; Hussein, H. Mucosal Immunity and Gut Microbiota in Dogs with Chronic Enteropathy. *Res. Vet. Sci.* **2019**, *122*, 156–164. [CrossRef]

29. Shaoul, R.; Okada, Y.; Cutz, E.; Marcon, M.A. Colonic Expression of Muc2, Muc5ac, and Tff1 in Inflammatory Bowel Disease in Children. *J. Pediatr. Gastroenterol. Nutr.* **2004**, *38*, 488–493. [CrossRef]
30. Wright, N.A.; Poulsom, R.; Stamp, G.; Van Noorden, S.; Sarraf, C.; Elia, G.; Ahnen, D.; Jeffery, R.; Longcroft, J.; Pike, C.; et al. Trefoil Peptide Gene Expression in Gastrointestinal Epithelial Cells in Inflammatory Bowel Disease. *Gastroenterology* **1993**, *104*, 12–20. [CrossRef]
31. Schmitz, S.; Hill, S.; Werling, D.; Allenspach, K. Expression of Trefoil Factor Genes in the Duodenum and Colon of Dogs with Inflammatory Bowel Disease and Healthy Dogs. *Vet. Immunol. Immunopathol.* **2013**, *151*, 168–172. [CrossRef] [PubMed]
32. Van der Heyden, S.; Vercauteren, G.; Daminet, S.; Paepe, D.; Chiers, K.; Polis, I.; Waelbers, T.; Hesta, M.; Schauvliege, S.; Wegge, B.; et al. Expression of P-Glycoprotein in the Intestinal Epithelium of Dogs with Lymphoplasmacytic Enteritis. *J. Comp. Pathol.* **2011**, *145*, 199–206. [CrossRef] [PubMed]
33. Allenspach, K.; Bergman, P.J.; Sauter, S.; Gröne, A.; Doherr, M.G.; Gaschen, F. P-Glycoprotein Expression in Lamina Propria Lymphocytes of Duodenal Biopsy Samples in Dogs with Chronic Idiopathic Enteropathies. *J. Comp. Pathol.* **2006**, *134*, 1–7. [CrossRef] [PubMed]
34. Danese, S.; Fiocchi, C. Endothelial Cell-Immune Cell Interaction in Ibd. *Dig. Dis.* **2016**, *34*, 43–50. [CrossRef]
35. Rodrigues, S.F.; Granger, D.N. Blood Cells and Endothelial Barrier Function. *Tissue Barriers* **2015**, *3*, e978720. [CrossRef]
36. Sands, B.E.; Kaplan, G.G. The Role of Tnfα in Ulcerative Colitis. *J. Clin. Pharmacol.* **2007**, *47*, 930–941. [CrossRef]
37. Kolaczkowska, E.; Kubes, P. Neutrophil Recruitment and Function in Health and Inflammation. *Nat. Rev. Immunol.* **2013**, *13*, 159–175. [CrossRef]
38. Langer, H.F.; Chavakis, T. Leukocyte—Endothelial Interactions in Inflammation. *J. Cell. Mol. Med.* **2009**, *13*, 1211–1220. [CrossRef]
39. Petri, W.A., Jr.; Miller, M.; Binder, H.J.; Levine, M.M.; Dillingham, R.; Guerrant, R.L. Enteric Infections, Diarrhea, and Their Impact on Function and Development. *J. Clin. Investig.* **2008**, *118*, 1277–1290. [CrossRef]
40. Kathrani, A.; Schmitz, S.; Priestnall, S.L.; Smith, K.C.; Werling, D.; Garden, O.A.; Allenspach, K. Cd11c+ Cells Are Significantly Decreased in the Duodenum, Ileum and Colon of Dogs with Inflammatory Bowel Disease. *J. Comp. Pathol.* **2011**, *145*, 359–366. [CrossRef]
41. German, A.J.; Helps, C.R.; Hall, E.J.; Day, M.J. Cytokine Mrna Expression in Mucosal Biopsies from German Shepherd Dogs with Small Intestinal Enteropathies. *Dig. Dis. Sci.* **2000**, *45*, 7–17. [CrossRef] [PubMed]
42. Jergens, A.E.; Sonea, I.M.; O'Connor, A.M.; Kauffman, L.K.; Grozdanic, S.D.; Ackermann, M.R.; Evans, R.B. Intestinal Cytokine Mrna Expression in Canine Inflammatory Bowel Disease: A Meta-Analysis with Critical Appraisal. *Comp. Med.* **2009**, *59*, 153–162. [PubMed]
43. Osada, H.; Ogawa, M.; Hasegawa, A.; Nagai, M.; Shirai, J.; Sasaki, K.; Shimoda, M.; Itoh, H.; Kondo, H.; Ohmori, K. Expression of Epithelial Cell-Derived Cytokine Genes in the Duodenal and Colonic Mucosae of Dogs with Chronic Enteropathy. *J. Vet. Med. Sci.* **2017**, *79*, 393–397. [CrossRef] [PubMed]
44. O'Sullivan, S.; Gilmer, J.F.; Medina, C. Matrix Metalloproteinases in Inflammatory Bowel Disease: An Update. *Mediat. Inflamm.* **2015**, *2015*, 964131. [CrossRef]
45. Gao, Q.; Meijer, M.J.; Kubben, F.J.; Sier, C.F.; Kruidenier, L.; van Duijn, W.; van den Berg, M.; van Hogezand, R.A.; Lamers, C.B.; Verspaget, H.W. Expression of Matrix Metalloproteinases-2 and -9 in Intestinal Tissue of Patients with Inflammatory Bowel Diseases. *Dig. Liver Dis.* **2005**, *37*, 584–592. [CrossRef]
46. Kirkegaard, T.; Hansen, A.; Bruun, E.; Brynskov, J. Expression and Localisation of Matrix Metalloproteinases and Their Natural Inhibitors in Fistulae of Patients with Crohn's Disease. *Gut* **2004**, *53*, 701–709. [CrossRef]
47. Garg, P.; Vijay-Kumar, M.; Wang, L.; Gewirtz, A.T.; Merlin, D.; Sitaraman, S.V. Matrix Metalloproteinase-9-Mediated Tissue Injury Overrides the Protective Effect of Matrix Metalloproteinase-2 During Colitis. *Am. J. Physiol. Gastrointest. Liver Physiol.* **2009**, *296*, G175–G184. [CrossRef]
48. Hogan, S.P. Functional Role of Eosinophils in Gastrointestinal Inflammation. *Immunol. Allergy Clin. N. Am.* **2009**, *29*, 129–140. [CrossRef]
49. Kim, J.H.; Lee, S.Y.; Bak, S.M.; Suh, I.B.; Lee, S.Y.; Shin, C.; Shim, J.J.; In, K.H.; Kang, K.H.; Yoo, S.H. Effects of Matrix Metalloproteinase Inhibitor on Lps-Induced Goblet Cell Metaplasia. *Am. J. Physiol. Cell. Mol. Physiol.* **2004**, *287*, L127–L133. [CrossRef]
50. Lubbe, W.J.; Zhou, Z.Y.; Fu, W.; Zuzga, D.; Schulz, S.; Fridman, R.; Muschel, R.J.; Waldman, S.A.; Pitari, G.M. Tumor Epithelial Cell Matrix Metalloproteinase 9 Is a Target for Antimetastatic Therapy in Colorectal Cancer. *Clin. Cancer Res.* **2006**, *12*, 1876–1882. [CrossRef]
51. Hanifeh, M.; Rajamäki, M.M.; Syrjä, P.; Mäkitalo, L.; Kilpinen, S.; Spillmann, T. Identification of Matrix Metalloproteinase-2 and -9 Activities within the Intestinal Mucosa of Dogs with Chronic Enteropathies. *Acta Vet. Scand.* **2018**, *60*, 16. [CrossRef] [PubMed]
52. Mancho, C.; Sainz, Á.; García-Sancho, M.; Villaescusa, A.; Rodríguez-Franco, F. Evaluation of Perinuclear Antineutrophilic Cytoplasmic Antibodies in Sera from Dogs with Inflammatory Bowel Disease or Intestinal Lymphoma. *Am. J. Vet. Res.* **2011**, *72*, 1333–1337. [CrossRef] [PubMed]
53. Benvenuti, E.; Pierini, A.; Gori, E.; Lucarelli, C.; Lubas, G.; Marchetti, V. Neutrophil-to-Lymphocyte Ratio (Nlr) in Canine Inflammatory Bowel Disease (Ibd). *Vet. Sci.* **2020**, *7*, 141. [CrossRef] [PubMed]
54. Becher, A.; Suchodolski, J.S.; Steiner, J.M.; Heilmann, R.M. Blood Neutrophil-to-Lymphocyte Ratio (Nlr) as a Diagnostic Marker in Dogs with Chronic Enteropathy. *J. Vet. Diagn. Investig.* **2021**, *33*, 516–527. [CrossRef]

55. Hanifeh, M.; Sankari, S.; Rajamäki, M.M.; Syrjä, P.; Kilpinen, S.; Suchodolski, J.S.; Heilmann, R.M.; Guadiano, P.; Lidbury, J.; Steiner, J.M.; et al. S100a12 Concentrations and Myeloperoxidase Activities Are Increased in the Intestinal Mucosa of Dogs with Chronic Enteropathies. *BMC Vet. Res.* **2018**, *14*, 125. [CrossRef]

56. Heilmann, R.M.; Nestler, J.; Schwarz, J.; Grützner, N.; Ambrus, A.; Seeger, J.; Suchodolski, J.S.; Steiner, J.M.; Gurtner, C. Mucosal Expression of S100a12 (Calgranulin C) and S100a8/A9 (Calprotectin) and Correlation with Serum and Fecal Concentrations in Dogs with Chronic Inflammatory Enteropathy. *Vet. Immunol. Immunopathol.* **2019**, *211*, 64–74. [CrossRef]

57. Khaki-Khatibi, F.; Qujeq, D.; Kashifard, M.; Moein, S.; Maniati, M.; Vaghari-Tabari, M. Calprotectin in Inflammatory Bowel Disease. *Clin. Chim. Acta* **2020**, *510*, 556–565. [CrossRef]

58. Walsham, N.E.; Sherwood, R.A. Fecal Calprotectin in Inflammatory Bowel Disease. *Clin. Exp. Gastroenterol.* **2016**, *9*, 21–29. [CrossRef]

59. Petryszyn, P.; Staniak, A.; Wolosianska, A.; Ekk-Cierniakowski, P. Faecal Calprotectin as a Diagnostic Marker of Inflammatory Bowel Disease in Patients with Gastrointestinal Symptoms: Meta-Analysis. *Eur. J. Gastroenterol. Hepatol.* **2019**, *31*, 1306–1312. [CrossRef]

60. Heilmann, R.M.; Jergens, A.E.; Ackermann, M.R.; Barr, J.W.; Suchodolski, J.S.; Steiner, J.M. Serum Calprotectin Concentrations in Dogs with Idiopathic Inflammatory Bowel Disease. *Am. J. Vet. Res.* **2012**, *73*, 1900–1907. [CrossRef]

61. Heilmann, R.M.; Berghoff, N.; Mansell, J.; Grützner, N.; Parnell, N.K.; Gurtner, C.; Suchodolski, J.S.; Steiner, J.M. Association of Fecal Calprotectin Concentrations with Disease Severity, Response to Treatment, and Other Biomarkers in Dogs with Chronic Inflammatory Enteropathies. *J. Vet. Intern. Med.* **2018**, *32*, 679–692. [CrossRef] [PubMed]

62. Félix, A.P.; Souza, C.M.M.; de Oliveira, S.G. Biomarkers of Gastrointestinal Functionality in Dogs: A Systematic Review and Meta-Analysis. *Anim. Feed. Sci. Technol.* **2022**, *283*, 115183. [CrossRef]

63. Okabe, Y.; Medzhitov, R. Tissue Biology Perspective on Macrophages. *Nat. Immunol.* **2016**, *17*, 9. [CrossRef] [PubMed]

64. Mantovani, A.; Biswas, S.K.; Galdiero, M.R.; Sica, A.; Locati, M. Macrophage Plasticity and Polarization in Tissue Repair and Remodelling. *J. Pathol.* **2013**, *229*, 176–185. [CrossRef]

65. Bain, C.C.; Bravo-Blas, A.; Scott, C.L.; Perdiguero, E.G.; Geissmann, F.; Henri, S.; Malissen, B.; Osborne, L.C.; Artis, D.; Mowat, A.M. Constant Replenishment from Circulating Monocytes Maintains the Macrophage Pool in the Intestine of Adult Mice. *Nat. Immunol.* **2014**, *15*, 929–937. [CrossRef]

66. Hume, D.A. Macrophages as Apc and the Dendritic Cell Myth. *J. Immunol.* **2008**, *181*, 5829–5835. [CrossRef]

67. Mowat, A.M.; Bain, C.C. Mucosal Macrophages in Intestinal Homeostasis and Inflammation. *J. Innate Immun.* **2011**, *3*, 550–564. [CrossRef]

68. Zhou, Z.; Ding, M.; Huang, L.; Gilkeson, G.; Lang, R.; Jiang, W. Toll-Like Receptor-Mediated Immune Responses in Intestinal Macrophages; Implications for Mucosal Immunity and Autoimmune Diseases. *Clin. Immunol.* **2016**, *173*, 81–86. [CrossRef]

69. Bain, C.C.; Schridde, A. Origin, Differentiation, and Function of Intestinal Macrophages. *Front. Immunol.* **2018**, *9*, 2733. [CrossRef]

70. Kamada, N.; Hisamatsu, T.; Okamoto, S.; Sato, T.; Matsuoka, K.; Arai, K.; Nakai, T.; Hasegawa, A.; Inoue, N.; Watanabe, N.; et al. Abnormally Differentiated Subsets of Intestinal Macrophage Play a Key Role in Th1-Dominant Chronic Colitis through Excess Production of Il-12 and Il-23 in Response to Bacteria. *J. Immunol.* **2005**, *175*, 6900. [CrossRef]

71. Xavier, R.J.; Podolsky, D.K. Unravelling the Pathogenesis of Inflammatory Bowel Disease. *Nature* **2007**, *448*, 427–434. [CrossRef] [PubMed]

72. Nolte, A.; Junginger, J.; Baum, B.; Hewicker-Trautwein, M. Heterogeneity of Macrophages in Canine Histiocytic Ulcerative Colitis. *Innate Immun.* **2017**, *23*, 228–239. [CrossRef] [PubMed]

73. German, A.J.; Hall, E.J.; Day, M.J. Immune Cell Populations within the Duodenal Mucosa of Dogs with Enteropathies. *J. Vet. Intern. Med.* **2001**, *15*, 14–25. [CrossRef] [PubMed]

74. German, A.J.; Hall, E.J.; Kelly, D.F.; Watson, A.D.J.; Day, M.J. An Immunohistochemical Study of Histiocytic Ulcerative Colitis in Boxer Dogs. *J. Comp. Pathol.* **2000**, *122*, 163–175. [CrossRef]

75. Manchester, A.C.; Hill, S.; Sabatino, B.; Armentano, R.; Carroll, M.; Kessler, B.; Miller, M.; Dogan, B.; McDonough, S.P.; Simpson, K.W. Association between Granulomatous Colitis in French Bulldogs and Invasive Escherichia Coli and Response to Fluoroquinolone Antimicrobials. *J. Vet. Intern. Med.* **2013**, *27*, 56–61. [CrossRef]

76. Wagner, A.; Junginger, J.; Lemensieck, F.; Hewicker-Trautwein, M. Immunohistochemical Characterization of Gastrointestinal Macrophages/Phagocytes in Dogs with Inflammatory Bowel Disease (Ibd) and Non-Ibd Dogs. *Vet. Immunol. Immunopathol.* **2018**, *197*, 49–57. [CrossRef]

77. Dandrieux, J.R.; Martinez Lopez, L.M.; Stent, A.; Jergens, A.; Allenspach, K.; Nowell, C.J.; Firestone, S.M.; Kimpton, W.; Mansfield, C.S. Changes in Duodenal Cd163-Positive Cells in Dogs with Chronic Enteropathy after Successful Treatment. *Innate Immun.* **2018**, *24*, 400–410. [CrossRef]

78. Luckschander, N.; Hall, J.A.; Gaschen, F.; Forster, U.; Wenzlow, N.; Hermann, P.; Allenspach, K.; Dobbelaere, D.; Burgener, I.A.; Welle, M. Activation of Nuclear Factor-Kb in Dogs with Chronic Enteropathies. *Vet. Immunol. Immunopathol.* **2010**, *133*, 228–236. [CrossRef]

79. Eissa, S.; Abdulkarim, H.; Dasouki, M.; Al Mousa, H.; Arnout, R.; Al Saud, B.; Rahman, A.A.; Zourob, M. Multiplexed Detection of Dock8, Pgm3 and Stat3 Proteins for the Diagnosis of Hyper-Immunoglobulin E Syndrome Using Gold Nanoparticles-Based Immunosensor Array Platform. *Biosens. Bioelectron.* **2018**, *117*, 613–619. [CrossRef]

80. Filippone, R.T.; Sahakian, L.; Apostolopoulos, V.; Nurgali, K. Eosinophils in Inflammatory Bowel Disease. *Inflamm. Bowel Dis.* **2019**, *25*, 1140–1151. [CrossRef]

81. Acharya, K.R.; Ackerman, S.J. Eosinophil Granule Proteins: Form and Function. *J. Biol. Chem.* **2014**, *289*, 17406–17415. [CrossRef] [PubMed]

82. Impellizzeri, G.; Marasco, G.; Eusebi, L.H.; Salfi, N.; Bazzoli, F.; Zagari, R.M. Eosinophilic Colitis: A Clinical Review. *Dig. Liver Dis.* **2019**, *51*, 769–773. [CrossRef] [PubMed]

83. Cerquetella, M.; Spaterna, A.; Laus, F.; Tesei, B.; Rossi, G.; Antonelli, E.; Villanacci, V.; Bassotti, G. Inflammatory Bowel Disease in the Dog: Differences and Similarities with Humans. *World J. Gastroenterol.* **2010**, *16*, 1050–1056. [CrossRef] [PubMed]

84. Junginger, J.; Schwittlick, U.; Lemensieck, F.; Nolte, I.; Hewicker-Trautwein, M. Immunohistochemical Investigation of Foxp3 Expression in the Intestine in Healthy and Diseased Dogs. *Vet. Res.* **2012**, *43*, 23. [CrossRef]

85. Sattasathuchana, P.; Steiner, J.M. Canine Eosinophilic Gastrointestinal Disorders. *Anim. Health Res. Rev.* **2014**, *15*, 76–86. [CrossRef]

86. Dainese, R.; Galliani, E.A.; De Lazzari, F.; D'Incà, R.; Mariné-Barjoan, E.; Vivinus-Nebot, M.H.; Hébuterne, X.; Sturniolo, G.C.; Piche, T. Role of Serological Markers of Activated Eosinophils in Inflammatory Bowel Diseases. *Eur. J. Gastroenterol. Hepatol.* **2012**, *24*, 393–397. [CrossRef]

87. Sattasathuchana, P.; Allenspach, K.; Lopes, R.; Suchodolski, J.S.; Steiner, J.M. Evaluation of Serum 3-Bromotyrosine Concentrations in Dogs with Steroid-Responsive Diarrhea and Food-Responsive Diarrhea. *J. Vet. Intern. Med.* **2017**, *31*, 1056–1061. [CrossRef]

88. Bastan, I.; Ge, X.N.; Dileepan, M.; Greenberg, Y.G.; Guedes, A.G.; Hwang, S.H.; Hammock, B.D.; Washabau, R.J.; Rao, S.P.; Sriramarao, P. Inhibition of Soluble Epoxide Hydrolase Attenuates Eosinophil Recruitment and Food Allergen-Induced Gastrointestinal Inflammation. *J. Leukoc. Biol.* **2018**, *104*, 109–122. [CrossRef]

89. Bastan, I.; Rendahl, A.K.; Seelig, D.; Day, M.J.; Hall, E.J.; Rao, S.P.; Washabau, R.J.; Sriramarao, P. Assessment of Eosinophils in Gastrointestinal Inflammatory Disease of Dogs. *J. Vet. Intern. Med.* **2018**, *32*, 1911–1917. [CrossRef]

90. Bischoff, S.C. Mast Cells in Gastrointestinal Disorders. *Eur. J. Pharmacol.* **2016**, *778*, 139–145. [CrossRef]

91. De Zuani, M.; Dal Secco, C.; Frossi, B. Mast Cells at the Crossroads of Microbiota and Ibd. *Eur. J. Immunol.* **2018**, *48*, 1929–1937. [CrossRef] [PubMed]

92. Locher, C.; Tipold, A.; Welle, M.; Busato, A.; Zurbriggen, A.; Griot-Wenk, M.E. Quantitative Assessment of Mast Cells and Expression of Ige Protein and Mrna for Ige and Interleukin 4 in the Gastrointestinal Tract of Healthy Dogs and Dogs with Inflammatory Bowel Disease. *Am. J. Vet. Res.* **2001**, *62*, 211–216. [CrossRef] [PubMed]

93. Kleinschmidt, S.; Meneses, F.; Nolte, I.; Hewicker-Trautwein, M. Distribution of Mast Cell Subtypes and Immune Cell Populations in Canine Intestines: Evidence for Age-Related Decline in T Cells and Macrophages and Increase of Iga-Positive Plasma Cells. *Res. Vet. Sci.* **2008**, *84*, 41–48. [CrossRef] [PubMed]

94. Kleinschmidt, S.; Meneses, F.; Nolte, I.; Hewicker-Trautwein, M. Characterization of Mast Cell Numbers and Subtypes in Biopsies from the Gastrointestinal Tract of Dogs with Lymphocytic-Plasmacytic or Eosinophilic Gastroenterocolitis. *Vet. Immunol. Immunopathol.* **2007**, *120*, 80–92. [CrossRef]

95. Berghoff, N.; Steiner, J.M. Laboratory Tests for the Diagnosis and Management of Chronic Canine and Feline Enteropathies. *Vet. Clin. N. Am. Small Anim. Pract.* **2011**, *41*, 311–328. [CrossRef]

96. Berghoff, N.; Hill, S.; Parnell, N.K.; Mansell, J.; Suchodolski, J.S.; Steiner, J.M. Fecal and Urinary N-Methylhistamine Concentrations in Dogs with Chronic Gastrointestinal Disease. *Vet. J.* **2014**, *201*, 289–294. [CrossRef]

97. Bouma, G.; Strober, W. The Immunological and Genetic Basis of Inflammatory Bowel Disease. *Nat. Rev. Immunol.* **2003**, *3*, 521–533. [CrossRef]

98. Smyth, M.J.; Godfrey, D.I. Nkt Cells and Tumor Immunity—A Double-Edged Sword. *Nat. Immunol.* **2000**, *1*, 459–460. [CrossRef]

99. Tanaka, J.; Saga, K.; Kido, M.; Nishiura, H.; Akamatsu, T.; Chiba, T.; Watanabe, N. Proinflammatory Th2 Cytokines Induce Production of Thymic Stromal Lymphopoietin in Human Colonic Epithelial Cells. *Dig. Dis. Sci.* **2010**, *55*, 1896–1904. [CrossRef]

100. Wilson, S.B.; Delovitch, T.L. Janus-Like Role of Regulatory Inkt Cells in Autoimmune Disease and Tumour Immunity. *Nat. Rev. Immunol.* **2003**, *3*, 211–222. [CrossRef]

101. Saubermann, L.J.; Beck, P.; De Jong, Y.P.; Pitman, R.S.; Ryan, M.S.; Kim, H.S.; Exley, M.; Snapper, S.; Balk, S.P.; Hagen, S.J.; et al. Activation of Natural Killer T Cells by Alpha-Galactosylceramide in the Presence of Cd1d Provides Protection against Colitis in Mice. *Gastroenterology* **2000**, *119*, 119–128. [CrossRef] [PubMed]

102. Heller, F.; Fuss, I.J.; Nieuwenhuis, E.E.; Blumberg, R.S.; Strober, W. Oxazolone Colitis, a Th2 Colitis Model Resembling Ulcerative Colitis, Is Mediated by Il-13-Producing Nk-T Cells. *Immunity* **2002**, *17*, 629–638. [CrossRef]

103. Steel, A.W.; Mela, C.M.; Lindsay, J.O.; Gazzard, B.G.; Goodier, M.R. Increased Proportion of Cd16+ Nk Cells in the Colonic Lamina Propria of Inflammatory Bowel Disease Patients, but Not after Azathioprine Treatment. *Aliment. Pharmacol. Ther.* **2011**, *33*, 115–126. [CrossRef] [PubMed]

104. Takayama, T.; Kamada, N.; Chinen, H.; Okamoto, S.; Kitazume, M.T.; Chang, J.; Matuzaki, Y.; Suzuki, S.; Sugita, A.; Koganei, K.; et al. Imbalance of Nkp44(+)Nkp46(−) and Nkp44(−)Nkp46(+) Natural Killer Cells in the Intestinal Mucosa of Patients with Crohn's Disease. *Gastroenterology* **2010**, *139*, 1995–2004.e15. [CrossRef]

105. Fathollahi, A.; Aslani, S.; Mostafaei, S.; Rezaei, N.; Mahmoudi, M. The Role of Killer-Cell Immunoglobulin-Like Receptor (Kir) Genes in Susceptibility to Inflammatory Bowel Disease: Systematic Review and Meta-Analysis. *Inflamm. Res.* **2010**, *67*, 727–736. [CrossRef]

106. Vadstrup, K.; Bendtsen, F. Anti-Nkg2d Mab: A New Treatment for Crohn's Disease? *Int. J. Mol. Sci.* **2017**, *18*, 1997. [CrossRef]

107. Baumgarth, N.; Tung, J.W.; Herzenberg, L.A. Inherent Specificities in Natural Antibodies: A Key to Immune Defense against Pathogen Invasion. *Springer Semin. Immunopathol.* **2005**, *26*, 347–362. [CrossRef]

108. de Veer, M.J.; Kemp, J.M.; Meeusen, E.N. The Innate Host Defence against Nematode Parasites. *Parasite Immunol.* **2007**, *29*, 1–9. [CrossRef]

109. Bunker, J.J.; Erickson, S.A.; Flynn, T.M.; Henry, C.; Koval, J.C.; Meisel, M.; Jabri, B.; Antonopoulos, D.A.; Wilson, P.C.; Bendelac, A. Natural Polyreactive Iga Antibodies Coat the Intestinal Microbiota. *Science* **2017**, *358*, eaan6619. [CrossRef]

110. Pietzsch, J.; Hoppmann, S. Human S100a12: A Novel Key Player in Inflammation? *Amino Acids* **2009**, *36*, 381–389. [CrossRef]

111. Foell, D.; Wittkowski, H.; Vogl, T.; Roth, J. S100 Proteins Expressed in Phagocytes: A Novel Group of Damage-Associated Molecular Pattern Molecules. *J. Leukoc. Biol.* **2007**, *81*, 28–37. [CrossRef] [PubMed]

112. Heilmann, R.M.; Grellet, A.; Allenspach, K.; Lecoindre, P.; Day, M.J.; Priestnall, S.L.; Toresson, L.; Procoli, F.; Grützner, N.; Suchodolski, J.S.; et al. Association between Fecal S100a12 Concentration and Histologic, Endoscopic, and Clinical Disease Severity in Dogs with Idiopathic Inflammatory Bowel Disease. *Vet. Immunol. Immunopathol.* **2014**, *158*, 156–166. [CrossRef]

113. Cabrera-García, A.I.; Protschka, M.; Alber, G.; Kather, S.; Dengler, F.; Müller, U.; Steiner, J.M.; Heilmann, R.M. Dysregulation of Gastrointestinal Rage (Receptor for Advanced Glycation End Products) Expression in Dogs with Chronic Inflammatory Enteropathy. *Vet. Immunol. Immunopathol.* **2021**, *234*, 110216. [CrossRef] [PubMed]

114. Cabrera-García, A.I.; Suchodolski, J.S.; Steiner, J.M.; Heilmann, R.M. Association between Serum Soluble Receptor for Advanced Glycation End-Products (Rage) Deficiency and Severity of Clinicopathologic Evidence of Canine Chronic Inflammatory Enteropathy. *J. Vet. Diagn. Investig.* **2020**, *32*, 664–674. [CrossRef] [PubMed]

115. Cabrera-García, A.I.; Protschka, M.; Kather, S.; Dengler, F.; Alber, G.; Müller, U.; Steiner, J.; Heilmann, R. Dysregulation of Gastrointestinal Rage (Receptor for Advanced Glycation End Products) Expression in a Spontaneous Animal Model of Inflammatory Bowel Disease. *Inflamm. Bowel Dis.* **2021**, *27*, S3. [CrossRef]

116. Allenspach, K.; Mochel, J.P. Current Diagnostics for Chronic Enteropathies in Dogs. *Vet. Clin. Pathol.* **2022**, *50*, 18–28. [CrossRef] [PubMed]

117. O'Neill, L.A.J. How Toll-like Receptors Signal: What We Know and What We Don't Know. *Curr. Opin. Immunol.* **2006**, *18*, 3–9. [CrossRef]

118. Fritz, J.H.; Ferrero, R.L.; Philpott, D.J.; Girardin, S.E. Nod-Like Proteins in Immunity, Inflammation and Disease. *Nat. Immunol.* **2006**, *7*, 1250–1257. [CrossRef]

119. McMahon, L.A.; House, A.K.; Catchpole, B.; Elson-Riggins, J.; Riddle, A.; Smith, K.; Werling, D.; Burgener, I.A.; Allenspach, K. Expression of Toll-Like Receptor 2 in Duodenal Biopsies from Dogs with Inflammatory Bowel Disease Is Associated with Severity of Disease. *Vet. Immunol. Immunopathol.* **2010**, *135*, 158–163. [CrossRef]

120. Burgener, I.A.; König, A.; Allenspach, K.; Sauter, S.N.; Boisclair, J.; Doherr, M.G.; Jungi, T.W. Upregulation of Toll-Like Receptors in Chronic Enteropathies in Dogs. *J. Vet. Intern. Med.* **2008**, *22*, 553–560. [CrossRef]

121. Kaser, A.; Pasaniuc, B. Ibd Genetics: Focus on (Dys) Regulation in Immune Cells and the Epithelium. *Gastroenterology* **2014**, *146*, 896–899. [CrossRef] [PubMed]

122. Kathrani, A.; House, A.; Catchpole, B.; Murphy, A.; German, A.; Werling, D.; Allenspach, K. Polymorphisms in the Tlr4 and Tlr5 Gene Are Significantly Associated with Inflammatory Bowel Disease in German Shepherd Dogs. *PLoS ONE* **2010**, *5*, e15740. [CrossRef] [PubMed]

123. Kathrani, A.; Lee, H.; White, C.; Catchpole, B.; Murphy, A.; German, A.; Werling, D.; Allenspach, K. Association between Nucleotide Oligomerisation Domain Two (Nod2) Gene Polymorphisms and Canine Inflammatory Bowel Disease. *Vet. Immunol. Immunopathol.* **2014**, *161*, 32–41. [CrossRef] [PubMed]

124. Kathrani, A.; Werling, D.; Allenspach, K. Canine Breeds at High Risk of Developing Inflammatory Bowel Disease in the South-Eastern Uk. *Vet. Rec.* **2011**, *169*, 635. [CrossRef] [PubMed]

125. Kathrani, A.; Holder, A.; Catchpole, B.; Alvarez, L.; Simpson, K.; Werling, D.; Allenspach, K. Tlr5 Risk-Associated Haplotype for Canine Inflammatory Bowel Disease Confers Hyper-Responsiveness to Flagellin. *PLoS ONE* **2012**, *7*, e30117. [CrossRef] [PubMed]

126. Cheroutre, H.; Lambolez, F.; Mucida, D. The Light and Dark Sides of Intestinal Intraepithelial Lymphocytes. *Nat. Rev. Immunol.* **2011**, *11*, 445–456. [CrossRef]

127. Wang, L.; Zhu, L.; Qin, S. Gut Microbiota Modulation on Intestinal Mucosal Adaptive Immunity. *J. Immunol. Res.* **2019**, *2019*, 4735040. [CrossRef]

128. Tatiya-aphiradee, N.; Chatuphonprasert, W.; Jarukamjorn, K. Immune Response and Inflammatory Pathway of Ulcerative Colitis. *J. Basic Clin. Physiol. Pharmacol.* **2019**, *30*, 1–10. [CrossRef]

129. Degasperi, G.R. Mucosal Immunology in the Inflammatory Bowel Diseases. In *Biological Therapy for Inflammatory Bowel Disease*; IntechOpen: London, UK, 2019. [CrossRef]

130. Heilmann, R.M.; Suchodolski, J.S. Is Inflammatory Bowel Disease in Dogs and Cats Associated with a Th1 or Th2 Polarization? *Vet. Immunol. Immunopathol.* **2015**, *168*, 131–134. [CrossRef]

131. Korn, T.; Bettelli, E.; Oukka, M.; Kuchroo, V.K. Il-17 and Th17 Cells. *Annu. Rev. Immunol.* **2009**, *27*, 485–517. [CrossRef]

132. Choy, M.C.; Visvanathan, K.; De Cruz, P. An Overview of the Innate and Adaptive Immune System in Inflammatory Bowel Disease. *Inflamm. Bowel Dis.* **2017**, *23*, 2–13. [CrossRef] [PubMed]

133. Maillard, M.H.; Snapper, S.B. Cytokines and chemokines in mucosal homeostasis. In *Inflammatory Bowel Disease—Translating Basic Science Bowel Disease—Translating Basic Science into Clinical Practice*; Targan, S.R., Shanahan, F., Karp, L.C., Eds.; Wiley-Blackwell: Hoboken, NJ, USA, 2010; pp. 119–156. [CrossRef]
134. Marafini, I.; Sedda, S.; Dinallo, V.; Monteleone, G. Inflammatory Cytokines: From Discoveries to Therapies in Ibd. *Expert Opin. Biol. Ther.* **2019**, *19*, 1207–1217. [CrossRef] [PubMed]
135. Neurath, M.F.; Finotto, S.; Glimcher, L.H. The Role of Th1/Th2 Polarization in Mucosal Immunity. *Nat. Med.* **2002**, *8*, 567–573. [CrossRef] [PubMed]
136. Ridyard, A.E.; Nuttall, T.J.; Else, R.W.; Simpson, J.W.; Miller, H.R.P. Evaluation of Th1, Th2 and Immunosuppressive Cytokine Mrna Expression within the Colonic Mucosa of Dogs with Idiopathic Lymphocytic–Plasmacytic Colitis. *Vet. Immunol. Immunopathol.* **2002**, *86*, 205–214. [CrossRef]
137. Hammad, H.; Lambrecht, B.N. Barrier Epithelial Cells and the Control of Type 2 Immunity. *Immunity* **2015**, *43*, 29–40. [CrossRef]
138. Boden, E.K.; Lord, J.D. Cd4 T Cells in Ibd: Crossing the Line? *Dig. Dis. Sci.* **2017**, *62*, 2208–2210. [CrossRef]
139. Ohta, H.; Takada, K.; Sunden, Y.; Tamura, Y.; Osuga, T.; Lim, S.Y.; Murakami, M.; Sasaki, N.; Wickramasekara Rajapakshage, B.K.; Nakamura, K.; et al. Cd4+ T Cell Cytokine Gene and Protein Expression in Duodenal Mucosa of Dogs with Inflammatory Bowel Disease. *J. Vet. Med. Sci.* **2014**, *76*, 409–414. [CrossRef]
140. Schmitz, S.; Garden, O.A.; Werling, D.; Allenspach, K. Gene Expression of Selected Signature Cytokines of T Cell Subsets in Duodenal Tissues of Dogs with and without Inflammatory Bowel Disease. *Vet. Immunol. Immunopathol.* **2012**, *146*, 87–91. [CrossRef]
141. Kinjo, T.; Azuma, Y.; Nishiyama, K.; Fujimoto, Y.; Miki, M.; Kuramoto, N. Intestinal Il-17 Expression in Canine Inflammatory Bowel Disease. *Int. J. Vet. Health Sci. Res.* **2017**, *5*, 171–175. [CrossRef]
142. Ohta, H.; Takada, K.; Torisu, S.; Yuki, M.; Tamura, Y.; Yokoyama, N.; Osuga, T.; Lim, S.Y.; Murakami, M.; Sasaki, N.; et al. Expression of Cd4+ T Cell Cytokine Genes in the Colorectal Mucosa of Inflammatory Colorectal Polyps in Miniature Dachshunds. *Vet. Immunol. Immunopathol.* **2013**, *155*, 259–263. [CrossRef]
143. Maeda, S.; Ohno, K.; Fujiwara-Igarashi, A.; Uchida, K.; Tsujimoto, H. Changes in Foxp3-Positive Regulatory T Cell Number in the Intestine of Dogs with Idiopathic Inflammatory Bowel Disease and Intestinal Lymphoma. *Vet. Pathol.* **2016**, *53*, 102–112. [CrossRef] [PubMed]
144. Maeda, S.; Ohno, K.; Uchida, K.; Nakashima, K.; Fukushima, K.; Tsukamoto, A.; Nakajima, M.; Fujino, Y.; Tsujimoto, H. Decreased Immunoglobulin a Concentrations in Feces, Duodenum, and Peripheral Blood Mononuclear Cells of Dogs with Inflammatory Bowel Disease. *J. Vet. Intern. Med.* **2013**, *27*, 47–55. [CrossRef] [PubMed]
145. Kolodziejska-Sawerska, A.; Rychlik, A.; Depta, A.; Wdowiak, M.; Nowicki, M.; Kander, M. Cytokines in Canine Inflammatory Bowel Disease. *Pol. J. Vet. Sci.* **2013**, *16*, 165–171. [CrossRef] [PubMed]
146. Yang, X.O.; Panopoulos, A.D.; Nurieva, R.; Chang, S.H.; Wang, D.; Watowich, S.S.; Dong, C. Stat3 Regulates Cytokine-Mediated Generation of Inflammatory Helper T Cells. *J. Biol. Chem.* **2007**, *282*, 9358–9363. [CrossRef] [PubMed]
147. Sugimoto, K. Role of Stat3 in Inflammatory Bowel Disease. *World J. Gastroenterol.* **2008**, *14*, 5110–5114. [CrossRef]
148. Neufert, C.; Pickert, G.; Zheng, Y.; Wittkopf, N.; Warntjen, M.; Nikolae, A.; Ouyang, W.; Neurath, M.F.; Becker, C. Activation of Epithelial Stat3 Regulates Intestinal Homeostasis. *Cell Cycle* **2010**, *9*, 652–655. [CrossRef] [PubMed]
149. Manz, A.; Allenspach, K.; Kummer, S.; Richter, B.; Walter, I.; Macho-Maschler, S.; Tichy, A.; Burgener, I.A.; Luckschander-Zeller, N. Upregulation of Signal Transducer and Activator of Transcription 3 in Dogs with Chronic Inflammatory Enteropathies. *J. Vet. Intern. Med.* **2021**, *35*, 1288–1296. [CrossRef]
150. Madakamutil, L.T.; Christen, U.; Lena, C.J.; Wang-Zhu, Y.; Attinger, A.; Sundarrajan, M.; Ellmeier, W.; von Herrath, M.G.; Jensen, P.; Littman, D.R.; et al. Cd8αα-Mediated Survival and Differentiation of Cd8 Memory T Cell Precursors. *Science* **2004**, *304*, 590. [CrossRef]
151. van Wijk, F.; Cheroutre, H. Mucosal T Cells in Gut Homeostasis and Inflammation. *Expert Rev. Clin. Immunol.* **2010**, *6*, 559–566. [CrossRef]
152. McVay, L.D.; Li, B.; Biancaniello, R.; Creighton, M.A.; Bachwich, D.; Lichtenstein, G.; Rombeau, J.L.; Carding, S.R. Changes in Human Mucosal Γδ T Cell Repertoire and Function Associated with the Disease Process in Inflammatory Bowel Disease. *Mol. Med.* **1997**, *3*, 183–203. [CrossRef]
153. Haas, E.; Rütgen, B.C.; Gerner, W.; Richter, B.; Tichy, A.; Galler, A.; Bilek, A.; Thalhammer, J.G.; Saalmüller, A.; Luckschander-Zeller, N. Phenotypic Characterization of Canine Intestinal Intraepithelial Lymphocytes in Dogs with Inflammatory Bowel Disease. *J. Vet. Intern. Med.* **2014**, *28*, 1708–1715. [CrossRef] [PubMed]
154. Kubinak, J.L.; Petersen, C.; Stephens, W.Z.; Soto, R.; Bake, E.; O'Connell, R.M.; Round, J.L. Myd88 Signaling in T Cells Directs Iga-Mediated Control of the Microbiota to Promote Health. *Cell Host Microbe* **2015**, *17*, 153–163. [CrossRef] [PubMed]
155. Mizoguchi, A.; Bhan, A.K. Immunobiology of B Cells in Inflammatory Bowel Disease. In *Crohn's Disease and Ulcerative Colitis: From Epidemiology and Immunobiology to a Rational Diagnostic and Therapeutic Approach*; Baumgart, D.C., Ed.; Springer International Publishing: Cham, Switzerland, 2017; pp. 111–117. [CrossRef]
156. Scott, M.G.; Nahm, M.H.; Macke, K.; Nash, G.S.; Bertovich, M.J.; MacDermott, R.P. Spontaneous Secretion of Igg Subclasses by Intestinal Mononuclear Cells. Differences between Ulcerative Colitis, Crohn's Disease, and Controls. *Clin. Exp. Immunol.* **1986**, *66*, 209–215.

157. Silva, F.A.R.; Rodrigues, B.L.; Ayrizono, M.d.L.S.; Leal, R.F. The Immunological Basis of Inflammatory Bowel Disease. *Gastroenterol. Res. Pract.* **2016**, *2016*, 2097274. [CrossRef] [PubMed]

158. Ahluwalia, B.; Moraes, L.; Magnusson, M.K.; Öhman, L. Immunopathogenesis of Inflammatory Bowel Disease and Mechanisms of Biological Therapies. *Scand. J. Gastroenterol.* **2018**, *53*, 379–389. [CrossRef]

159. Hevia, A.; López, P.; Suárez, A.; Jacquot, C.; Urdaci, M.C.; Margolles, A.; Sánchez, B. Association of Levels of Antibodies from Patients with Inflammatory Bowel Disease with Extracellular Proteins of Food and Probiotic Bacteria. *BioMed Res. Int.* **2014**, *2014*, 351204. [CrossRef]

160. Macpherson, A.; Khoo, U.Y.; Forgacs, I.; Philpott-Howard, J.; Bjarnason, I. Mucosal Antibodies in Inflammatory Bowel Disease Are Directed against Intestinal Bacteria. *Gut* **1996**, *38*, 365. [CrossRef]

161. Galler, A.; Rütgen, B.C.; Haas, E.; Saalmüller, A.; Hirt, R.A.; Gerner, W.; Schwendenwein, I.; Richter, B.; Thalhammer, J.G.; Luckschander-Zeller, N. Immunophenotype of Peripheral Blood Lymphocytes in Dogs with Inflammatory Bowel Disease. *J. Vet. Intern. Med.* **2017**, *31*, 1730–1739. [CrossRef]

162. Stonehewer, J.; Simpson, J.W.; Else, R.W.; MacIntyre, N. Evaluation of B and T Lymphocytes and Plasma Cells in Colonic Mucosa from Healthy Dogs and from Dogs with Inflammatory Bowel Disease. *Res. Vet. Sci.* **1998**, *65*, 59–63. [CrossRef]

163. Planer, J.D.; Peng, Y.; Kau, A.L.; Blanton, L.V.; Ndao, I.M.; Tarr, P.I.; Warner, B.B.; Gordon, J.I. Development of the Gut Microbiota and Mucosal Iga Responses in Twins and Gnotobiotic Mice. *Nature* **2016**, *534*, 263–266. [CrossRef]

164. Reboldi, A.; Arnon, T.I.; Rodda, L.B.; Atakilit, A.; Sheppard, D.; Cyster, J.G. Iga Production Requires B Cell Interaction with Subepithelial Dendritic Cells in Peyer's Patches. *Science* **2016**, *352*, aaf4822. [CrossRef] [PubMed]

165. Maeda, S.; Ohno, K.; Fujiwara-Igarashi, A.; Tomiyasu, H.; Fujino, Y.; Tsujimoto, H. Methylation of Tnfrsf13b and Tnfrsf13c in Duodenal Mucosa in Canine Inflammatory Bowel Disease and Its Association with Decreased Mucosal Iga Expression. *Vet. Immunol. Immunopathol.* **2014**, *160*, 97–106. [CrossRef] [PubMed]

166. Castigli, E.; Wilson, S.A.; Scott, S.; Dedeoglu, F.; Xu, S.; Lam, K.-P.; Bram, R.J.; Jabara, H.; Geha, R.S. Taci and Baff-R Mediate Isotype Switching in B Cells. *J. Exp. Med.* **2005**, *201*, 35–39. [CrossRef] [PubMed]

167. Estruch, J.J.; Barken, D.; Bennett, N.; Krawiec, D.K.; Ogilvie, G.K.; Powers, B.E.; Polansky, B.J.; Sueda, M.T. Evaluation of Novel Serological Markers and Autoantibodies in Dogs with Inflammatory Bowel Disease. *J. Vet. Intern. Med.* **2020**, *34*, 1177–1186. [CrossRef] [PubMed]

168. Soontararak, S.; Chow, L.; Johnson, V.; Coy, J.; Webb, C.; Wennogle, S.; Dow, S. Humoral Immune Responses against Gut Bacteria in Dogs with Inflammatory Bowel Disease. *PLoS ONE* **2019**, *14*, e0220522. [CrossRef]

169. Netea, M.G.; Domínguez-Andrés, J.; Barreiro, L.B.; Chavakis, T.; Divangahi, M.; Fuchs, E.; Joosten, L.A.B.; van der Meer, J.W.M.; Mhlanga, M.M.; Mulder, W.J.M.; et al. Defining Trained Immunity and Its Role in Health and Disease. *Nat. Rev. Immunol.* **2020**, *20*, 375–388. [CrossRef]

170. Konstantinidis, A.O.; Pardali, D.; Adamama-Moraitou, K.K.; Gazouli, M.; Dovas, C.I.; Legaki, E.; Brellou, G.D.; Savvas, I.; Jergens, A.E.; Rallis, T.S.; et al. Colonic Mucosal and Serum Expression of Micrornas in Canine Large Intestinal Inflammatory Bowel Disease. *BMC Vet. Res.* **2020**, *16*, 69. [CrossRef]

171. Tamura, Y.; Ohta, H.; Yokoyama, N.; Lim, S.Y.; Osuga, T.; Morishita, K.; Nakamura, K.; Yamasaki, M.; Takiguchi, M. Evaluation of Selected Cytokine Gene Expression in Colonic Mucosa from Dogs with Idiopathic Lymphocytic-Plasmacytic Colitis. *J. Vet. Med. Sci.* **2014**, *76*, 1407–1410. [CrossRef]

Article

A Preliminary Study of Modulen IBD Liquid Diet in Hospitalized Dogs with Protein-Losing Enteropathy

Aarti Kathrani * and Gina Parkes

Department of Clinical Science and Services, Royal Veterinary College, Hertfordshire AL9 7TA, UK; gparkes@rvc.ac.uk
* Correspondence: akathrani@rvc.ac.uk

Simple Summary: Modulen IBD is an oral liquid food that results in remission rates similar to immunosuppressive drugs in children with inflammatory bowel disease. This diet has not been previously investigated in dogs. We aimed to describe the use of Modulen IBD in hospitalized dogs with inflammatory protein-losing enteropathy (PLE) when combined with whey powder and a multivitamin/mineral blend to ensure this was complete and balanced for dogs. Five dogs hospitalized for PLE that had an esophagostomy feeding tube placed were eligible and prospectively enrolled. All dogs received Modulen IBD without concurrent immunosuppressive drugs and tolerated tube feedings. All dogs had resolution of anorexia, three had stable or improved serum albumin concentrations, four had improved or normalized serum globulin concentrations, and four dogs had improved or normalized serum cholesterol concentrations 2–3 days after initiating the diet. In conclusion, the Modulen IBD liquid diet was well-tolerated in-hospital and resolved anorexia in all dogs and helped to improve selected biochemical parameters in some dogs. Further studies are needed to assess the long-term effects of feeding this diet to dogs with inflammatory PLE.

Citation: Kathrani, A.; Parkes, G. A Preliminary Study of Modulen IBD Liquid Diet in Hospitalized Dogs with Protein-Losing Enteropathy. *Animals* **2022**, *12*, 1594. https://doi.org/10.3390/ani12121594

Academic Editor: Anshan Shan

Received: 3 May 2022
Accepted: 18 June 2022
Published: 20 June 2022

Publisher's Note: MDPI stays neutral with regard to jurisdictional claims in published maps and institutional affiliations.

Abstract: Modulen IBD is an enteral liquid diet that can induce remission rates similar to glucocorticoids in children with inflammatory bowel disease. The Modulen IBD liquid diet has not been previously investigated in dogs. Our study aimed to describe the use of the Modulen IBD liquid diet in hospitalized dogs with inflammatory protein-losing enteropathy (PLE), including its tolerance and effects on appetite and gastrointestinal signs, and laboratory parameters during hospitalization. Of the 14 dogs hospitalized for PLE that had an esophagostomy feeding tube placed at the time of endoscopy, 5 were eligible and prospectively enrolled. The Modulen IBD liquid diet was supplemented with whey powder isolate and a multivitamin/mineral blend to ensure the diet was complete and balanced for canine adult maintenance and had a macronutrient profile desirable for PLE. All five dogs tolerated tube feedings with the Modulen IBD liquid diet, allowing an increase of 75 to 100% of the resting energy requirement (RER) by day 3 to 4. The diet was administered without glucocorticoid in all five dogs. All five of these dogs had a resolution of anorexia allowing the voluntary intake of a commercial hydrolyzed protein diet prior to the use of glucocorticoids. Of these five dogs, three (60%) had stable or improved serum albumin concentrations (median % increase: 10.3, range: 0–31.1), four (80%) had improved or normalized serum globulin concentrations (median % increase: 12.9, range: 5.1–66.2) and four (80%) had improved or normalized serum cholesterol concentrations (median % increase: 31.5, range: 4.8–63) 2–3 days after initiating the diet. However, there were no significant differences in these selected biochemical parameters pre- and post-feeding with the diet ($p > 0.080$). In conclusion, the Modulen IBD liquid diet, fed via an esophagostomy feeding tube was well-tolerated in-hospital and resolved anorexia in all dogs and helped to improve selected biochemical parameters in some dogs. Further studies are needed to assess the long-term effects of feeding this diet on the rate of serum albumin increase and remission in dogs with inflammatory PLE.

Keywords: nutrition; canine; gastrointestinal; diarrhea

1. Introduction

Approximately 50% of dogs with protein-losing enteropathy (PLE) due to chronic enteropathy (CE) or lymphangiectasia have a poor prognosis due to failure to respond to standard treatments, including immunosuppressive therapy [1,2]. It is therefore critical to develop more efficacious treatments for the management of dogs with inflammatory PLE. Exclusive enteral nutrition alone has been shown to induce remission in 80% of pediatric patients with Crohn's disease (CD) [3], a form of human inflammatory bowel disease, and similar remission rates have been seen in adults with newly diagnosed CD [4]. Indeed, several studies and a meta-analysis comparing exclusive enteral nutrition to corticosteroids in pediatric patients with active CD found no significant difference in remission rates between the two forms of therapy [5–7]. Although the mechanism of action of exclusive enteral nutrition is unknown, it has been shown to improve mucosal healing (74% versus 33% with corticosteroids [8]), decrease inflammatory cytokines [9], alter the gut microbiota [10], improve nutritional status, and increase serum albumin concentrations compared to corticosteroids [11,12]. To our knowledge, exclusive enteral nutrition with a liquid diet has never been investigated in canine gastrointestinal diseases. As these diets result in greater intestinal mucosal healing and serum albumin concentrations compared to treatment with corticosteroids, studies assessing their effects in dogs with inflammatory PLE are warranted to determine if they can help to improve the response to treatment and prognosis of these dogs.

The Modulen IBD liquid diet is a powdered formulation that is used for the dietary management of active CD, as a sole source of nutrition or in the remission phase of CD, as a nutritional support. The diet can be administered as an oral sip feed or via an enteral feeding tube. The diet is 100% casein-based, using whole protein, and is supplemented with medium-chain triglycerides and carbohydrates in the form of glucose syrup, resulting in low osmolality for good tolerance. The diet contains transforming growth factor beta, a naturally occurring anti-inflammatory factor. The Modulen IBD liquid diet has been shown in multiple studies to decrease intestinal inflammation and promote mucosal healing in humans with CD [13–16]. To our knowledge, this diet has never been investigated in dogs.

Our study aimed to describe the use of the Modulen IBD liquid diet fed exclusively via an esophagostomy tube in hospitalized dogs with inflammatory PLE. We aimed to determine the tolerance to this diet, its effects on appetite and gastrointestinal signs, and selected laboratory parameters during hospitalization in these dogs. Determining the short-term effect of this diet in dogs with inflammatory PLE will help to provide data to justify the potential cost and complexity of longer-term studies utilizing this diet. We hypothesized that exclusive enteral nutrition with the Modulen IBD liquid diet in hospitalized dogs with inflammatory PLE would be well-tolerated and result in improvement in appetite, gastrointestinal signs, and select biochemical parameters, including serum albumin, globulin, and cholesterol concentrations, during the hospitalization period.

2. Materials and Methods

2.1. Eligibility Criteria

This study was performed at the Queen Mother Hospital for Animals, Royal Veterinary College, U.K from January 2020 to October 2020, inclusive. Dogs presenting to the Internal Medicine Service with gastrointestinal signs that had biochemical results consistent with PLE (panhypoproteinemia with low or low-normal serum cholesterol concentration) were eligible for inclusion, provided they had thorough and appropriate investigations to rule out all other known causes, such as hypoadrenocorticism and gastrointestinal parasites. Gastrointestinal histopathologic diagnosis and clinical indication for the placement of an esophagostomy feeding tube (the presence of anorexia for at least 3 to 5 days) was also required for inclusion in the study. Exclusion criteria included the presence of concurrent disease that may require additional dietary strategies, such as pancreatitis and hepatic or renal disease, if the dog was growing, pregnant or lactating, and gastrointestinal histopathology that subsequently showed neoplasia or changes consistent with infection rather than chronic inflammation.

The dogs were enrolled at the time of gastrointestinal endoscopy and placement of an esophagostomy feeding tube. The dogs were treated with the Modulen IBD liquid diet without the use of concurrent glucocorticoids. All dogs were started on the treatment the same day after the collection of gastrointestinal biopsy specimens via endoscopy, and treatment was continued until the histopathology results were returned 3 to 5 days later.

2.2. Formulation of the Diet

The diet was specifically formulated for each dog by a board-certified veterinary nutritionist (A.K.), taking into consideration the dog's body weight at admission, biochemical results including electrolytes and cholesterol, and diet history including fat tolerance, using a computer software program (www.BalanceIT.com, accessed dates: January 2020 to October 2020). The diet was formulated to ensure it was complete and balanced and that all essential nutrients met the minimum requirements and/or recommended allowances per the National Research Council Nutrient Requirements 2006 for canine adult maintenance on an energy basis.

As the Modulen IBD liquid diet is formulated for humans and not for dogs, two additional ingredients were required to be added to this diet to make it suitable for canine feeding. Firstly, whey powder was required to ensure that the diet met the protein requirements for canine adult maintenance and that the fat content did not exceed 35% on a metabolizable energy basis, and secondly, a multivitamin and mineral blend (BalanceIT Canine Supplement) was required to ensure all the essential vitamins and minerals for canine adult maintenance were met. Although the exact amounts of these three ingredients varied for each dog, at least 75% of the calories were provided by the Modulen IBD liquid diet, and overall, the caloric distribution was in the region of 30.7–33.2% for protein, 32.7–33.9% for fat, and 34.0–35.3% for carbohydrate on a metabolizable energy basis for all five dog formulations.

2.3. Preparation of the Diet

The diet was prepared using cleaned equipment and bottled water according to the package instructions. All three ingredients (the Modulen IBD liquid diet, whey powder, and multivitamin and mineral blend) were weighed using a gram scale and combined in a clean bowl. Bottled water was added in 25 to 50 mL increments until the slurry reached a consistency that was considered to easily pass via the esophagostomy feeding tube. The final volume of the slurry was measured, and the kcal/mL was calculated. The slurry was prepared daily and divided into four syringed feeds and refrigerated until needed.

2.4. In-Hospital Feeding and Monitoring

Feeding commenced the evening of the procedure of the gastrointestinal endoscopy and esophagostomy feeding tube placement once the dog was suitably recovered from the general anesthetic. The feeding was commenced on day 1 as 25% of the resting energy requirement (RER) using the exponential equation of 70 multiplied by the dog's bodyweight in kilograms raised to the power of 0.75, divided into 4 feeds over 24 h. This was increased by 25% of the RER increments every 24 h if the dog tolerated the diet (showed no signs of regurgitation, vomiting, or lip smacking) or exacerbation of gastrointestinal signs, such as diarrhea. If the dog showed any signs of intolerance or worsening of signs, then the amount of slurry offered was maintained for another 24 h and the signs were reassessed or decreased by 25% of the RER depending on the severity of the intolerance or worsening of the signs. If the dog was not tolerating 25% of the RER, then this was decreased to 10% of the RER and reassessed 24 h later. All feeds were exclusively completed with the Modulen IBD liquid diet. No other treats or foods were administered and multiple signs were placed on the dog's kennel and treatment sheet to prevent inadvertent feeding. Once the feeds reached 75% to 100% of the RER, Purina Pro Plan Veterinary Diets HA Canine dry food was offered to gauge appetite, and voluntary oral intake was recorded.

The slurry was administered via esophagostomy tube manually at a rate of approximately 1 mL per minute. The slurry was warmed for 10–15 min at room temperature, followed by gently warming in a water bath of warm water prior to administration, if needed. The esophagostomy feeding tube was flushed with 5 mL of tap water before and after each feed.

2.5. Assessment of Tolerance to Diet, Effect on Gastrointestinal Signs, and Biochemical Parameters

Tolerance of the diet was closely assessed by monitoring the dog for any signs of nausea (hypersalivating or lip smacking), regurgitation, vomiting, or restlessness. This was assessed throughout hospitalization when the dog was receiving the Modulen IBD liquid diet. Written notes following every feed were made and recorded, even if the dog showed no signs.

Throughout the hospitalization period, stool quality and the presence of any gastrointestinal signs were recorded. Appetite was assessed only after the dog was receiving at least 75–100% of the RER of the Modulen IBD liquid diet by offering Purina Pro Plan Veterinary Diets HA Canine dry food at this stage and recording voluntary intake.

Blood was collected by venipuncture 2–3 days following the initiation of the Modulen IBD liquid diet for a standard biochemistry panel to allow for comparisons of serum albumin, globulin, cholesterol, and C-reactive protein (CRP) concentrations to admission (pre-Modulen IBD liquid diet) values.

The Modulen IBD liquid diet was discontinued once the dog was voluntarily consuming Purina Pro Plan Veterinary Diets HA Canine dry food reliably.

2.6. Statistical Analysis

The related-samples Wilcoxon signed rank test was used to determine if there were any significant differences in serum albumin, globulin, cholesterol, and CRP concentrations 2–3 days following initiation of the Modulen IBD liquid diet compared to admission (pre-Modulen IBD liquid diet) concentrations for the five dogs. This was performed using the IBM SPSS (Statistical Product and Service Solutions) version 26 statistical software program. Statistical significance was placed at $p < 0.05$. The median, range, and percentage change were calculated using Microsoft Excel for Mac Version 16.57.

2.7. Ethical Approval and Owner Consent

The Royal Veterinary College granted ethical approval for the study (URN 2019 1912–3). Written informed consent for participation in the study was obtained from all owners of the dogs.

3. Results

3.1. Study Dogs

Fourteen dogs were eligible for inclusion; however, seven were excluded: two due to concurrent liver disease, two due to the primary investigator not being informed of the case for enrollment, one due to concurrent pancreatitis and liver disease, one due to duodenal small cell lymphoma, and one due to acute blood loss from a gastrointestinal ulcer as the cause for the panhypoproteinemia. Two dogs were also excluded as they were started on the Modulen IBD diet with concurrent glucocorticoids, as their serum albumin concentration was considered severe (<12 g/L). Five dogs met the inclusion criteria.

The five dogs had a median (range) age of 5.8 years (3–10.9) and consisted of two intact males, two neutered males, and one female neutered dog. The breeds consisted of an English bulldog, a cocker spaniel, a crossbreed, a beagle, and a Yorkshire terrier. The presenting clinical signs included vomiting and diarrhea in four dogs and diarrhea in one. The median duration of gastrointestinal signs was 2 months with a range of 1 week to 3 years. The median (range) body weight was 9.05 kg (7.5–26.1) and the median (range) body condition score was 2/9 (range 2–5). Prior to referral, two dogs received one injection of dexamethasone one day prior to referral and four dogs received antimicrobials (metronidazole ($n = 4$) and oxytetracycline ($n = 1$)), though none of these medications were ongoing or continued at the time of referral. Two dogs previously received diet trials with a commercial hydrolyzed protein diet.

The following investigations were performed for all five dogs: complete blood count, serum biochemistry panel, serum vitamin B12, transabdominal ultrasound, and upper and lower gastrointestinal endoscopy. Four of the five dogs also had urinalysis, basal cortisol, and fecal parasitology performed. Four dogs were empirically dewormed prior to referral [milbemycin oxime and praziquantel ($n = 1$), milbemycin oxime ($n = 1$), praziquantel ($n = 1$), and fenbendazole ($n = 1$)], two had fecal cultures performed, one had serum pancreatic lipase immunoreactivity, and one had a bile acid stimulation test.

The median (range) serum albumin and vitamin B12 concentrations were 19.4 g/dL (14.8–22.1) and 361 ng/L (190–459), respectively. All 5 dogs were diagnosed with chronic inflammatory enteropathy; 3 had lymphoplasmacytic and neutrophilic duodenitis and 2 lymphoplasmacytic duodenitis, 3 had evidence of lymphangiectasia on duodenal histopathology. Four dogs had ileal histopathology performed; 2 had lymphoplasmacytic and neutrophilic ileitis and 2 lymphoplasmacytic ileitis, 2 had evidence of lymphangiectasia or lacteal dilation on ileal histopathology. All dogs had colonic histopathology performed: four had lymphoplasmacytic and neutrophilic colitis and one had plasmacytic colitis. With regards to the neutrophilic infiltration, for the duodenum, this was moderate in three; for the ileum, this was moderate in two; and for the colon, this was mild in one, mild to moderate in one, moderate in one, and moderate to marked in one. One dog was subsequently diagnosed with primary hyperparathyroidism at the same visit based on ionized hypercalcemia and high-normal parathyroid hormone (PTH) concentration (ionized calcium 1.66 mmol/L, reference range 1.25–1.45; PTH 4.0 pmol/L, reference range 0.5–5.8; and PTH-related protein 0.0 pmol/L, reference range 0.0–1.0). The median (range) canine chronic enteropathy clinical activity index [17] (CCECAI) for all dogs was 14 (9–18).

Treatment at discharge included Purina Pro Plan Veterinary Diets HA Canine dry food and prednisolone for four out of the five dogs, and for the one remaining dog, it included a combination of the Purina HA diet with Hill's Prescription Diet z/d Canine dry food as the dog eventually preferred this combination during hospitalization. For one out of the four dogs treated with diet and prednisolone, treatment also included metronidazole.

3.2. Diet Formulation

The diet for all five dogs was formulated to ensure the minimum requirements and/or recommended allowances of all essential nutrients met those established by the National Research Council 2006 profiles for canine adult maintenance on an energy basis. The diet formulation for all five dogs was comprised of the same three ingredients: Modulen IBD liquid diet, commercial whey powder isolate, and BalanceIT Canine Supplement. The Modulen IBD liquid diet provided between 77–80% of the total calories for all five formulations. The median (range) percentage of protein, fat, and carbohydrate on a metabolizable energy basis for all five formulations was 32% (30.7–33.2), 33.3% (32.7–33.9), and 34.7% (34.0–35.3), respectively.

3.3. Tolerance to Diet

All Modulen IBD liquid diet feeds at 25% of the RER were well-tolerated in three of the five dogs. For one of the dogs, the first 25% RER feed was associated with lip smacking, while the rest of the feeds were well-tolerated. For the one remaining dog, regurgitation into the mouth followed by swallowing was seen for one of the feeds; however, this resolved after the dog was started on omeprazole and ondansetron.

Two out of the five dogs tolerated all Modulen IBD liquid feeds at 50% of the RER. For one of the dogs, two feeds were associated with lip smacking, which resolved when administration was slowed. A second dog had a small amount of regurgitation following the first feed at 50% RER and none after this. This same dog also showed some lip smacking with another 50% RER feed, which settled when administration was slowed. The one remaining dog belched once during one 50% RER feed and tolerated the rest of the feedings well.

Four out of five dogs tolerated all Modulen IBD liquid feeds at 75% of the RER. The one remaining dog showed lip smacking with two feeds at 75% RER; however, on both occasions, this resolved once administration was slowed.

Only three out of the five dogs received the Modulen IBD liquid diet at 100% of the RER, as the remainder of the dogs had very good to ravenous appetites and ate Purina Pro Plan Veterinary Diets HA Canine dry food well when offered at the 75% RER Modulen feeds. Of these three dogs, two tolerated all feeds at 100% of the RER. The one remaining dog demonstrated intermittent lip smacking with one feed, which was reduced when administration was slowed.

3.4. Effect of Diet on Appetite and Gastrointestinal Signs during Hospitalization

All five dogs presented with anorexia, necessitating the placement of an esophagostomy feeding tube at the time of the gastrointestinal endoscopy. All five dogs voluntarily ate Purina Pro Plan Veterinary Diets HA Canine dry food prior to the commencement of glucocorticoid and after receiving the Modulen IBD liquid diet at 75% of the RER ($n = 3$) or 100% of the RER ($n = 2$). For all five dogs, appetite was reported to be very good to ravenous. Once all five dogs voluntarily consumed the Purina Pro Plan Veterinary Diets HA Canine dry food diet reliably, they received this diet exclusively, except for one dog, which received this diet with Hill's Prescription Diet z/d Canine dry food, as the dog eventually preferred this combination during hospitalization.

The dogs were hospitalized for a median (range) of 5 days (4–8) following initiation of the Modulen IBD liquid diet. None of the five dogs had any episodes of vomiting during hospitalization after initiating the Modulen IBD liquid diet. Only two of the five dogs had episodes of regurgitation that were independent of feeding during hospitalization after initiating the Modulen IBD liquid diet. One dog had three episodes of regurgitation: once when feedings were at 25% of the RER and twice when at 50% of the RER, and the second dog had a small amount of regurgitation once when feeding was at 50% of the RER.

For the five dogs, the effect of feeding the Modulen IBD liquid diet on stool frequency and consistency during hospitalization was as follows: for one dog, only one stool was passed when the dog was receiving 75% of the RER, and this was reported to be soft formed and light brown in color with no blood, mucus, straining, or excess volume. A second dog passed no stools during hospitalization after the diet was initiated. For a third dog, when receiving 25–50% of the RER feeds, watery stool with fresh blood and mucus was being passed up to eight times per day. This dog was then reported to pass no stools once feeds reached 75 to 100% of the RER. A fourth dog passed liquid bloody stool three times when receiving 25% of the RER feeds, which increased to six times per day when receiving 50% of the RER feeds, but then improved from projectile to smaller amounts and returned to a normal color with mucus, but no blood, at 75% of the RER feeds. The frequency at this time also improved to three times per day and the consistency further improved to soft and semi-formed. For the remaining dog, the diet did not help to improve frequency or consistency of the stool, and the dog continued to pass stool between seven to nine times per day when receiving 25% to 100% of the RER feeds. The consistency varied from watery to jelly-like and back to watery during this time. The stool was initially green, then bloody with mucus. This one dog received metronidazole in-hospital, as fecal consistency and frequency were not improving, and this medication was continued after discharge from the hospital.

3.5. Effect of Diet on Selected Biochemical Parameters

Serum biochemistry was performed on all dogs 2–3 days after initiating the Modulen IBD liquid feeds. For the five dogs, three (60%) had stable or improved serum albumin concentrations (median % increase: 10.3, range: 0–31.1), four (80%) had improved or normalized serum globulin concentrations (median % increase: 12.9, range: 5.1–66.2), four

(80%) dogs had improved or normalized serum cholesterol concentrations (median % increase: 31.5, range: 4.8–63), and three had improved serum CRP concentrations (median % decrease: 7.4, range: 7.4–80.9).

The related-samples Wilcoxon signed rank test showed no significant differences in the selected biochemical parameters, both pre- and post-Modulen IBD liquid diet, for all dogs ($p > 0.080$). The results are presented in Table 1.

Table 1. Selected serum biochemical parameters for the five dogs diagnosed with inflammatory protein-losing enteropathy before (pre) and 2–3 days after (post) receiving the Modulen IBD liquid diet exclusively in-hospital without concurrent glucocorticoids. The percentage change (%) that were deemed to be beneficial are presented in bold. The *p*-value represents the results from the related-samples Wilcoxon signed rank test. CRP = C-reactive protein. RR = laboratory reference range. The * dog was subsequently diagnosed with primary hyperparathyroidism based on ionized calcium and serum parathyroid hormone (PTH) and PTH-related protein concentrations at the same visit.

Dog	Albumin (g/L) RR 26.3–38.2			Globulin (g/L) RR 23.4–42.2			Cholesterol (mmol/L) RR 3.2–6.2			CRP (mg/L)		
	Pre	Post	%	Pre	Post	%	Pre	Post	%	Pre	Post	%
1	19.8	19.4	−2.0	22.2	25.9	**+16.7**	3.00	3.80	**+26.7**	74.0	68.5	**−7.4**
2	15.5	17.1	**+10.3**	17.8	18.7	**+5.1**	2.35	1.55	−34.0	13.1	2.5	**−80.9**
3	21.7	21.7	**0.0**	23.0	25.1	**+9.1**	2.56	3.49	**+36.3**	18.6	18.8	+1.1
4 *	16.0	14.4	−10.0	22.8	22.0	−3.5	2.70	2.83	**+4.8**	4.30	71.8	+1569.8
5	14.8	19.4	**+31.1**	14.5	24.1	**+66.2**	2.00	3.26	**+63.0**	42.0	38.9	−7.4
p-value		0.581			0.080			0.176			0.686	

4. Discussion

The Modulen IBD liquid diet is a prescription diet intended for humans with inflammatory bowel disease (IBD), with several studies proving its therapeutic efficacy [13–16]. However, to our knowledge, this diet has not been previously investigated clinically in dogs. Therefore, our study sought to assess the tolerance to this diet and the short-term effect on gastrointestinal signs and selected biochemical parameters during hospitalization in dogs with inflammatory PLE.

Inflammatory PLE in dogs has a guarded prognosis, with disease-associated death reported in approximately half of the cases [1,2,18]. Therefore, a more efficacious treatment is needed to help improve the prognosis in these dogs. Diet is integral to the treatment of dogs with inflammatory PLE, with multiple studies showing its efficacy [2,19–22]. However, it is unknown what dietary composition and macronutrients are ideal for treatment. So far, most of the studies have centered on the effect of low-fat diets in this disease [23,24]. Therefore, further studies investigating the effects of specific diets to determine the most efficacious are needed. As in canine inflammatory PLE, diet forms an integral part of treatment in humans with IBD [25]. The Modulen IBD liquid diet is designed to be fed as the sole source of nutrition to humans with IBD, and several studies have shown that this diet, when fed exclusively, is able to resolve intestinal inflammation and induce remission in children and adults with IBD [13–16,26]. The possible causes for the reduction in intestinal inflammation include its effects on the microbiota [15,16], mucosal healing [13,15], reduction of proinflammatory cytokines [27], and improving growth and body composition [13,14,28,29]. Therefore, this specific diet was deemed as an ideal candidate to investigate further in dogs with inflammatory PLE.

As formulated, the Modulen IBD liquid diet is relatively high in fat at 42% on a metabolizable energy basis. However, of the 23 g of fat in this diet, 6 g come from medium chain triglycerides. Medium chain triglycerides have been shown to modulate intestinal inflammation and cause less damage than long chain triglycerides in an animal model of ileitis [30], and therefore might help to reduce intestinal inflammation in canine inflammatory PLE. The Modulen IBD liquid diet also contains transforming growth factor beta

(TGF-beta). This cytokine has anti-inflammatory properties in the intestinal tract [31] and plays a crucial role in maintaining tolerance against self-antigens and those derived from food and commensal bacteria [32]. Interestingly, duodenal TGF-beta mRNA expression has been shown to be significantly lower in dogs with IBD than in healthy dogs [33]. As the direct therapeutic effect of TGF beta within the Modulen IBD diet remains to be proven [34], studies are needed to specifically ascertain if this cytokine has any beneficial effects on the gastrointestinal tract in dogs with PLE.

As being fed only the Modulen IBD liquid diet does not meet the daily requirements of all essential nutrients for canine maintenance, whey powder and a multivitamin and mineral blend were added to this diet to ensure it was complete and balanced for adult dogs. All dogs were closely monitored for worsening gastrointestinal signs when the diet was being introduced, as this may have indicated an intolerance to the whey powder or multivitamin and mineral blend. For the five dogs in our study, as the gastrointestinal signs generally improved, it was assumed that they tolerated the ingredients present in the diet. However, future studies using this diet and additional ingredients should take into account that dogs may be intolerant to some ingredients, particularly the whey powder, and therefore these dogs should be closely monitored for any indications of this so that the diet can be discontinued.

Our study showed that the Modulen IBD liquid diet was well-tolerated in all dogs, with only a few episodes of lip smacking or regurgitation in a few of the dogs; however, the former resolved or improved when the speed of administration was slowed. One study in dogs showed vomiting as a common complication following initiation of enteral feeding in dogs with hemorrhagic diarrhea [35]. As our study showed that this potential common complication was absent with the Modulen IBD liquid diet, this diet may be more advantageous compared to other enteral liquid diets fed to dogs. The largest effect of the Modulen IBD liquid diet was on the dogs' appetites whilst hospitalized. All dogs were anorexic prior to hospitalization, necessitating the placement of an esophagostomy feeding tube. However, for all five dogs, their appetites increased, allowing the significant voluntary intake of Purina Pro Plan Veterinary Diets HA Canine dry food after receiving the Modulen IBD liquid diet at 75 or 100% of their RER (on day 3 to 4 of initiating the diet). The possible reason for this increase in hunger could be the positive effects of the diet in reducing intestinal inflammation and promoting mucosal healing. The ability of the Modulen IBD liquid diet to improve voluntary oral intake in dogs with inflammatory PLE has many advantages. Firstly, this allows the enteral feeding tube to be removed earlier, and therefore the effort of feeding via the tube and any complications associated with the tube being in place can be avoided. Secondly, more therapeutic options can be provided via the mouth, such as hydrolyzed protein diets, which may be difficult to blendarize and pass through a feeding tube due to their consistency. In addition, the potential lower palatability of hydrolyzed protein diets may make oral feeding difficult when appetite is reduced, hence potentially decreasing compliance with these diets, and therefore their effect, which has been proven in multiple studies [36–39]. Therefore, ensuring voluntary oral intake returns as soon as possible in these dogs is advantageous, and thus may make the Modulen IBD liquid diet more favorable.

For some dogs in our study, the Modulen IBD liquid diet resulted in an improvement in stool consistency and frequency during hospitalization. Interestingly, human formulations of enteral liquid diets have anecdotally been avoided in dogs due to the concern that the increased osmolality may worsen stool consistency. However, this may be less of a concern with the Modulen IBD liquid diet, as the osmolality is lower due to the higher fat content. The reasons for the improvement in stool consistency in our study dogs are unknown but can likely be attributed to the direct effect of the diet on the intestinal mucosa. Further, the TGF-beta in the diet's formulation might have helped to improve intestinal inflammation and mucosal healing, resulting in less diarrhea. Similarly, the higher medium chain triglyceride content might have also helped to improve intestinal inflammation. In one dog, the frequency and consistency of the diarrhea did not improve following the initiation of the diet. As inflammatory PLE in dogs is heterogenous with regards to its pathogenesis, this may explain the varied effect of this diet on the gastrointestinal signs in this dog. Further studies will help to determine which subset of dogs with inflammatory PLE is likely to benefit most from this diet, as well as the mechanisms for any beneficial effects on stool consistency and frequency.

The presence of initial transient green stool has been noted as an effect of the Modulen IBD liquid diet in humans [40] and is due to a buildup of biliverdin, which is a likely consequence of the cessation of the normal microbial breakdown of biliverdin to stercobilin. Interestingly, in our study, this was seen convincingly in one dog. In this dog the occurrence of green stool was also present transiently at the initiation of the diet. This suggests similar effects of this diet in both species, and therefore any beneficial effects of this diet in humans with IBD, such as reduction in intestinal inflammation and increase in remission rates, may also cross over to dogs with inflammatory PLE.

The Modulen IBD liquid diet has been shown to increase serum albumin concentrations faster than corticosteroids in children with IBD [12]. A faster increase in serum albumin is desirable in dogs with inflammatory PLE so that complications such as ascites and thromboembolism can be avoided. Additionally, serum albumin concentration has been shown to be a prognostic indicator in dogs with chronic enteropathy [17], and one study showed that the normalization of plasma albumin concentrations within 50 days of initial treatment of PLE was associated with a longer survival time [41]. Therefore, treatment aimed at increasing serum albumin concentrations faster is advantageous and desirable in dogs with inflammatory PLE. In our study, we performed serum biochemistry 2–3 days after initiating the diet. In the five dogs in our study, the diet resulted in stable or improved serum albumin concentrations in three dogs 2–3 days after starting the diet. A more marked improvement may have been seen had the dogs been receiving 100% of the RER of the diet at the time of blood collection. In addition, the relatively longer half-life of serum albumin of approximately 7 days in dogs [42] and the relatively little acceleration of albumin synthesis in humans with gastrointestinal protein loss [43] may mean follow-up biochemistry following longer feeding times would potentially result in the effects of the diet being more apparent. In one dog, the serum albumin concentration decreased by at least 10%; however, this dog also had concurrent primary hyperparathyroidism that was subsequently diagnosed at the same visit, and therefore the impact of changing dietary calcium levels on subsequent serum albumin concentrations is unknown. This further highlights the need to take into account comorbidities when considering feeding with the Modulen IBD liquid diet. One of the dogs in our study had at least a 30% improvement in serum albumin following the diet. This dog eventually went into biochemical and clinical remission with dietary therapy alone, and therefore, in this dog, the Modulen IBD liquid diet may have had more of an impact. Multiple studies have shown that there is a subset of dogs with inflammatory PLE that may enter remission with dietary therapy alone [2,19,20]. Thus, further studies should aim to investigate the effects of the Modulen IBD liquid diet in this subgroup.

The Modulen IBD liquid diet appeared to have the greatest effect on serum globulin concentrations 2–3 days after feeding this diet. Serum globulins are made up of three components: alpha and beta globulins, which are synthesized by the liver, and gamma globulins, which are synthesized by plasma cells [44]. An increase in serum globulins in our study may reflect decreased loss via the gastrointestinal tract, although serum globulin may also increase with inflammation. However, in our study, as the serum CRP followed similar trends of reduction with the increase in serum globulin, this suggests that the inflammation is an unlikely cause for the increase seen in our study. The measurement of fecal alpha-1 proteinase inhibitor before and after receiving the diet in these dogs may have helped to definitively determine the mechanism for the serum albumin and globulin changes seen [45]. In our study, the serum globulin concentration 2–3 days after receiving the Modulen diet did not reach statistical significance; however, this might have been due to our study being underpowered, as only five dogs were included in the statistical analysis.

Therefore, our study justifies the need for studies utilizing a larger number of dogs with inflammatory PLE. A small number of dogs were enrolled in our pilot study in an effort to reduce the number of animals used in a research study with an unknown outcome.

The Modulen IBD liquid diet also had a positive effect of increasing serum cholesterol concentrations 2–3 days after initiating feeding. Although this was apparent for four out of the five dogs, one dog had a decrease despite improvements in serum albumin and globulin and normalization of serum CRP. This dog had lymphangiectasia on intestinal histopathology, and therefore one possibility for the decreased serum cholesterol following feeding of this diet might have been due to the higher fat content resulting in increased loss of lymph and, therefore, cholesterol via the gastrointestinal tract. However, given that the CRP had normalized in this dog and that the other dogs with lymphangiectasia or lacteal dilation on intestinal histopathology still had increased concentrations of cholesterol following the diet suggests that there might be another reason for this dog's decrease. As stated above, the heterogenous pathogenesis of inflammatory PLE may mean that not all dogs with this condition will respond equally to dietary management, and therefore further studies are needed to help identify which subset is most likely to respond to this strategy.

Although the Modulen IBD liquid diet did not significantly reduce inflammation across the five dogs, as measured by CRP, this is likely due to dog four showing a major increase. As mentioned above, dog four was subsequently diagnosed with primary hyperparathyroidism at the same visit, which may have influenced subsequent CRP concentrations [46]. However, another possibility could be that this dog is a dietary non-responder. There are children with Crohn's disease who are non-responders to this diet, though they are relatively small in proportion at 10–23% of those treated [47]. Therefore, it is possible that with a larger study, the percentage change in CRP will become significant, as the presence of a non-responder is likely to make a bigger difference in a small study.

The results of our study necessitate further studies assessing the longer-term effects of the Modulen IBD liquid diet in a larger number of dogs with inflammatory PLE. Studies assessing whether this diet helps to increase biochemical and clinical remission are needed. In addition, the mechanism of action of this diet and the reason for its efficacy, as well as identifying which subset of dogs within this condition are most likely to respond, are needed.

In addition to the limitations mentioned above, additional limitations of our study include: the Modulen IBD liquid diet was not fed as a complete diet, as this diet as fed is not complete and balanced for dogs. However, at least 75% of the total calories were provided from this diet, and therefore it is suspected that the efficacy would have been maintained. As our study was assessing short-term feeding, data on the effect of the diet on body weight was not taken into account. Therefore, future studies assessing the longer-term outcome should take this into consideration, especially as malnutrition is prevalent in this disease and the Modulen IBD liquid diet has been shown to improve body composition in children with IBD receiving this diet [13,14,28,29]. Unfortunately, no standardized scoring system for the feces was used to allow more objective data on any changes after feeding the diet. Similarly, repeat CCECAI scores were not assessed during hospitalization to allow for a measure of objective change. Hence, future studies should also account for this. Finally, our study did not include a control diet to help compare the specific effects of the Modulen IBD liquid diet against, and, as such, future studies should also focus on a group of dogs with inflammatory PLE receiving a control diet, such as a therapeutic gastrointestinal highly digestible diet.

5. Conclusions

In conclusion, the Modulen IBD liquid diet fed via an esophagostomy feeding tube was well-tolerated in-hospital, and resolved anorexia in all dogs with inflammatory PLE and helped to improve selected biochemical parameters in some dogs. Further studies are needed to assess the long-term effects of feeding this diet on the rate of serum albumin increase and remission in dogs with inflammatory PLE.

Author Contributions: Conceptualization, A.K.; methodology, A.K.; formal analysis, A.K.; investigation, A.K. and G.P.; resources, A.K.; data curation, A.K. and G.P.; writing—original draft preparation, A.K.; writing—review and editing, G.P.; funding acquisition, A.K. All authors have read and agreed to the published version of the manuscript.

Funding: This research received external funding in the form of a gift from Nestle Purina.

Institutional Review Board Statement: The animal study protocol was approved by the Ethics Committee of the Royal Veterinary College (URN 2019 1912-3).

Informed Consent Statement: Written informed consent for participation into the study was obtained from all owners of the dogs.

Data Availability Statement: All data is provided in the manuscript.

Acknowledgments: The authors thank Ian Sanderson, Camilla Hindar, Sophie Broughton, Deidre Mullowney, Sarah Tayler, and Emma Hugill for their help with the study.

Conflicts of Interest: The authors declare no conflict of interest.

References

1. Kathrani, A.; Sanchez-Vizcaino, F.; Hall, E.J. Association of chronic enteropathy activity index, blood urea concentration, and risk of death in dogs with protein-losing enteropathy. *J. Vet. Intern. Med.* **2019**, *33*, 536–543. [CrossRef] [PubMed]
2. Allenspach, K.; Rizzo, J.; Jergens, A.E.; Chang, Y.M. Hypovitaminosis D is associated with negative outcome in dogs with protein losing enteropathy: A retrospective study of 43 cases. *BMC Vet. Res.* **2017**, *13*, 96. [CrossRef] [PubMed]
3. Ashton, J.J.; Gavin, J.; Beattie, R.M. Exclusive enteral nutrition in Crohn's disease: Evidence and practicalities. *Clin. Nutr.* **2019**, *38*, 80–89. [CrossRef] [PubMed]
4. Gonzalez-Huix, F.; de Leon, R.; Fernandez-Banares, F.; Esteve, M.; Cabre, E.; Acero, D.; Abad-Lacruz, A.; Figa, M.; Guilera, M.; Planas, R.; et al. Polymeric enteral diets as primary treatment of active Crohn's disease: A prospective steroid controlled trial. *Gut* **1993**, *34*, 778–782. [CrossRef] [PubMed]
5. Dziechciarz, P.; Horvath, A.; Shamir, R.; Szajewska, H. Meta-analysis: Enteral nutrition in active Crohn's disease in children. *Aliment. Pharm.* **2007**, *26*, 795–806. [CrossRef]
6. Heuschkel, R.B. Enteral nutrition in children with Crohn's disease. *J. Pediatr. Gastroenterol. Nutr.* **2000**, *31*, 575. [CrossRef]
7. Heuschkel, R.B.; Menache, C.C.; Megerian, J.T.; Baird, A.E. Enteral nutrition and corticosteroids in the treatment of acute Crohn's disease in children. *J. Pediatr. Gastroenterol. Nutr.* **2000**, *31*, 8–15. [CrossRef]
8. Borrelli, O.; Cordischi, L.; Cirulli, M.; Paganelli, M.; Labalestra, V.; Uccini, S.; Russo, P.M.; Cucchiara, S. Polymeric diet alone versus corticosteroids in the treatment of active pediatric Crohn's disease: A randomized controlled open-label trial. *Clin. Gastroenterol. Hepatol.* **2006**, *4*, 744–753. [CrossRef]
9. Nahidi, L.; Day, A.S.; Lemberg, D.A.; Leach, S.T. Differential effects of nutritional and non-nutritional therapies on intestinal barrier function in an in vitro model. *J. Gastroenterol.* **2012**, *47*, 107–117. [CrossRef]
10. Leach, S.T.; Mitchell, H.M.; Eng, W.R.; Zhang, L.; Day, A.S. Sustained modulation of intestinal bacteria by exclusive enteral nutrition used to treat children with Crohn's disease. *Aliment. Pharm.* **2008**, *28*, 724–733. [CrossRef]
11. Kang, Y.; Park, S.; Kim, S.; Kim, S.Y.; Koh, H. Therapeutic efficacy of exclusive enteral nutrition with specific polymeric diet in pediatric crohn's disease. *Pediatr. Gastroenterol. Hepatol. Nutr.* **2019**, *22*, 72–79. [CrossRef] [PubMed]
12. Navas-Lopez, V.M.; Blasco-Alonso, J.; Lacasa Maseri, S.; Giron Fernandez-Crehuet, F.; Serrano Nieto, M.J.; Vicioso Recio, M.I.; Sierra Salinas, C. Exclusive enteral nutrition continues to be first line therapy for pediatric Crohn's disease in the era of biologics. *An. Pediatr* **2015**, *83*, 47–54. [CrossRef]
13. Matuszczyk, M.; Meglicka, M.; Landowski, P.; Czkw)anianc, E.; Sordyl, B.; Szymanska, E.; Kierkus, J. Oral exclusive enteral nutrition for induction of clinical remission, mucosal healing, and improvement of nutritional status and growth velocity in children with active Crohn's disease—A prospective multicentre trial. *Prz. Gastroenterol.* **2021**, *16*, 346–351. [CrossRef] [PubMed]
14. Agin, M.; Yucel, A.; Gumus, M.; Yuksekkaya, H.A.; Tumgor, G. The effect of enteral nutrition support rich in TGF-beta in the treatment of inflammatory bowel disease in childhood. *Medicina* **2019**, *55*, 620. [CrossRef]
15. Pigneur, B.; Lepage, P.; Mondot, S.; Schmitz, J.; Goulet, O.; Dore, J.; Ruemmele, F.M. Mucosal healing and bacterial composition in response to enteral nutrition vs steroid-based induction therapy-A randomised prospective clinical trial in children with crohn's disease. *J. Crohns. Colitis.* **2019**, *13*, 846–855. [CrossRef]
16. Lionetti, P.; Callegari, M.L.; Ferrari, S.; Cavicchi, M.C.; Pozzi, E.; de Martino, M.; Morelli, L. Enteral nutrition and microflora in pediatric Crohn's disease. *JPEN J. Parenter. Enter. Nutr.* **2005**, *29* (Suppl. 4), S173–S175. [CrossRef]
17. Allenspach, K.; Wieland, B.; Grone, A.; Gaschen, F. Chronic enteropathies in dogs: Evaluation of risk factors for negative outcome. *J. Vet. Intern. Med.* **2007**, *21*, 700–708. [CrossRef]
18. Craven, M.D.; Washabau, R.J. Comparative pathophysiology and management of protein-losing enteropathy. *J. Vet. Intern. Med.* **2019**, *33*, 383–402. [CrossRef]

19. Economu, L.; Chang, Y.M.; Priestnall, S.L.; Kathrani, A. The effect of assisted enteral feeding on treatment outcome in dogs with inflammatory protein-losing enteropathy. *J. Vet. Intern. Med.* **2021**, *35*, 1297–1305. [CrossRef]
20. Simmerson, S.M.; Armstrong, P.J.; Wunschmann, A.; Jessen, C.R.; Crews, L.J.; Washabau, R.J. Clinical features, intestinal histopathology, and outcome in protein-losing enteropathy in Yorkshire Terrier dogs. *J. Vet. Intern. Med.* **2014**, *28*, 331–337. [CrossRef]
21. Nagata, N.; Ohta, H.; Yokoyama, N.; Teoh, Y.B.; Nisa, K.; Sasaki, N.; Osuga, T.; Morishita, K.; Takiguchi, M. Clinical characteristics of dogs with food-responsive protein-losing enteropathy. *J. Vet. Intern. Med.* **2020**, *34*, 659–668. [CrossRef] [PubMed]
22. Wennogle, S.A.; Stockman, J.; Webb, C.B. Prospective evaluation of a change in dietary therapy in dogs with steroid-resistant protein-losing enteropathy. *J. Small Anim Pract.* **2021**, *62*, 756–764. [CrossRef]
23. Rudinsky, A.J.; Howard, J.P.; Bishop, M.A.; Sherding, R.G.; Parker, V.J.; Gilor, C. Dietary management of presumptive protein-losing enteropathy in Yorkshire terriers. *J. Small Anim. Pract.* **2017**, *58*, 103–108. [CrossRef] [PubMed]
24. Okanishi, H.; Yoshioka, R.; Kagawa, Y.; Watari, T. The clinical efficacy of dietary fat restriction in treatment of dogs with intestinal lymphangiectasia. *J. Vet. Intern. Med.* **2014**, *28*, 809–817. [CrossRef] [PubMed]
25. Narula, N.; Dhillon, A.; Zhang, D.; Sherlock, M.E.; Tondeur, M.; Zachos, M. Enteral nutritional therapy for induction of remission in Crohn's disease. *Cochrane Database Syst. Rev.* **2018**, *4*, CD000542. [CrossRef] [PubMed]
26. Triantafillidis, J.K.; Mantzaris, G.; Stamataki, A.; Asvesttis, K.; Malgarinos, G.; Gikas, A. Complete remission of severe scleritis and psoriasis in a patient with active Crohn's disease using Modulen IBD as an exclusive immunomodulating diet. *J. Clin. Gastroenterol.* **2008**, *42*, 550–551. [CrossRef] [PubMed]
27. Fell, J.M. Control of systemic and local inflammation with transforming growth factor beta containing formulas. *JPEN J. Parenter. Enter. Nutr.* **2005**, *29* (Suppl. 4), S126–S128; discussion S129–S133, S184–S188. [CrossRef]
28. Gerasimidis, K.; Talwar, D.; Duncan, A.; Moyes, P.; Buchanan, E.; Hassan, K.; O'Reilly, D.; McGrogan, P.; Edwards, C.A. Impact of exclusive enteral nutrition on body composition and circulating micronutrients in plasma and erythrocytes of children with active Crohn's disease. *Inflamm. Bowel. Dis.* **2012**, *18*, 1672–1681. [CrossRef]
29. Hartman, C.; Berkowitz, D.; Weiss, B.; Shaoul, R.; Levine, A.; Adiv, O.E.; Shapira, R.; Fradkin, A.; Wilschanski, M.; Tamir, A.; et al. Nutritional supplementation with polymeric diet enriched with transforming growth factor-beta 2 for children with Crohn's disease. *Isr. Med. Assoc. J.* **2008**, *10*, 503–507.
30. Tsujikawa, T.; Ohta, N.; Nakamura, T.; Satoh, J.; Uda, K.; Ihara, T.; Okamoto, T.; Araki, Y.; Andoh, A.; Sasaki, M.; et al. Medium-chain triglycerides modulate ileitis induced by trinitrobenzene sulfonic acid. *J. Gastroenterol. Hepatol.* **1999**, *14*, 1166–1172. [CrossRef]
31. Hahm, K.B.; Im, Y.H.; Parks, T.W.; Park, S.H.; Markowitz, S.; Jung, H.Y.; Green, J.; Kim, S.J. Loss of transforming growth factor beta signalling in the intestine contributes to tissue injury in inflammatory bowel disease. *Gut* **2001**, *49*, 190–198. [CrossRef]
32. Sanjabi, S.; Oh, S.A.; Li, M.O. Regulation of the immune response by TGF-beta: From conception to autoimmunity and infection. *Cold Spring Harb Perspect. Biol.* **2017**, *9*. [CrossRef]
33. Maeda, S.; Ohno, K.; Uchida, K.; Nakashima, K.; Fukushima, K.; Tsukamoto, A.; Nakajima, M.; Fujino, Y.; Tsujimoto, H. Decreased immunoglobulin A concentrations in feces, duodenum, and peripheral blood mononuclear cells of dogs with inflammatory bowel disease. *J. Vet. Intern. Med.* **2013**, *27*, 47–55. [CrossRef] [PubMed]
34. Di Caro, S.; Fragkos, K.C.; Keetarut, K.; Koo, H.F.; Sebepos-Rogers, G.; Saravanapavan, H.; Barragry, J.; Rogers, J.; Mehta, S.J.; Rahman, F. Enteral nutrition in adult crohn's disease: Toward a paradigm shift. *Nutrients* **2019**, *11*, 2222. [CrossRef] [PubMed]
35. Will, K.; Nolte, I.; Zentek, J. Early enteral nutrition in young dogs suffering from haemorrhagic gastroenteritis. *J. Vet. Med. A Physiol. Pathol. Clin. Med.* **2005**, *52*, 371–376. [CrossRef] [PubMed]
36. Allenspach, K.; Culverwell, C.; Chan, D. Long-term outcome in dogs with chronic enteropathies: 203 cases. *Vet. Rec.* **2016**, *178*, 368. [CrossRef] [PubMed]
37. Mandigers, P.J.; Biourge, V.; van den Ingh, T.S.; Ankringa, N.; German, A.J. A randomized, open-label, positively-controlled field trial of a hydrolyzed protein diet in dogs with chronic small bowel enteropathy. *J. Vet. Intern. Med.* **2010**, *24*, 1350–1357. [CrossRef]
38. Walker, D.; Knuchel-Takano, A.; McCutchan, A.; Chang, Y.M.; Downes, C.; Miller, S.; Stevens, K.; Verheyen, K.; Phillips, A.D.; Miah, S.; et al. A comprehensive pathological survey of duodenal biopsies from dogs with diet-responsive chronic enteropathy. *J. Vet. Intern. Med.* **2013**, *27*, 862–874. [CrossRef]
39. Wang, S.; Martins, R.; Sullivan, M.C.; Friedman, E.S.; Misic, A.M.; El-Fahmawi, A.; De Martinis, E.C.P.; O'Brien, K.; Chen, Y.; Bradley, C.; et al. Diet-induced remission in chronic enteropathy is associated with altered microbial community structure and synthesis of secondary bile acids. *Microbiome* **2019**, *7*, 126. [CrossRef]
40. Walton, C.; Montoya, M.P.; Fowler, D.P.; Turner, C.; Jia, W.; Whitehead, R.N.; Griffiths, L.; Waring, R.H.; Ramsden, D.B.; Cole, J.A.; et al. Enteral feeding reduces metabolic activity of the intestinal microbiome in Crohn's disease: An observational study. *Eur. J. Clin. Nutr.* **2016**, *70*, 1052–1056. [CrossRef]
41. Nakashima, K.; Hiyoshi, S.; Ohno, K.; Uchida, K.; Goto-Koshino, Y.; Maeda, S.; Mizutani, N.; Takeuchi, A.; Tsujimoto, H. Prognostic factors in dogs with protein-losing enteropathy. *Vet. J.* **2015**, *205*, 28–32. [CrossRef]
42. Morris, M.A.; Preddy, L. Glycosylation accelerates albumin degradation in normal and diabetic dogs. *Biochem. Med. Metab. Biol.* **1986**, *35*, 267–270. [CrossRef]

43. Wochner, R.D.; Weissman, S.M.; Waldmann, T.A.; Houston, D.; Berlin, N.I. Direct measurement of the rates of synthesis of plasma proteins in control subjects and patients with gastrointestinal protein loss. *J. Clin. Investig.* **1968**, *47*, 971–982. [CrossRef] [PubMed]
44. Busher, J.T. Serum albumin and globulin. In *Clinical Methods: The History, Physical, and Laboratory Examinations*; Walker, H.K., Hall, W.D., Hurst, J.W., Eds.; Butterworths: Boston, MA, USA, 1990.
45. Murphy, K.F.; German, A.J.; Ruaux, C.G.; Steiner, J.M.; Williams, D.A.; Hall, E.J. Fecal alpha1-proteinase inhibitor concentration in dogs with chronic gastrointestinal disease. *Vet. Clin. Pathol./Am. Soc. Vet. Clin. Pathol.* **2003**, *32*, 67–72. [CrossRef] [PubMed]
46. Emam, A.A.; Mousa, S.G.; Ahmed, K.Y.; Al-Azab, A.A. Inflammatory biomarkers in patients with asymptomatic primary hyperparathyroidism. *Med. Princ Pract.* **2012**, *21*, 249–253. [CrossRef] [PubMed]
47. Swaminath, A.; Feathers, A.; Ananthakrishnan, A.N.; Falzon, L.; Li Ferry, S. Systematic review with meta-analysis: Enteral nutrition therapy for the induction of remission in paediatric Crohn's disease. *Aliment. Pharm.* **2017**, *46*, 645–656. [CrossRef] [PubMed]

Article

Risk Factors and Clinical Presentation in Dogs with Increased Serum Pancreatic Lipase Concentrations—A Descriptive Analysis

Harry Cridge [1,*]**, Nicole Scott** [2] **and Jörg M. Steiner** [2]

[1] Department of Small Animal Clinical Sciences, Michigan State University College of Veterinary Medicine, East Lansing, MI 48824, USA
[2] Gastrointestinal Laboratory, Department of Small Animal Clinical Sciences, College of Veterinary Medicine and Biomedical Sciences, Texas A&M University, College Station, TX 77843, USA; nicolescott1320@tamu.edu (N.S.); jsteiner@cvm.tamu.edu (J.M.S.)
* Correspondence: cridgeh1@msu.edu

Simple Summary: Over the past several years, there has been an increasing importance placed on clinical presentation as part of a diagnosis of pancreatitis, and yet there is comparatively little data investigating the full array of clinical signs that may be seen in dogs with pancreatitis. Our study showed that dogs with increased pancreatic lipase immunoreactivity concentrations could display a wide range of clinical signs, which may be related to pancreatitis or a concurrent disease. Non-specific clinical signs, such as anorexia, were prevalent. Additionally, overt abdominal pain was infrequently reported, and veterinarians should be cautious in ruling out pancreatitis due to a lack of abdominal pain alone. Additionally, limited data is available on potential risk factors for pancreatitis in dogs; this information could be important in the development of disease prevention strategies. In our study, the most common concurrent disease was hepatobiliary abnormalities. Additional studies are needed to determine whether this is a causative or associative relationship. Drug use reflected common prescribing practices, and anti-epileptic drug use was low despite prior studies documenting clear associations between phenobarbital and potassium bromide and drug-associated pancreatitis. Adult maintenance diets, in addition to human foods and dog treats, were commonly fed prior to the development of an increased pancreatic lipase concentration.

Abstract: Limited data exist regarding the full array of clinical signs seen in dogs with pancreatitis and potential risk factors for the disease. Laboratory submissions from the Gastrointestinal Laboratory at Texas A&M University were retrospectively reviewed for dogs with an increased serum pancreatic lipase immunoreactivity (cPLI) concentration ($\geq$400 µg/L), and an internet-based survey was distributed to the attending veterinarian and/or technician on each case. The survey contained questions related to (i) clinical signs, (ii) prior gastrointestinal upset, (iii) comorbidities, (iv) pre-existing medical therapies, and (v) dietary history. One hundred and seventy (170) survey responses were recorded. The top three clinical signs reported were inappetence (62%), diarrhea (53%), and vomiting (49%). Abdominal pain was noted in only 32% of dogs, likely associated with poor pain detection. Additionally, the majority of dogs (71%) had prior episodes of gastrointestinal upset within the past 12 months, lending support for the commonality of recurrent acute pancreatitis, or acute on chronic disease. Hepatobiliary abnormalities (24%) were the most common concurrent disease, and endocrine disorders were seen in a low proportion of respondents (5–8%). Adult maintenance diets (65%), dog treats (40%), and human foods (29%) were commonly consumed by dogs prior to the discovery of increased cPLI concentration.

Keywords: canine; pancreatitis; cPLI; etiology; clinical signs

Citation: Cridge, H.; Scott, N.; Steiner, J.M. Risk Factors and Clinical Presentation in Dogs with Increased Serum Pancreatic Lipase Concentrations—A Descriptive Analysis. *Animals* **2022**, *12*, 1581. https://doi.org/10.3390/ani12121581

Academic Editors: Aarti Kathrani and Romy M. Heilmann

Received: 20 May 2022
Accepted: 16 June 2022
Published: 19 June 2022

Publisher's Note: MDPI stays neutral with regard to jurisdictional claims in published maps and institutional affiliations.

1. Introduction

Pancreatitis is the most common disorder of the exocrine pancreas in dogs [1]. Histopathology is traditionally considered the gold standard diagnostic modality; however, it is not routinely used due to its invasive nature and inherent limitations, including the potential to miss localized lesions and challenges associated with determining the significance of subclinical histopathologic lesions [2,3]. Given this, there is a reliance on the use of a clinical gold standard, which comprises a combined assessment of history, clinical signs, diagnostic imaging, and serum pancreatic lipase immunoreactivity (cPLI) concentration. Many studies have investigated diagnostic imaging and cPLI assays; however, comparatively little data exists on clinical presentation. The largest study that focused on clinical signs of pancreatitis was published almost 25 years ago and utilized a histopathologic gold standard [4]. Other studies utilizing various diagnostic criteria and population sizes have reported varied frequencies of abdominal pain, vomiting, and diarrhea in pancreatitis [5–7]. Published data frequently represents referral-level populations. There is, therefore, a critical need to further investigate the range and frequency of clinical signs in dogs with increased serum cPLI concentrations across a multitude of practice types.

Additionally, pancreatitis in dogs is commonly considered idiopathic in nature, but several risk factors have been proposed [8]. Identification of risk factors would allow for the development of disease prevention strategies. Therefore, the aims of our study were to (i) document demographic data of dogs with an increased serum cPLI concentration, (ii) document the clinical presentation of dogs with an increased serum cPLI concentration, (iii) investigate the prevalence of comorbidities and pre-existing drug therapies in the study population, and (iv) document the dietary history of dogs with an increased serum cPLI concentration.

2. Materials and Methods

2.1. Case Identification and Survey Distribution

The submission records of a non-commercial laboratory (Gastrointestinal Laboratory, Texas A&M University, College Station, TX, USA) were retrospectively reviewed for dogs with an increased serum cPLI concentration ($\geq$400 µg/L; measured by Spec cPL, Idexx Laboratories, Westbrook, ME, USA) between 1 November 2021, and 25 February 2022. This assay is not included in routine biochemical profiles and is specifically requested by the attending veterinarian on each case. Repeated measures were excluded. One thousand five hundred and thirty-one (1531) dogs were identified. Internet-based surveys (Qualtrics Version 16.0, Proto, UT, USA) were distributed via email, across four batches. Batching was used to ensure that surveys were distributed close to the time of sample collection to reduce recall bias. Surveys were completed by a licensed veterinarian or a licensed veterinary technician at the practice that submitted the serum sample. Multiple practice types were surveyed, including primary care clinics, emergency clinics, and specialty referral hospitals across the United States of America.

2.2. Data Handling and Analysis

Demographic data (i.e., breed, age, sex) were compiled from the laboratory records for all dogs with an increased serum cPLI concentration ($\geq$400 µg/L) during the search period (n = 1531). Then data from the returned internet surveys (n = 170) were also reviewed under the following categories: (i) clinical signs, (ii) prior episodes of gastrointestinal upset, (iii) comorbidities, (iv) pre-existing medical therapies and nutraceuticals, and (v) dietary history. The survey utilized is available for review (Supplementary Material 1). Qualtrics (Qualtrics Version 16.0, Proto, UT, USA) was used for data compilation and calculation of percentages. Median and ranges were also calculated, and missing data were handled in accordance with the recommendations of an expert consultant on survey-based research. The denominator used for the calculation of percentages in question sub-analyses was based on the number of veterinarians that answered "yes" to at least one part of the question. This assumes that some veterinarians only entered "yes" rather than also utilizing the "no" and "unknown" columns, where available. Breed related odds ratios (OR) were calculated using a two-sided Fisher's exact test compared to a control population composed of the latest publicly available breed/litter registration data from the American Kennel Club (2009) [9]. Odds ratios were calculated against the total number of dogs from other breeds. Woolf logit interval was used to compute 95% confidence intervals (CI). Only significant odds ratios were reported. Only breeds with $\geq$ 5 dogs represented were included in the analysis. Analysis was performed using commercially available software (GraphPad Prism Version 9.0, GraphPad Software Inc., San Diego, CA, USA). Significance for all statistical comparisons was set at $p < 0.05$.

3. Results

3.1. Survey Population

A total of 1531 dogs were identified with a serum cPLI concentration (as measured by Spec cPL) $\geq$ 400 µg/L during the study period. Primary care practices represented 37% of samples, whereas 59% came from emergency or referral hospitals. Practice type could not be identified in 4% of samples. The median cPLI concentration was 763 µg/L (range, 400—$\geq$ 2000 µg/L). Mixed breed dogs represented almost a third of dogs (451/1486, 30.3%). The most common breeds were Yorkshire Terriers (100/1486, 6.7%), Boxers (51/1486, 3.4%), and Labrador Retrievers (51/1486, 3.4%). Detailed demographic data are shown in Table 1. Body weight was unavailable for review.

The following breeds were at increased risk of an increased serum cPLI concentration: Greyhound (OR: 28.5, 95% CI: 13.92–58.47, $p \leq 0.001$), Pointer (OR: 8.12, 95% CI: 3.84–17.17, $p \leq 0.001$), Fox terrier (OR: 6.82, 95% CI: 3.65–12.77, $p \leq 0.001$), Eskimo (OR: 6.67, 95% CI: 2.75–16.16, $p = 0.001$), Black and tan coonhound (OR: 3.40, 95% CI: 1.52–7.61, $p = 0.01$), Staffordshire terrier (OR: 2.70, 95% CI: 1.21–6.04, $p = 0.03$), Cardigan Welsh corgi (OR: 2.59, 95% CI: 1.08–6.26, $p = 0.047$), Cavalier King Charles spaniel (OR: 2.51, 95% CI: 1.85–3.40, $p \leq 0.001$), Soft-coated wheaten terrier (OR: 2.48, 95% CI: 1.24–4.99, $p = 0.02$), Brussels griffon (OR: 2.44, 95% CI: 1.09–5.44, $p = 0.04$), and Miniature pinscher (OR: 2.33, 95% CI: 1.57–3.46, $p \leq 0.001$) (See Table 2).

The following breeds were at decreased risk of an increased cPLI concentration: Beagle (OR: 0.19, 95% CI: 0.11–0.33, $p \leq 0.001$), Labrador retriever (OR: 0.24, 95% CI: 0.18–0.32, $p \leq 0.001$), Bulldog (OR: 0.26, 95% CI: 0.15–0.43, $p \leq 0.001$), Golden retriever (OR: 0.32, 95% CI: 0.21–0.48, $p \leq 0.001$), German shepherd dog (OR: 0.35, 95% CI: 0.25–0.50, $p \leq 0.001$), Australian shepherd (OR: 0.35, 95% CI: 0.15–0.84, $p = 0.01$), Siberian husky (OR: 0.36, 95% CI: 0.17–0.75, $p \leq 0.001$), German shorthair pointer (OR: 0.41, 95% CI: 0.22–0.77, $p \leq 0.001$), Rottweiler (OR: 0.43, 95% CI: 0.25–0.74, $p \leq 0.001$), Poodle (OR: 0.46, 95% CI: 0.29–0.71, $p = \leq 0.001$), Pug (OR: 0.50, 95% CI: 0.28–0.88, $p = 0.01$), and Shih Tzu (OR: 0.64, 95% CI: 0.43–0.94, $p = 0.02$) (see Table 2). Non-significant odds ratios for breeds are not reported.

One-hundred and seventy (170) surveys were returned, indicating an 11.1% positive return rate. The results from the returned questionnaires are further described below.

Table 1. Demographic details on the survey population. Note: Age was unavailable for 65 dogs. Sex status was unavailable for 31 dogs. Breed was unavailable for 45 dogs. Rounding resulted in a total > 100% for gender.

Variable	Dogs with cPLI $\geq$ 400 µg/L
Age (years), median (range)	10.25 (0.25–19)
Sex	79/1500 (5.3%) male intact
	669/1500 (44.6%) male neutered
	31/1500 (2.1%) female intact
	721/1500 (48.1%) female spayed

Table 1. *Cont.*

Variable	Dogs with cPLI $\geq$ 400 µg/L
Breed	Mixed breed (451) Yorkshire terrier (100) Boxer (51) Labrador retriever (51) Dachshund (46) Cavalier King Charles spaniel (43) Chihuahua (41) German shepherd (34) Miniature pinscher (30) Pomeranian (28) Miniature poodle (27) Cocker spaniel (26) Shih Tzu (26) Maltese (25) Miniature schnauzer (25) Golden retriever (23) French bulldog (20) Jack Russel terrier (20) Standard poodle (20) Other breeds with less than 20 dogs represented: Boston terrier (18), Great dane (16), Australian shepherd (15), Bulldog (14), Havanese (14), Beagle (13), Rottweiler (13), Shetland sheepdog (13), Border collie (12), Pug (12), Bichon frise (11), Papillon (11), Fox terrier (10), German shorthaired pointer (10), Greyhound (8), Soft-coated wheaten terrier (8), Basset hound (7), Bernese mountain dog (7), Pointer (7), Siberian husky (7), Bull terrier (6), Cairn terrier (6), Black and tan coonhound (6), Brussels griffon (6), Staffordshire terrier (Pit Bull) (6), West highland white terrier (6), Australian heeler (5), Belgian malinois (5), Eskimo (5), Lhasa apso (5), Toy poodle (5), Silky terrier (5), American staffordshire terrier (5), Vizsla (5), Cardigan eelsh corgi (5), Chesapeake bay retriever (4), Foxhound (4), Newfoundland (4), Shiba inu (4), Whippet (4), Catahoula leopard dog (3), English springer spaniel (3), Great pyrenees (3), Irish setter (3), Italian greyhound (3), Portuguese water dog (3), Border terrier (2), Boykin spaniel (2), Chinese crested (2), Collie (2), Doberman pinscher (2), English cocker spaniel (2), Keeshond (2), Manchester terrier (2), Pekingese (2), Redbone coonhound (2), Rhodesian ridgeback (2), Samoyed (2), Schipperke (2), Scottish terrier (2), Swiss mountain dog (2), Weimaraner (2), Pembroke welsh corgi (2), Affenpinscher (1), Airedale terrier (1), Akita (1), Alaskan malamute (1), Australian kelpie (1), Basenji (1), Belgian tervuren (1), Blue tick coonhound (1), Bouvier des flandres (1), Brittany spaniel (1), English setter (1), Mastiff (1), Norwegian elkhound (1), Norwich terrier (1), Plott hound (1), Curly-coated retriever (1), Saluki (1), Shar-pei (1), Tibetan terrier (1), and Welsh terrier (1).

3.2. Clinical Signs and Prior Episodes

The most common clinical sign reported was inappetence (92/148 responses, 62%), followed by diarrhea (78/148 responses, 53%) and vomiting (72/148 responses, 49%). Only approximately one-third of dogs had perceived abdominal pain or discomfort (48/148 responses, 32%). Complete data are displayed in Table 3.

Table 2. Breeds with a statistically significant increase or decrease in risk for an increased serum cPLI concentration.

Risk	Breed [Odds Ratio]
Increased Risk	Greyhound (OR: 28.5)
	Pointer (OR: 8.12)
	Fox terrier (OR: 6.82)
	Eskimo (OR: 6.67)
	Black and tan coonhound (OR: 3.40)
	Staffordshire terrier (OR: 2.70)
	Cardigan Welsh corgi (OR: 2.59)
	Cavalier King Charles spaniel (OR: 2.51)
	Soft-coated wheaten terrier (OR: 2.48)
	Brussels griffon (OR: 2.44)
	Miniature pinscher (OR: 2.33)
Decreased Risk	Beagle (OR: 0.19)
	Labrador retriever (OR: 0.24)
	Bulldog (OR: 0.26)
	Golden retriever (OR: 0.32)
	German shepherd dog (OR: 0.35)
	Australian shepherd (OR: 0.35)
	Siberian husky (OR: 0.36)
	German shorthair pointer (OR: 0.41)
	Rottweiler (OR: 0.43)
	Poodle (OR: 0.46)
	Pug (OR: 0.50)
	Shih Tzu (OR: 0.64)

Table 3. Clinical signs reported in returned survey data from dogs with a serum cPLI $\geq$ 400 µg/L.

Clinical Sign	No. of Dogs Affected
Inappetence	92/148 (62%)
Diarrhea	78/148 (53%)
Vomiting	77/148 (49%)
Lethargy	67/148 (45%)
Nausea	52/148 (35%)
Abdominal pain or discomfort	48/148 (32%)
Regurgitation	15/148 (10%)
Other clinical signs	55/148 (37%)

Other stated clinical signs included weight loss (22 dogs), polydipsia and polyuria (11 dogs), neurologic compromise (5 dogs), respiratory compromise (3 dogs), borborygmus (5 dogs), and trembling (1 dog). The majority of dogs (105/148, 71%) had displayed prior episodes of gastrointestinal upset in the 12-month period immediately preceding serum sample collection.

3.3. Comorbidities

The survey also enquired about the presence or absence of comorbidities as reported and diagnosed by the attending veterinarian. The three most common comorbidities were hepatobiliary abnormalities (35/146, 24%), kidney disease (27/146, 18%), and hypothyroidism (11/146, 8%). Complete data are displayed in Table 4.

Table 4. Comorbidities reported in returned survey data from dogs with a serum cPLI concentration ≥ 400 µg/L.

Disease	No. of Dogs Affected
Hepatobiliary abnormalities (non-specified)	35/146 (24%)
Kidney disease (non-specified)	27/146 (18%)
Hypothyroidism	11/146 (8%)
Hyperadrenocorticism	8/146 (5%)
Diabetes mellitus	7/146 (5%)
Hyperlipidemia	6/146 (4%)
Tick-borne infections	3/146 (2%)
Foreign body (within 3 months)	1/146 (1%)
Other medical or surgical disorders	91/146 (62%)

Other stated comorbidities included chronic enteropathies (26 dogs), allergic skin disease (16 dogs), cardiac disease (12), urogenital disorders (11 dogs), seizures/other neurological disease (10 dogs), immune-mediated disease (5 dogs), cancer (5 dogs), orthopedic disease (5 dogs), hyperparathyroidism (2 dogs), systemic hypertension (2 dogs), and colorectal polyps (1 dog).

3.4. Pre-Existing Medical Therapies and Nutraceuticals

The most common pre-existing medical therapies received by dogs within 3 months of sample collection included supplements/nutraceuticals (51/143 responses, 36%), antibiotics (37/143 responses, 26%), and corticosteroids (24/143 responses, 17%). Complete data are displayed in Table 5. The most commonly stated nutraceuticals included joint supplements, followed by probiotics and anti-oxidants. The most common antibiotic reported to be administered was metronidazole. The most common anti-seizure medication reported to be administered was phenobarbital. The most common immunosuppressive agent, other than corticosteroids, reported to be administered was oclacitinib.

Table 5. Pre-existing medical therapies and nutraceuticals reported in survey data from dogs with a serum cPLI concentration ≥ 400 µg/L.

Therapeutics	No. of Dogs
Supplements/Nutraceuticals	51/143 (36%)
Antibiotics	37/143 (26%)
Corticosteroids	24/143 (17%)
Non-steroidal anti-inflammatories (NSAIDs) or NSAID-like drugs	18/143 (13%)
Immunosuppressives (other than corticosteroids)	12/143 (8%)
Anti-seizure medications	3/143 (2%)
Chemotherapy drugs	1/143 (1%)

3.5. Dietary History (Prior to Sample Collection)

The top three nutrition sources fed to dogs prior to sample collection in the study were adult maintenance diets (80/123 responses, 65%), dog treats (49/123 responses, 40%), and human foods (36/123 responses, 29%). Of the dogs that received other therapeutic diets, renal/urinary diets were fed to 10 of the dogs. Complete data are displayed in Table 6. As dogs likely received more than one nutritional source, the predominant diet was also enquired about during the survey. The predominant diet type is shown in Table 7.

Table 6. Pre-existing diets reported in survey data from dogs with a serum cPLI concentration ≥ 400 μg/L. Note: some dogs did receive more than one diet type.

Dietary Type	No. of Dogs
Adult maintenance diet (over-the-counter)	80/123 (65%)
Commercial dog treats	49/123 (40%)
Human foods (table foods)	36/123 (29%)
Low-fat diet (based on label description)	23/123 (19%)
Other therapeutic diet(s)	18/123 (15%)
Gastrointestinal therapeutic diet (non-low fat)	18/123 (15%)
Homemade diet	16/123 (13%)
Hydrolyzed diet	9/123 (7%)
Limited-ingredient novel protein diet	8/123 (7%)

Table 7. Predominant diet type reported in survey data from dogs with a serum cPLI concentration ≥ 400 μg/L.

Predominant Diet Type	No. of Dogs
Adult maintenance diet (over-the-counter)	53/131 (40%)
Low-fat diet (based on label description)	22/131 (17%)
Gastrointestinal therapeutic diet (non-low fat)	15/131 (11%)
Other therapeutic diets	14/131 (11%)
Homemade diet	11/131 (8%)
Hydrolyzed diet	10/131 (8%)
Human foods (table foods)	3/131 (2%)
Commercial dog treats	2/131 (2%)
Limited-ingredient novel protein diet	1/131 (1%)

4. Discussion

In this descriptive survey-based study, we investigated the demographic data, clinical presentation, and potential risk factors for an increased serum pancreatic lipase immunoreactivity (cPLI) concentration ≥ 400 μg/L in a large cohort of dogs from multiple practice types. Increased serum cPLI concentration was utilized to determine study eligibility as it is an objective and highly specific biomarker of acute pancreatitis in dogs [10–12]. While other diagnostic criteria such as ultrasonographic features were initially considered in study planning, they were ultimately deemed more subjective than cPLI concentration, particularly in the face of a multi-practice evaluation. Ultrasound is well-known to be affected by operator experience, patient discomfort, and the information provided to the sonographer at the time of evaluation [13–15]. Clinical signs were not eligible for study inclusion due to the objectives of the study.

Mixed breed dogs represented almost one-third of dogs with an increased serum cPLI concentration. The top three affected breeds were Yorkshire terriers, Boxers, and Labrador retrievers. Mixed breed dogs and Labradors likely represent the common occurrence of this breed in the general dog population. Indeed, Labrador retrievers had a decreased risk of having an increased serum cPLI concentration relative to other breeds. Yorkshire terriers have also been identified as one of the most common breeds affected by pancreatitis in other studies, suggesting a breed-predisposition; however, a significant odds ratio was not detected in our study [4]. Boxers and Cavalier King Charles spaniels are reportedly predisposed to chronic pancreatitis, and this likely represents their over-representation in our study, particularly in the face of a high prevalence of prior clinical signs of gastrointestinal upset [16]. Cavalier King Charles spaniels had an almost 2.5× increased risk of having an increased serum cPLI concentration compared to other breeds. Other potentially predisposed breeds include the Black and tan coonhound, Brussels griffon, Cardigan Welsh corgi, Eskimo, Fox terrier, Greyhound, Miniature pinscher, Pointer, Soft-coated wheaten terrier, and Staffordshire terrier. Additional studies are needed on the causes of potential breed predispositions. No gender appeared to be over-represented in this study group, and most dogs were middle-aged to older, with a median age of 10.25 years, although a wide range of ages was reported.

The most common clinical sign reported in dogs with an increased serum Spec cPL concentration in this study was inappetence. This is important because it is often an underestimated clinical sign and has a broad list of differential diagnoses. Pancreatitis should therefore be considered an appropriate differential diagnosis for dogs with non-specific inappetence. The second most common clinical sign was diarrhea. The third most common sign was vomiting, but this was only seen in half of the study respondents. The prevalence of vomiting may also be related to the distribution of lesions within the pancreas [7]. Abdominal pain, which is commonly considered a cornerstone of acute pancreatitis, was only detected in around one-third of dogs in our study, which was a surprising finding given the prevalence (58–67%) reported in prior studies and anecdotal perception of dogs with pancreatitis [4–6]. Interestingly, these studies involved referral populations or fatal cases of pancreatitis, and the prevalence of abdominal pain has been reported to be as low as 15% in a study utilizing multiple practice types and a clinical diagnosis utilizing a point-of-care pancreatic lipase assay [7]. We therefore suspect that our data more accurately reflect the true prevalence of detected abdominal pain in dogs with increased serum cPLI concentrations. The percentage in this study likely reflects veterinarian and owner perception, rather than what is truly present. More studies are needed on pain perception in dogs with increased serum cPLI concentrations. It is the authors' opinion that analgesics should be administered to all dogs with pancreatitis, even in the absence of overt abdominal pain, as abdominal discomfort is likely under-recognized. Other clinical signs commonly reported in the study population included lethargy, diarrhea, and nausea. Weight loss, polyuria/polydipsia, neurologic complications, respiratory compromise, borborygmi, and trembling were also intermittently reported. These clinical signs may represent a primary presentation of pancreatitis or may result from a comorbidity. The high proportion of animals with prior episodes of gastrointestinal upset suggests the presence of recurrent episodes of pancreatitis, acute on chronic disease, or pancreatitis occurring as a comorbidity to other intestinal disorders. Acute on chronic disease is a distinct potential given the high frequency of histopathologic evidence of chronic pancreatitis found in necropsy studies, which is otherwise undetected [16]. Long-term monitoring of cPLI concentrations and other biomarkers/imaging findings in dogs following an episode of pancreatitis is urgently required to further investigate this potential and its frequency.

Many diseases have been identified as potential risk factors for pancreatitis in dogs, including endocrinopathies, hypertriglyceridemia, infectious agents, and other miscellaneous disorders [8]. Hyperadrenocorticism (HAC) was reported in the history of 12/101 dogs with pancreatitis in one retrospective study, and in another study, dogs with HAC had higher cPLI concentrations than a control group of dogs [17,18]. Similar results have also been seen with other methodologies of pancreatic lipase quantification, e.g., 1-2-diglyceride and 1-2,o-dilauryl-rac-glycero-glutaric acid-(6′-methylresorufin) ester [19]. One study also documented a 4.5× increased risk of HAC in dogs with pancreatitis compared to those without pancreatitis [20]. In our study, a small, yet important, proportion of dogs with increased cPLI concentrations ($\geq$400 µg/L) had a known history of hyperadrenocorticism. The exact pathophysiologic mechanism is not yet known [20]. Diabetes mellitus (DM) has also been reported in association with pancreatitis [20–23]. In our study, a small number of dogs with an increased cPLI concentration ($\geq$400 µg/L) had a known history of diabetes mellitus. While the relationship between pancreatitis and DM is multidirectional, pancreatitis often leads to DM [24]. This is further supported by evidence of glucose intolerance in dogs with presumed chronic pancreatitis [25]. Hypertriglyceridemia may also mediate the relationship between DM and pancreatitis [20]. In our study, hypothyroidism was noted in a small number of dogs with an increased cPLI concentration (>400 µg/L). Hypothyroidism has also been documented in 4/101 dogs in another retrospective study of pancreatitis in dogs [17]. Hyperlipidemia is thought to mediate any potential relationship between hypothyroidism and pancreatitis [22]. Hypothyroidism is also a common disease in middle-aged to older dogs, as with the signalment of dogs in this study. Other diseases reported in dogs with an increased serum cPLI concentration included hepatobiliary abnormalities, renal disease, hyperlipidemia, and tick-borne disease. The frequency of reported hepatobiliary abnormalities was somewhat unexpected but likely reflects reactive hepatic changes, post hepatic cholestasis, or less commonly functional cholestasis secondary to the cytokine response in pancreatitis, rather than pre-existing disease [26,27]. Reference criteria for the diagnosis of hepatic disease, including histopathology, were not provided. Thus, biochemical abnormalities alone may have been utilized to suggest the presence of hepatic disease. A similar caveat exists for renal disease. Hepatic lipase does not cross-react with the Spec cPL assay and is therefore unlikely to be the cause of the cPLI elevations in this study [28]. Timing of hepatic disease/biochemical abnormality relative to that of pancreatitis in each dog was unknown. Another study did, however, note a high prevalence of ultrasonographic evidence of pancreatitis (30.2%) in dogs with cholangitis or cholangiohepatitis [29]. Further research into this potential relationship is needed. The relationship between renal disease and pancreatitis is the source of much debate. Given the molecular weight of pancreatic lipase and a lack of a consistent relationship between pancreatic lipase and creatinine concentration, it is suspected that changes in GFR alone are not responsible for increased Spec cPL concentrations in dogs with renal disease [26,30,31]. Another study investigating DGGR lipase also noted a high prevalence of hyperlipasemia in dogs with naturally occurring renal disease [32]. Additional research is therefore needed to further investigate the relationship between naturally occurring renal disease and pancreatitis. Renal and urinary diets are often higher in dietary fat than standard adult maintenance diets, and this could be a contributing factor in this relationship; indeed, 10 dogs in our study were receiving such diets [33]. However, dietary fat level was not definitively assessed, and as such, this remains speculative. The renal disease reported may also represent acute kidney injury secondary to acute pancreatitis [34]. The low frequency of tick-borne disease likely reflects an incidental association. The risk factors studied in this project were determined based on a temporal association with the onset of pancreatitis. Given the data type, it is important to consider that these risk factors may not always represent a causative relationship and may instead be incidental or represent shared risk factors for disease. Additionally, the direction of a relationship cannot always be definitively determined.

Drug-associated pancreatitis (DAP) is an important etiology in humans with pancreatitis, and the Badalov classification system is used to identify potential at-risk drugs [35]. A recent review article highlighted the current dearth of knowledge on this topic in dogs and called for further reports of potential DAP [26]. Our data provides significant additional information regarding drug histories in dogs with increased serum cPLI concentrations. The top three most common types of medications used over a 3-month period prior to sample collection were antibiotics, corticosteroids, and non-steroidal anti-inflammatory drugs or related pharmaceuticals, e.g., grapiprant. Supplements and nutraceutics were also prescribed to over one-third of the study population. These are anecdotally among the most commonly prescribed drug classes, and as such, this finding may represent an incidental association. The most common antibiotic reported to have been administered was metronidazole, which is commonly used in the management of diarrhea (as seen in a large proportion of this study population), although this is associated with controversy and is no longer routinely recommended due to negative effects on the intestinal microbiome [36,37]. Corticosteroids have previously been implicated in the development of pancreatitis; however, the role of exogenous steroids in the etiology of pancreatitis is perpetually evolving [26]. Administration of corticosteroids to healthy dogs does not result in significant changes in cPLI concentrations or does not commonly (1/6) increase serum cPLI concentrations into a diagnostic range for pancreatitis [38,39]. Indeed, corticosteroids are now being used with increasing frequency in the management of acute pancreatitis [40,41]. In contrast with those results, our study showed a moderate frequency of corticosteroid use prior to sampling in dogs with increased Spec cPL concentrations, and this may reflect the presence of concurrent disease treated with corticosteroids or secondary pancreatic inflammation influenced by concurrent disease [8]. We cannot, however, rule out the potential that corticosteroids caused pancreatitis in some of the dogs in our study. Other drugs frequently implicated in the development of pancreatitis, such as phenobarbital and potassium bromide, were infrequently reported in our study population. This likely is related to a low frequency of anti-epileptic drug use in our population. Future studies should investigate drug latency periods and re-exposure data to provide further definitive evidence of DAP in dogs and to help develop a DAP classification scheme for use in veterinary patients. Latency is defined as the interval between starting a medication and pancreatitis induction, with a consistent latency period providing supportive evidence of DAP.

High-fat diets have long been suggested as a risk factor for pancreatitis in dogs based on early studies showing that high-fat diets induce or worsen the severity of pancreatitis in dogs [42,43]. However, limited data exist on naturally occurring disease, and a recent review concluded that there was insufficient evidence to make a definitive conclusion on the effects of high levels of dietary fat on the development of pancreatitis [8]. Adult maintenance diets were fed to the majority of respondents in the current study; however, dog treats and human foods were also fed to a significant proportion of the test population. Other diets commonly utilized in the management of gastrointestinal disorders were fed at a lower frequency, including therapeutic (not labeled low fat) gastrointestinal diets, hydrolyzed diets, and limited-ingredient novel protein diets. Diets labeled as low fat were fed to approximately one-fifth of the test population and may reflect the high frequency of prior gastrointestinal signs in the test population and subsequent prescribing practices of veterinarians. It is important to note that the level of dietary fat was not definitively assessed in this study. In a prior study of healthy dogs, there was no difference in Spec cPL concentrations between those dogs fed a maintenance diet (4.01 g of fat/100 kcal) and one labeled as a low-fat diet (1.55 g of fat/100 kcal) [44]. The high proportion of human foods and dog treats reflects anecdote and the results from a prior study, in which access to unusual food items, table scraps, and trash increased the risk of pancreatitis in dogs [45]. As this was a survey study, labeled claims as assessed by a licensed veterinarian or veterinary nurse/technician were utilized to determine which diets were low fat. The details required to calculate dietary fat on an energy basis were unavailable.

Additional limitations of our study are predominantly related to the nature of survey-based research. This study is an association study, and thus the level of evidence for a causative relationship should be considered weak. A control population and regression analysis would be required to determine true risk factors for pancreatitis. Although the survey was addressed to the veterinarian who submitted the sample and their nursing team, we cannot exclude the potential that someone else within the practice completed the survey based on the medical record. This could be a potential source of bias. Additionally, recall bias may have been introduced between the time of serum sampling and survey completion. The survey completion rate was 11.1%, and as such, data were not available from the majority of dogs with an increased serum cPLI concentration during the study period. Potential reasons for failure to complete the survey may include time constraints, lack of documented data in the medical record, or other factors. Additionally, dogs with milder clinical presentations (i.e., subclinical or mild disease) may not have had serum samples submitted to the laboratory for quantification of cPLI. Thus, as with other research, our study may have selected for more severe clinical presentations.

5. Conclusions

Dogs with increased pancreatic lipase immunoreactivity concentrations display a wide array of clinical signs. Of note, abdominal pain was infrequently reported and was likely under-detected. Pancreatitis should not be excluded based on a lack of perceived abdominal pain. Recurrent episodes of gastrointestinal upset are common in dogs with increased cPLI concentrations. Additional research on potential risk factors of pancreatitis, including potential DAP, is warranted.

Supplementary Materials: The following supporting information can be downloaded at: https://www.mdpi.com/article/10.3390/ani12121581/s1, Supplementary Material 1.

Author Contributions: Conceptualization, H.C. and J.M.S.; methodology, H.C.; formal analysis, H.C. and J.M.S.; investigation, N.S.; writing—original draft preparation, H.C.; writing—review and editing, H.C., N.S., J.M.S. All authors have read and agreed to the published version of the manuscript.

Funding: This research received no external funding.

Institutional Review Board Statement: Ethical review and approval were waived for this study due to retrospective identification of samples and subsequent retrieval of pre-existing data via a survey (IACUC Exemption approved by MSU IACUC committee on 23 August 2021).

Informed Consent Statement: Informed consent was provided by each veterinarian or veterinary technician (nurse) who completed the survey.

Data Availability Statement: Data are only available on request due to restrictions, e.g., privacy or ethical reasons. As the study identified animals with an increased serum cPLI concentration, no owner consent was signed, and personal data are therefore not publicly available.

Acknowledgments: The authors want to thank the staff of the Gastrointestinal Laboratory at Texas A&M University for technical assistance and for the distribution of surveys. The authors also wish to thank Debra Rusz, from the Michigan State University Office of Survey Research (OSR), for her expertise and guidance in data analysis. The authors also wish to thank the veterinarians and veterinary technicians (nurses) for taking the time to complete the surveys and further our knowledge of pancreatitis in dogs.

Conflicts of Interest: The authors declare no conflict of interest directly relevant to this study.

References

1. Xenoulis, P.G. Diagnosis of pancreatitis in dogs and cats. *J. Small Anim. Pract.* **2015**, *56*, 13–26. [CrossRef] [PubMed]
2. Newman, S.; Steiner, J.; Woosley, K.; Barton, L.; Ruaux, C.; Williams, D. Localization of pancreatic inflammation and necrosis in dogs. *J. Vet. Intern. Med.* **2004**, *18*, 488–493. [CrossRef] [PubMed]
3. Pratschke, K.M.; Ryan, J.; Mcalinden, A.; McLauchlan, G. Pancreatic surgical biopsy in 24 dogs and 19 cats: Postoperative complications and clinical relevance of histological findings. *J. Small Anim. Pract.* **2015**, *56*, 60–66. [CrossRef] [PubMed]
4. Hess, R.; Saunders, M.H.; Van Winkle, T.J.; Schofer, F.S.; Washabau, R.J. Clinical, clinicopathologic, radiographic, and ultrasonographic abnormalities in dogs with fatal acute pancreatitis: 70 cases (1986–1995). *J. Am. Vet. Med. Assoc.* **1998**, *213*, 665–670. [PubMed]
5. Kuzi, S.; Mazor, R.; Segev, G.; Nivy, R.; Mazaki-Tovi, M.; Chen, H.; Rimer, D.; Duneyevitz, A.; Yas, E.; Lavy, E.; et al. Prognostic markers and assessment of a previously published clinical severity index in 109 hospitalised dogs with acute presentation of pancreatitis. *Vet. Rec.* **2020**, *187*, e13. [CrossRef]
6. Keany, K.M.; Fosgate, G.T.; Perry, S.M.; Stroup, S.T.; Steiner, J.M. Serum concentrations of canine pancreatic lipase immunoreactivity and c-reactive protein for monitoring disease progression in dogs with acute pancreatitis. *J. Vet. Intern. Med.* **2021**, *35*, 2187–2195. [CrossRef]
7. Berman, C.; Lobetti, R.; Lindquist, E. A comparison of ultrasonographic and clinical findings in 293 dogs with acute pancreatitis: Different clinical presentation with left limb, right limb, or diffuse involvement of the pancreas. *J. S. Afr. Vet. Assoc.* **2020**, *91*, a2022. [CrossRef]
8. Cridge, H.; Lim, S.Y.; Algül, H.; Steiner, J.M. New insights into the etiology, risk factors, and pathogenesis of pancreatitis in dogs: Potential impacts on clinical practice. *J. Vet. Intern. Med.* **2022**, *363*, 847–864. [CrossRef]
9. American Kennel Club: Dog Registration Statistics (1991–2008). Year. Available online: http://images.akc.org/pdf/archives/AKCregstats_1991--2008.pdf (accessed on 6 June 2022).
10. McCord, K.; Morley, P.S.; Armstrong, J.; Simpson, K.; Rishniw, M.; Forman, M.A.; Biller, D.; Parnell, N.; Arnell, K.; Hill, S.; et al. A multi-institutional study evaluating the diagnostic utility of the spec cpl and snap cpl in clinical acute pancreatitis in 84 dogs. *J. Vet. Intern. Med.* **2012**, *26*, 888–896. [CrossRef]
11. Haworth, M.D.; Hosgood, G.; Swindells, K.L.; Mansfield, C.S. diagnostic accuracy of the snap and spec canine pancreatic lipase tests for pancreatitis in dogs presenting with clinical signs of acute abdominal disease. *J. Vet. Emerg. Crit. Care* **2014**, *24*, 135–143. [CrossRef]

12. Cridge, H.; MacLeod, A.G.; Pachtinger, G.E.; Mackin, A.J.; Sullivant, A.M.; Thomason, J.T.; Archer, T.M.; Lunsford, K.V.; Rosenthal, K.; Wills, R.W. Evaluation of SNAP cPL, Spec cPL, VetScan cPL rapid test and Precision PSL assays for the diagnosis of clinical pancreatitis in dogs. *J. Vet. Intern. Med.* **2018**, *32*, 658–664. [CrossRef] [PubMed]

13. Washabau, R. Pancreas. In *Canine and Feline Gastroenterology*; Elsevier Saunders: St. Louis, MO, USA, 2013; pp. 799–848.

14. Cridge, H.; Sullivant, A.M.; Wills, R.W.; Lee, A.M. Association between abdominal ultrasound findings, the specific canine pancreatic lipase assay, clinical severity indices, and clinical diagnosis in dogs with pancreatitis. *J. Vet. Intern. Med.* **2020**, *34*, 636–643. [CrossRef] [PubMed]

15. Hammes, K.; Kook, P.H. Effects of medical history and clinical factors on serum lipase activity and ultrasonographic evidence of pancreatitis: Analysis of 234 dogs. *J. Vet. Intern. Med.* **2022**, *36*, 946–965. [CrossRef] [PubMed]

16. Watson, P.J.; Roulois, A.J.A.; Scase, T.; Johnston, P.E.J.; Thompson, H.; Herrtage, M.E. Prevalence and breed distribution of chronic pancreatitis at post-mortem examination in first-opinion dogs. *J. Small Anim. Pract.* **2007**, *48*, 609–618. [CrossRef] [PubMed]

17. Hess, R.S.; Kass, P.H.; Shofer, F.S.; Van Winkle, T.J.; Washabau, R.J. Evaluation of risk factors for fatal acute pancreatitis in dogs. *J. Am. Vet. Med. Assoc.* **1999**, *214*, 46–51.

18. Mawby, D.I.; Whittemore, J.C.; Fecteau, K.A. Canine pancreatic-specific lipase concentrations in clinically healthy dogs and dogs with naturally occuring hyperadrenocorticism. *J. Vet. Intern. Med.* **2014**, *28*, 1244–1250. [CrossRef]

19. Linari, G.; Dondi, F.; Segatore, S.; Vasylyeva, K.; Linta, N.; Pietra, M.; O Leal, R.; Fracassi, F. Evaluation of 1,2-o-dilauryl-rac-glycero glutaric acid-(6′-methylresorufin) ester (DGGR) and 1,2-diglyceride lipase assays in dogs with naturally occurring hypercortisolism. *J. Vet. Diagn. Investig.* **2021**, *33*, 817–824. [CrossRef]

20. Kim, H.; Kang, J.H.; Heo, T.Y.; Kang, B.T.; Kim, G.; Chang, D.; Na, K.J.; Yang, M.P. Evaluation of hypertriglyceridemia as a mediator between endocrine diseases and pancreatitis in dogs. *J. Am. Anim. Hosp. Assoc.* **2019**, *55*, 92–100. [CrossRef]

21. Pápa, K.; Máthé, Á.; Abonyi-Tóth, Z.; Sterczer, A.; Psáder, R.; Hetyey, C.; Vajdovich, P.; Vörös, K. Occurrence, clinical features and outcome of canine pancreatitis (80 cases). *Acta Vet. Hung.* **2011**, *59*, 37–52. [CrossRef]

22. Cook, A.K.; Breitschwerdt, E.B.; Levine, J.F.; Bunch, S.E.; Linn, L.O. Risk factors associated with acute pancreatitis in dogs: 101 cases (1985–1990). *J. Am. Vet. Med. Assoc.* **1993**, *203*, 673–679.

23. Mattin, M.; O'Neill, D.G.; Church, D.B.; McGreevy, P.D.; Thomson, P.C.; Brodbelt, D.C. An epidemiological study of diabetes mellitus in dogs attending first opinion practice in the UK. *Vet. Rec.* **2014**, *174*, 349. [CrossRef] [PubMed]

24. Davison, L.J. Diabetes mellitus and pancreatitis—Cause or effect? *J. Small Anim. Pract.* **2015**, *56*, 50–59. [CrossRef] [PubMed]

25. Watson, P.; Herrtage, M. Use of glucagon stimulation tests to assess beta-cell function in dogs with chronic pancreatitis. *J. Nutr.* **2004**, *134*, 2081S–2083S. [CrossRef] [PubMed]

26. Cridge, H.; Twedt, D.C.; Marolf, A.J.; Sharkey, L.C.; Steiner, J.M. Advances in the diagnosis of acute pancreatitis in dogs. *J. Vet. Intern. Med.* **2021**, *35*, 2572–2587. [CrossRef] [PubMed]

27. Watson, P.J.; Roulois, A.J.A.; Scase, T.J.; Irvine, R.; Herrtage, M.E. Prevalence of hepatic lesions at post-mortem examination in dogs and association with pancreatitis. *J. Small Anim. Pract.* **2010**, *51*, 566–572. [CrossRef]

28. Lim, S.Y.; Xenoulis, P.G.; Stavroulaki, E.M.; Lidbury, J.A.; Suchodolski, J.S.; Carrière, F.; Steiner, J.M. The 1,2-o-dilauryl-rac-glycero-3-glutaric acid-(6′-methylresorufin) ester (dggr) lipase assay in cats and dogs is not specific for pancreatic lipase. *Vet. Clin. Pathol.* **2020**, *49*, 607–613. [CrossRef]

29. Peters, L.M.; Glanemann, B.; Garden, O.A.; Szladovits, B. Cytologic findings of 140 bile samples from dogs and cats and associated clinical pathologic data. *J. Vet. Intern. Med.* **2016**, *30*, 123–131. [CrossRef]

30. Hulsebosch, S.E.; Palm, C.A.; Segev, G.; Cowgill, L.D.; Kass, P.H.; Marks, S.L. Evaluation of canine pancreas-specific lipase activity, lipase activity, and trypsin-like immunoreactivity in an experimental model of acute kidney injury in dogs. *J. Vet. Intern. Med.* **2016**, *30*, 192–199. [CrossRef]

31. Steiner, J.M.; Williams, D.A. Purification of classical pancreatic lipase from dog pancreas. *Biochimie* **2002**, *84*, 1243–1251. [CrossRef]

32. Prümmer, J.K.; Howard, J.; Grandt, L.M.; Obrador de Aguilar, R.; Meneses, F.; Peters, L.M. Hyperlipasemia in critically ill dogs with and without acute pancreatitis: Prevalence, underlying diseases, predictors, and outcome. *J. Vet. Intern. Med.* **2020**, *34*, 2319–2329. [CrossRef]

33. Wingert, A.M.; Murray, O.A.; Lulich, J.P.; Hoelmer, A.M.; Merkel, L.K.; Furrow, E. Efficacy of medical dissolution for suspected struvite cystoliths in dogs. *J. Vet. Intern. Med.* **2021**, *35*, 2287–2295. [CrossRef] [PubMed]

34. Gori, E.; Lippi, I.; Guidi, G.; Perondi, F.; Pierini, A.; Marchetti, V. Acute pancreatitis and acute kidney injury in dogs. *Vet. J.* **2019**, *245*, 77–81. [CrossRef] [PubMed]

35. Badalov, N.; Baradarian, R.; Iswara, K.; Li, J.; Steinberg, W.; Tenner, S. Drug-induced acute pancreatitis: An evidence-based review. *Clin. Gastroenterol. Hepatol.* **2007**, *5*, 648–661. [CrossRef] [PubMed]

36. Langlois, D.K.; Koenigshof, A.M.; Mani, R. Metronidazole treatment of acute diarrhea in dogs: A randomized double blinded placebo-controlled clinical trial. *J. Vet. Intern. Med.* **2020**, *34*, 98–104. [CrossRef]

37. Pilla, R.; Gaschen, F.P.; Barr, J.W.; Olson, E.; Honneffer, J.; Guard, B.C.; Blake, A.B.; Villanueva, D.; Khattab, M.R.; AlShawaqfeh, M.K.; et al. Effects of metronidazole on the fecal microbiome and metabolome in healthy dogs. *J. Vet. Intern. Med.* **2020**, *34*, 1853–1866. [CrossRef]

38. Steiner, J.M.; Teague, S.R.; Lees, G.E.; Willard, M.D.; Williams, D.A.; Ruaux, C.G. Stability of canine pancreatic lipase immunoreactivity concentration in serum samples and effects of long-term administration of prednisone to dogs on serum canine pancreatic lipase immunoreactivity concentrations. *Am. J. Vet. Res.* **2009**, *70*, 1001–1005. [CrossRef]

39. Ohta, H.; Kojima, K.; Yokoyama, N.; Sasaki, N.; Kagawa, Y.; Hanazono, K.; Ishizuku, T.; Morishita, K.; Nakamura, K.; Takagi, S.; et al. Effects of immunosuppressive prednisolone therapy on pancreatic tissue and concentration of canine pancreatic lipase immunoreactivity in healthy dogs. *Can. J. Vet. Res.* **2018**, *82*, 278–286.

40. Okanishi, H.; Nagata, T.; Nakane, S.; Watari, T. comparison of initial treatment with and without corticosteroids for suspected acute pancreatitis in dogs. *J. Small Anim. Pract.* **2019**, *60*, 298–304. [CrossRef]

41. Bjørnkjær-Nielson, K.; Bjørnvad, C. Corticosteroid treatment for acute/Acute—on—Chronic experimental and naturally occurring pancreatitis in several species: A scoping review to inform possible use in dogs. *Acta Vet. Scand.* **2021**, *63*, 1–34. [CrossRef]

42. Lindsay, S.; Enenman, C.; Chaikoff, I. Pancreatitis accompanying hepatic disease in dogs fed a high fat, Low Protein Diet. *Medicine* **1948**, *45*, 635.

43. Haig, T. Experimental pancreatitis intensified by a high fat diet. *Surg. Gynecol. Obs.* **1970**, *131*, 914–918.

44. James, F.E.; Mansfield, C.S.; Steiner, J.M.; Williams, D.A.; Robertson, I.D. Pancreatic response in healthy dogs fed diets of various fat compositions. *Am. J. Vet. Res.* **2009**, *70*, 614–618. [CrossRef] [PubMed]

45. Lem, K.Y.; Fosgate, G.T.; Norby, B.; Steiner, J.M. Associations between dietary factors and pancreatitis in dogs. *J. Am. Vet. Med. Assoc.* **2008**, *233*, 1425–1431. [CrossRef] [PubMed]

*animals*

Article

A Descriptive Study on the Extent of Dietary Information Obtained during Consultations at a Veterinary Teaching Hospital

Andreina Schramm and Peter Hendrik Kook *

Clinic for Small Animal Internal Medicine, Vetsuisse Faculty, University of Zurich, 8057 Zurich, Switzerland; andreina.schramm@uzh.ch
* Correspondence: pkook@vetclinics.uzh.ch

Simple Summary: The majority of dogs with chronic idiopathic gastrointestinal (GI) disease are diet-responsive. We retrospectively evaluated what dietary information was gained when dogs were presented with chronic GI signs to either our Gastroenterology (GE) (n = 243) or Internal Medicine (IM) (n = 239) Service between 10/2017 and 01/2020. Referral documents mentioned the previously fed diet in 53/131 (40%) GE and 14/112 (13%) IM referrals. No dog had received more than one diet trial from the referring veterinarian. Diet trials had been performed in 127/199 (64%) GE and 56/156 (36%) IM dogs prior to presenting to our teaching hospital. The diet fed at the time of consultation could be named by 106/199 (53%) GE and 40/156 (26%) IM dog owners. Data on response to subsequent newly prescribed diets were available from 86 GE dogs and 88 IM dogs. A positive response to diet was noted in 50/86 (58%) GE and 26/88 (30%) IM dogs. A further 23/35 (66%) GE dogs and 12/21 (57%) IM dogs responded positively to a second diet trial, and 4/9 GE dogs (44%) and 6/7 (86%) IM dogs responded positively to a third diet trial. In conclusion, obtainable dietary information was infrequent. Positive response is still possible after initial diet failures. Further studies are now needed to determine if collecting complete dietary information at the time of consultations could lead to an increased percentage of diet-responsive disease.

Abstract: The majority of dogs with chronic idiopathic gastrointestinal (GI) disease respond to diet. So far, no study has assessed how much dietary information is obtained during consultations. We retrospectively evaluated what dietary information was available from dogs presenting to our Gastroenterology (GE), and Internal Medicine (IM) Service between 10/2017 and 01/2020. Data from 243 dogs presenting for first GE consultations were compared to 239 dogs presenting with chronic GI signs for first IM consultations. Referrals comprised 131 (54%) GE dogs and 112 (47%) IM dogs. Referral documents specified the previously fed diet in 53/131 (40%) GE and 14/112 (13%) IM dogs. No dog had received more than one previous diet trial for chronic GI signs. Irrespective of referral status, diet trials had been performed in 127/199 (64%) GE, and 56/156 (36%) IM dogs. The specific diet fed at the time of consultation could only be named by 106/199 (53%) GE and 40/156 (26%) IM dog owners. Data on response to subsequent newly prescribed diets were available from 86 GE dogs and 88 IM dogs. A positive response to diet was noted in 50/86 (58%) GE and 26/88 (30%) IM dogs. A further 23/35 (66%) GE dogs and 12/21 (57%) IM dogs responded positively to a second diet trial, and 4/9 GE dogs (44%) and 6/7 (86%) IM dogs responded positively to a third diet trial. In conclusion, overall dietary information gained from referring veterinarians and owners was often incomplete. More dietary information could be gained during GE consultations compared to IM consultations for chronic GI signs. A positive response to diet can still be seen after two diet failures. Further studies will help to ascertain if the percentage of diet-responsive GI disease increases when more complete dietary information is obtained at the time of consultations.

Keywords: food; history; response; owner; gastrointestinal; dog; referral

Citation: Schramm, A.; Kook, P.H. A Descriptive Study on the Extent of Dietary Information Obtained during Consultations at a Veterinary Teaching Hospital. *Animals* **2022**, *12*, 661. https://doi.org/10.3390/ani12050661

Academic Editors: Aarti Kathrani, Romy M. Heilmann and Edward J. Hall

Received: 25 November 2021
Accepted: 3 March 2022
Published: 6 March 2022

Publisher's Note: MDPI stays neutral with regard to jurisdictional claims in published maps and institutional affiliations.

1. Introduction

When assessing the response to therapy in dogs with chronic idiopathic gastrointestinal (GI) disease, food-responsive disease (FRD) has emerged as the most common entity, while antibiotic- and corticosteroid-responsive disease is less frequently documented [1]. This is best studied in dogs with chronic small intestinal enteropathy of varying severity [2–7], but also applies to dogs with chronic colitis [8], and dietary interventions are also successful when treating acute diarrhea [9,10]. Still, it is our long-term clinical experience that the majority of dog owners feeding commercial diets do not know what diet they have been feeding when asked during consultations. The situation is similar when assessing what dietary information is obtainable from referral documents. Not knowing the dog's current and previous diets understandably complicates the choice of a new diet and thus can compromise dietary success. Similarly, the literature has not investigated treatment success after a failed diet trial in dogs with GI disease and it is our experience that costly and invasive diagnostics such as endoscopy are not immediately indicated after one failed diet trial.

We hypothesized that information on diet (what diet was fed and response to diet) would be scarce in referral documents, and that the majority of dog owners would not be able to name diets that had been tried before or were fed at the time of consultation. We further hypothesized that less dietary information would be obtained from owners presenting their dogs to the Internal Medicine (IM) Service compared to owners presenting their dogs to a Gastroenterology (GE) Service, a subspecialty of our IM Service. An ancillary aim was to investigate how many dogs did in fact respond positively to diet after a first diet trial had failed.

2. Materials and Methods

2.1. Case Selection and Data Collection

All data were obtained from the hospital information system Vetera® (Vetera GmbH, 65344 Eltville am Rhein, Germany) at the Clinic for Small Animal Internal Medicine, Vetsuisse faculty, University of Zurich. Anamnestic information was recorded from dogs presented to the Gastroenterology (GE) Service, as well as to the general Internal Medicine (IM) Service for chronic GI signs.

Dogs presented for a first appointment between October 2017 and January 2020 were included. In order to be selected for the IM group, dogs had to present for chronic GI signs, and secondary GI disease (e.g., endocrine or renal disease) had to be ruled out. Because of the larger IM caseload, for each week the first 2 dogs presenting with chronic GI signs were selected for the IM group. Cases where it was apparent that a language barrier existed (for example when noted "owner speaks Russian, hardly any English") were excluded.

2.2. Data Handling and Analysis

Demographic data (breed, age, sex, body weight, and referral status (yes/no)) were compiled. At our hospital, owners can also schedule an appointment for a second opinion without having been referred. All pretreatments (dietary and medical) for chronic GI signs were recorded. Pretreatment was subdivided into the following categories: dietary, probiotics, antibiotics, corticosteroids, or immunosuppressive drugs other than corticosteroids

Available information on previous diet trials and current diet as well as information from re-checks at our hospital were extracted from the clinic system. In referred patients, the referral documents were screened for information on previous diet trials (number of diet trials, name of diets) and response to diet. Our records were checked to see if the following questions were asked: what diet(s) had been fed (name), how many diets were tried before consultation, for how long had they been tried, and what were the responses. Whether the following information could be obtained from our records was also noted: the name of the dog's current diet, if the diets were homemade or raw meat based (RMB), if the dog received any additional foods (e.g., cooked chicken, chicken hearts, cottage cheese) or other additives and supplements (e.g., vitamins, minerals, oils, fibers) or treats, and how many dogs received empirical dewormers.

It was further recorded if and what type of diet was prescribed by the attending clinician at our hospital. Diets were classified as highly-digestible (including low fat diets), hydrolyzed protein, or limited ingredient novel protein. Prescribed medical treatments were also recorded and categorized as treatment with probiotics, antibiotics, corticosteroids, and immunosuppressive drugs other than corticosteroids. If available, the same information was also recorded from first re-check consultations at our hospital.

The length of time the newly prescribed diet was fed and whether the response was positive was also recorded. Data on dietary compliance were recorded as compliant if the medical records mentioned that the owner was feeding the diet exclusively. Response to treatment was summarized as follows: 1. positive response to diet alone, 2. positive response to probiotics alone, 3. positive response to a combination of diet and probiotics, 4. positive response to antibiotics alone, or 5. positive response to corticosteroids alone.

Furthermore, medical files of all dogs were searched for information on additional diet trials performed at our hospital after the first re-check consultation. Length of diet trial and response was recorded. Due to the variation in the documentation of response to diet in our medical records, we summarized response to diet as either a "positive response to diet" or "negative response to diet". A positive response was defined as any improvement in GI signs, whereas negative response was defined as either no response or a worsening of signs. Therefore, unfortunately, this could not be further subdivided, for example into full or partial response. Microsoft® Excel (Version 16.35, for Mac) was used for data compilation, and calculation of median, ranges, and percentages.

3. Results

3.1. Study Population

A total of 243 dogs were included in the GE group and 239 dogs in the IM group. Demographic details can be found in Table 1. The most common breeds in the GE group were Crossbreeds (37/243, 15%), French Bulldogs (11/243, 5%), and Labrador Retrievers (10/243, 4%). While the most common dog breeds in the IM group were Crossbreeds (41/239, 17%), Poodles (13/239, 5%), and Labrador Retrievers (12/239, 5%).

Table 1. Demographic details on study population (GE = Gastroenterology, IM = Internal Medicine).

	GE Dogs (*n* = 243)	**IM Dogs (*n* = 239)**
Age (years), Median (range)	8 (1–19)	8 (1–18)
Sex	133/243 (55%) males 72/133 (54%) neutered 109/243 (45%) females 78/109 (72%) spayed	131/239 (55%) males 69/131 (53%) neutered 108/239 (45%) females 74/108 (69%) spayed
Body weight (kg), median (range)	11.1 (0.9–61)	13.6 (1.3–62.9)

3.2. Pretreatment

Information on pretreatment can be found in Table 2. Of those dogs that received probiotics, a high-dose (minimum of 200 billion bacteria) multi-strain probiotic [8] had been given to 7/23 (30%) of GE dogs and 2/25 (8%) of IM dogs. Treatment with empirical dewormers was recorded for 115 (47%) GE dogs, and 107 (45%) IM dogs.

Table 2. Pretreatment, summarized into categories (GE = Gastroenterology, IM = Internal Medicine).

	GE Dogs (*n* = 243)	**IM Dogs (*n* = 239)**
Probiotics	23/243 (9%)	25/239 (10%)
Corticosteroids	35/243 (14%)	26/239 (11%)
Other immunosuppressive drugs	5/243 (2%)	3/239 (1%)
Antibiotics	50/243 (21%)	49/239 (21%)

3.3. Dietary Information Available from Referring Veterinarians

Table 3 outlines information gained from referral documents. All referred cases had chronic (3 or more weeks) GI signs from the time the referral was made. None of the referral documents mentioned more than one diet trial for the GI signs.

Table 3. Diet-related information available from referring veterinarians for dogs presenting with chronic GI signs (GE = Gastroenterology, IM = Internal Medicine).

	GE Dogs	IM Dogs
Number of referred dogs	131/243 (54%)	112/239 (47%)
Previous diet mentioned in referral	53/131 (40%)	14/112 (13%)
Response to diet mentioned in referral	16/53 (30%)	9/14 (64%)

3.4. Dietary Information Gained during First Consultation

Table 4 outlines what dietary information could be gained during first consultations at our hospital. If more than one diet trial had been tried before consultation, no information on length of, or response to, those additional diet trials was available.

Table 4. Dietary information gained from owners of dogs with chronic GI signs during first consultation at our hospital (GE = Gastroenterology, IM = Internal Medicine).

	GE Dogs	IM Dogs
Clinician inquired on dog's diet	199/243 (82%)	156/239 (65%)
Previous diet trial performed	127/199 (64%)	56/156 (36%)
No. of previous diet trials (med., range)	1 (1–5)	1 (1–6)
Length of previous diet trial (med., range)	4 weeks (1–25)	2 weeks (1–8)
Response to previous diet trial recorded	109/127 (86%)	25/56 (45%)
Only manufacturer of diet could be named for current diet	122/199 (61%)	54/156 (35%)
Specific diet (manufacturer and product) currently fed named	106/199 (53%)	40/156 (26%)

Only 122 of 199 (61%) GE dog owners and 54/156 (35%) of IM dog owners could name the manufacturer (e.g., Royal Canin) of the current diet. The specific diet (manufacturer and product) currently fed could only be named by 106/199 (53%) GE dog owners and 40/156 (26%) IM dog owners. Available information on what types of diet were fed can be seen in Table 5.

Table 5. Type of diets fed at the time of first consultation at our hospital (GE = Gastroenterology, IM = Internal Medicine).

Type of Diet	GE Dogs *n* = 106	IM Dogs *n* = 40
Highly-digestible diet	40 (38%)	4 (10%)
Hydrolyzed protein diet	21 (20%)	1 (3%)
Limited ingredient novel protein diet	7 (7%)	1 (3%)
Other diet	37 (35%)	7 (18%)

3.5. Homemade and Raw Meat-Based Diets

The distribution of homemade diets and raw meat-based diets among diets fed at the time of first consultation are presented in Table 6.

Table 6. Homemade diets and raw meat-based diets (RMBD) fed at the time of first consultation to our hospital (GE = Gastroenterology, IM = Internal Medicine).

Type of Diet	GE Dogs *n* = 199	IM Dogs *n* = 156
Homemade diet	57/199 (29%)	31/156 (20%)
RMBD	24/199 (12%)	10/156 (6%)

3.6. Dietary Additives and Treats

Additional foods and additives were fed to 35/199 (18%) dogs in the GE group and 33/156 (21%) dogs in the IM group, while 19/199 (10%) GE dogs and 13/156 (8%) IM dogs received treats. Receiving "no treats" was specifically recorded in 4/199 (2%) GE dogs and in 1/156 (<1%) IM dogs.

The three most frequently used additional foods and additives in the GE group were: cooked chicken 5/35 (14%), cottage cheese 4/35 (11%), and Canikur® (Boehringer Ingelheim, Germany) (contains probiotics, clay, mannan-oligosaccharide) 2/35 (6%), and for the IM group: Cottage cheese 6/33 (18%), vitamins 3/33 (9%), and Bismutal® (Graeub AG, Switzerland) (contains carob, kaolin) 2/33 (6%).

3.7. Dietary and Medical Therapy Prescribed during First Consultation

A new diet (i.e., diet change) was prescribed to 118/243 (49%) GE dogs and 153/239 (64%) IM dogs. The type of diets prescribed can be seen in Table 7.

Table 7. Type of diets prescribed at first consultation at our hospital. Highly digestible diets included Royal Canin Veterinary Diet Sensitivity Control, Hill's Prescription Diet i/d, Hill's Prescription Diet i/d Sensitive, Hill's Prescription Diet i/d Low Fat. Hydrolyzed protein diets included Royal Canin Veterinary Diet Anallergenic, Hill's Prescription Diet z/d, Purina Pro Plan Veterinary Diets HA. Limited ingredient novel protein diets included Essendia Exclusion limited ingredient novel protein diets, Vet-concept limited ingredient novel protein (Sana) diets, Hill's Prescription Diet d/d. "Other diet" refers to any other diet. (GE = Gastroenterology, IM = Internal Medicine).

Type of Diet	GE Dogs n = 118	IM Dogs n = 153
Highly-digestible (incl. low-fat) diet	46 (39%)	95 (62%)
Hydrolyzed protein diet	36 (31%)	23 (15%)
Limited ingredient novel protein diet	17 (14%)	7 (5%)
Other diet	19 (16%)	28 (18%)

The medical treatments prescribed during first consultations at our hospital are listed in Table 8.

Table 8. Medical treatment prescribed at first consultation at our hospital. All high-dose multi-strain probiotics were SivoMixx® (Ormendes SA, 1008 Jouxtens-Mézery, Switzerland) (GE = Gastroenterology, IM = Internal Medicine).

Medical Treatment	GE Dogs n = 172	IM Dogs n = 224
High-dose multi-strain probiotics	105/172 (61%)	63/224 (28%)
Antibiotics	13/172 (8%)	45/224 (20%)
Corticosteroids	24/172 (14%)	19/224 (8%)
Other immunosuppressive drugs	6/17 (3%)	0 (0%)

3.8. Re-Check Consultations and Response to Diet and Medical Therapies

One hundred and fifty-nine of 243 (65%) GE dogs and 143/239 (60%) IM dogs were seen for a first re-check consultation after a median of 4 weeks (range 1–25 weeks) for GE dogs and 2 weeks (range 1–28 weeks) for IM dogs. Out of these, 92/159 (58%) GE dogs, and 89/143 (62%) IM dogs had received a new diet. The newly prescribed diet had been fed for a median of 3 weeks (range 1–24 weeks) in GE dogs and 2 weeks (range 1–22) in IM dogs.

Compliance to dietary and medical treatment was recorded for 60/92 (65%) GE dogs and 59/89 (66%) IM dogs. Compliance to dietary treatment alone was recorded for 9/12 (75%) GE dogs and 8/9 (89%) IM dogs.

Data on response to treatment were available for 143/159 (90%) GE dogs and 141/143 (99%) IM dogs. Data on response to new diet (and diet together with concurrent treatment) were available for 86/92 (93%) GE and 88/89 (99%) IM dogs and are presented in Table 9.

Table 9. Available data on positive response to diet change at first re-check at our hospital (GE = Gastroenterology, IM = Internal Medicine).

Positive Response to Treatment	GE Dogs (*n* = 86)	IM Dogs (*n* = 88)
Only diet change	19/86 (22%)	21/88 (24%)
Diet change with probiotics	31/86 (36%)	5/88 (6%)
Diet change with corticosteroids	0/86 (0%)	4/88 (5%)
Diet change with antibiotics	0/86 (0%)	6/88 (7%)

3.9. Positive Response to Second and Third Diet Trial

Thirty-five of 92 (38%) GE dogs and 21/59 (24%) IM dogs were prescribed a second diet. The second diet was fed for a median of 2 weeks in both groups (range GE 1–24 weeks; IM 1–22 weeks), 23 GE dogs (66%), and 12 IM dogs (57%) responded positively to the second diet. Data on a third diet trial were available in 9/35 (26%) GE dogs and 7/21 (33%) IM dogs. The third diet was fed for a median of 4 weeks (range 2–84) in GE dogs and 3 weeks (range 1–5 weeks) in IM dogs, 4 GE dogs (44%), and 6 IM dogs (86%) responded positively to the third diet. None of the dogs with a positive response to a second and third diet change received additional new treatments. Results can be seen in Table 10.

Table 10. Available responses to second and third diet change documented at our hospital. A positive response was defined as any improvement in GI signs, whereas a negative response was defined as either no response or a worsening of signs. (GE = Gastroenterology, IM = Internal Medicine).

Outcome after Second Diet Change	GE Dogs (*n* = 35)	IM Dogs (*n* = 21)
Positive response	23/35 (66%)	12/21 (57%)
Negative response	7/35 (20%)	4/21 (19%)
No data	5/35 (14%)	5/21 (24%)
Outcome after Third Diet Change	**GE Dogs (*n* = 9)**	**IM Dogs (*n* = 7)**
Positive response	4/9 (44%)	6/7 (86%)
Negative response	1/9 (11%)	0/7 (0%)
No data	4/9 (44%)	1(7 (14%)

3.10. Change in Diet Type from First to Second Diet Trial

Twenty-six of 35 (74%) GE dogs and 13/21 (62%) IM dogs changed diet type between first and second diet trial. For GE dogs, the three main changes were from highly-digestible to hydrolyzed protein diets: *n* = 5 (19%), hydrolyzed protein to other diets: *n* = 5 (19%), and other diets to highly-digestible diets: *n* = 4 (15%). For IM dogs, the main changes were highly-digestible to hydrolyzed protein diets: *n* = 5 (38%), and highly-digestible to limited ingredient novel protein diets: *n* = 3 (23%).

3.11. Change in Diet Type from Second to Third Diet Trial

Six of nine (67%) GE dogs and five of seven (71%) IM dogs changed diet type between the second and third diet trial. For GE dogs, the main change was from hydrolyzed protein to other diets: *n* = 3 (50%).

4. Discussion

In this retrospective study, we investigated how much dietary information (e.g., exact types of diets fed, length of diet trials) can be obtained from referring veterinarians and dog owners when presenting to our hospital for chronic GI signs. The amount of dietary information collected during consultations has not yet been addressed in veterinary medicine. We conducted this study to highlight an apparent contradiction: On the one hand, it has become increasingly clear that the majority of dogs with chronic idiopathic GI diseases can be controlled by diet alone [2,3,6,7]; on the other hand—despite the well-established importance of diet—many owners still do not know what they are feeding when they present their dogs for consultations. It is also our experience that not only dog owners but also referring veterinarians seem to underestimate the impact of diet on their patients' GI conditions.

At our institution, we offer a separate service within the Internal Medicine Service for patients presenting with GI problems. However, dogs with GI disease are also seen by the Internal Medicine Service as GE consultations are only available on 2 days per week compared to 5 days per week for IM consultations. This setting gave us the opportunity to compare dietary information taken at two services with different degrees of specialization and experience when treating dogs with GI disease. By doing so we could assess how results differ when dogs are seen by a larger group of veterinarians with differing interests in internal medicine versus a smaller service lead primarily by one internist focused on GE. We assumed that more dietary information would be retrievable from medical records if owners or referring veterinarians were seeking an appointment with the GE Service. Although our data confirmed this, diet was only mentioned in referral documents from 40% of cases referred to the GE Service and 13% referred to the IM Service. Referral documents including a diet history specification section might help to improve this. Unexpectedly, information on the actual response to previously tried diets was more frequently available in the IM group. However, this may have been due to selection bias as diet was overall only mentioned in 14/112 IM cases. We assume that one of the reasons for the low percentages of a specific mention of how dogs had responded to the previously tried diet, especially in the GE group, was that as the GI signs still persisted and necessitated a referral, a failed response was obvious and therefore mention of this was not warranted. The fact that none of the referral documents (GE and IM group) mentioned more than one previous diet trial may reflect the perception that failure to respond to one diet is synonymous with a general non-responsiveness to diet. Another possibility is that dog owners were not willing to pursue further diet trials, but rather wanted a second opinion or further diagnostic tests performed. In addition, dog owners might not perceive dietary treatment to be as effective as drug treatment in cases of GI disease. In order to better understand which factors motivate owners to try a diet for their dogs with GI signs, prospective survey studies are needed.

More owners in the GE group were asked about their dog's diet (82 vs. 65% IM group). Therefore, 18% of owners coming in for a GE consultation were not asked about their dog's diet. It is possible that nothing was recorded in cases where owners did not know what they were feeding. Alternatively, lacking data here may have been due to negligent documentation. Irrespective of referral status, approximately two-thirds in the GE group and only one-third of dogs with chronic GI signs seen by the IM Service had a previous diet trial recorded, and clearly more information on response to previous diet trials were recorded in the GE group (86%) vs. IM group (45%). Again, not recording response rates may have been synonymous with the tacit assumption that GI signs did not improve with the mentioned diet.

The most impressive result of this study is that only 106/199 (53%) dog owners seeking medical advice from the GE Service could in fact name the specific diet their dog was receiving at the time of consultation. Even worse, only 40/156 (26%) owners in the IM group knew the specific diet currently fed to their dog. Frequently, owners assumed they knew what they were feeding when they could only recall the manufacturer of the diet (e.g., Royal Canin); not realizing that one manufacturer can produce a multitude of different diets. It is not uncommon for the last author to sit together with dog owners in front of the computer screen searching for images of what their dog's food bag looks like in an attempt to identify the current diet—knowing so little about previously fed diets is probably one reason for the frequent selection of a highly digestible or hydrolyzed protein diet for the first diet trial, as more specific aspects such as dietary protein remain often unknown.

Approximately 20% of all dogs were fed additional foods besides their diet, and 10% received treats. These numbers are higher than what have been reported for healthy dogs from the U.S. and Australia [11]. This is likely because owners of sick dogs are keen to ensure that their dog consumes at least some food. It is possible that our results are still an underestimation, as we cannot guarantee with a retrospective study design that these questions were routinely asked, especially as treats are fed more often according to other studies [11,12]. Additional foods or treats besides the prescribed diet may be responsible for diet failure. However, only very few studies on chronic enteropathies in dogs have excluded treats in their study design [13–15]. The situation is similar when assessing studies on acute diarrhea in dogs [9].

Considerably more dogs in the GE group received high-dose multi-strain probiotics often together with a new diet compared to the IM group, which is likely related to an increased awareness of the benefits of probiotics in the GE Service. The combination of diet with a probiotic makes it difficult to assess which treatment the dog ultimately responded to. Explicit mention of exclusive positive response to probiotics was noted in only a handful of cases. Further differentiation between diet and probiotic responsiveness might be interesting for future studies on therapeutic outcome, as a recent study showed clinical, histological, and immunomodulatory benefits of a high-dose multi-strain probiotic in dogs with inflammatory bowel disease when all dogs remained on their pre-trial diets [16].

Approximately 150 dogs (=60%) were seen for a first re-check in both groups. The reason for longer time intervals between first consultation and re-check between groups was that the GE Service offers less consultation appointments compared to the IM Service. Compliance to diet, either as the sole treatment or more commonly in combination with other treatments, was recorded in two-thirds of cases in both groups (GE 69/104 (66%), IM 67/98 (68%)). This is comparable to dietary compliance reported in dogs with atopic skin disease (73% dietary compliance) [17]. Similar dietary compliance (69.6%) of owners was shown in a study on homemade diets in dogs [18]. Interestingly, no data on compliance to therapy could be recorded during later re-checks. Our compliance data imply that one-third of dogs were either not fed the newly prescribed diet exclusively or were not given medication as prescribed. However, it is also possible that clinicians were not asking and recording compliance data. Dogs treated with diet only, had a higher compliance (GE 75%, IM 89%); most probably as these dogs had milder disease and thus less inappetence. Understandably, lacking compliance can affect response rates to diet. So far, this factor has not been specifically considered in studies on dogs with GI disease. Dietary compliance is discussed as a key point of success for the therapy of chronic enteropathies [1,6,7,19,20], but specific documentation of dietary compliance is found very rarely in the literature [4,21,22]. Prospective studies are needed for a more precise characterization of dietary compliance.

When assessing all available data from all dogs receiving a new diet, including dogs that were additionally treated with probiotics, a positive response to diet was recorded in 50/86 (58%) GE dogs and 26/88 (30%) IM dogs. These numbers might be higher, had more owners known what they were feeding at first consultation because two-thirds of 35 dogs in the GE group still responded positively to a second diet trial after failing the first diet. However, additional studies would be needed to assess this. A positive response to a stringent diet after referral despite previous unsuccessful diet trials has been reported [13], but it is usually not specified in the literature how many diets were tried before dogs were designated as FRD or not. It is conceivable that this is one of the reasons why none of the referral letters contained information on more than one previous diet trial.

A second diet was tried less commonly in the IM group, which may have been due to different awareness of the importance of diet. When planning the study, we had expected more positive responses to a third diet change. We believe the comparably small data set on responses to a third diet trial was due to the fact that only consultations at the hospital and no telephone updates were included. However, when taking both groups into consideration, 10/16 (63%) dogs still responded to diet after having failed two previous diet trials.

When taking into account all positive responses to all three diet trials, then 77/86 (90%) GE dogs, and 44/88 (50%) IM dogs responded positively to diet. These figures must be viewed with caution, as a standardized and graduated assessment of response to diet was not available. In addition, numbers depend on the underlying disease process and how convincingly the relevance of diet has been communicated to owners in both services. The difference between services is certainly also influenced by the low rate of identified diets at first consultation in the IM group (26%).

The variability of diet trial duration in our study might have been due to owner specifics (i.e., differing patience or expectations regarding dietary success), differing recommendations of attending clinicians, and also in part due to limited availability of follow-up appointments.

The limitations of our study are the typical limitations of a retrospective study design. Whenever data were not available, we did not know if this was on behalf of the attending clinician not asking all required questions (e.g., compliance), negligent documentation, or owner unable to recall the specific details of the diet fed. Similar to large-scale studies on long-term outcome in dogs with CE [2], we graded the response to diet subjectively (improved versus not improved) and did not document specifically which clinical signs had improved and by how much. Our primary goals were to investigate how much dietary information could be gained from referring veterinarians and from owners of dogs with chronic idiopathic GI signs, and also how many dogs still improve clinically with diet after a failed diet trial. A full diagnostic work-up including endoscopy was not a prerequisite of this study. Therefore, results may differ when related to a more defined disease process in future studies, as diet can play a main role in the therapy or represent a supportive measure depending on the diagnosis. It would be interesting to compare our results to other institutions, but this is the first study evaluating the extent of obtainable dietary information during a veterinary consultation.

5. Conclusions

Dietary information gained from referring veterinarians and owners of dogs presented for chronic GI signs is often incomplete or lacking. Dietary information appeared more often complete when dogs were presented to the GE Service, but this was still just over half of the cases. A substantial percentage of dogs still improve clinically after having failed a first diet trial. Further studies will help to ascertain if the percentage of diet-responsive GI disease increases when more complete dietary information is obtained at the time of consultations.

Author Contributions: Conceptualization, P.H.K.; methodology, P.H.K.; validation, P.H.K.; formal analysis, A.S.; investigation, A.S.; data curation, A.S.; writing—original draft preparation, A.S.; writing—review and editing, P.H.K.; supervision, P.H.K.; project administration, P.H.K. All authors have read and agreed to the published version of the manuscript.

Funding: This research received no external funding.

Institutional Review Board Statement: Ethical review and approval were waived for this study as the study was retrospective.

Data Availability Statement: Data are only available on request due to restrictions, e.g., privacy or ethical reasons. As the study is retrospective, no owner consent was signed and personal data are therefore not publicly available.

Conflicts of Interest: The authors declare no conflict of interest.

References

1. Dandrieux, J.R.S. Inflammatory bowel disease versus chronic enteropathy in dogs: Are they one and the same? *J. Small Anim. Pract.* **2016**, *57*, 589–599. [CrossRef] [PubMed]
2. Allenspach, K.; Culverwell, C.; Chan, D. Long-term outcome in dogs with chronic enteropathies: 203 cases. *Vet. Rec.* **2016**, *178*, 368. [CrossRef]
3. Volkmann, M.; Steiner, J.M.; Fosgate, G.T.; Zentek, J.; Hartmann, S.; Kohn, B. Chronic Diarrhea in Dogs—Retrospective Study in 136 Cases. *J. Vet. Intern. Med.* **2017**, *31*, 1043–1055. [CrossRef] [PubMed]

4. Dandrieux, J.R.S.; Martinez Lopez, L.M.; Prakash, N.; Mansfield, C.S. Treatment response and long term follow up in nineteen dogs diagnosed with chronic enteropathy in Australia. *Aust. Vet. J.* **2019**, *97*, 301–307. [CrossRef] [PubMed]
5. Kawano, K.; Shimakura, H.; Nagata, N.; Masashi, Y.; Suto, A.; Suto, Y.; Uto, S.; Ueno, H.; Hasegawa, T.; Ushigusa, T.; et al. Prevalence of food-responsive enteropathy among dogs with chronic enteropathy in Japan. *J. Vet. Med. Sci.* **2016**, *78*, 1377–1380. [CrossRef] [PubMed]
6. Rudinsky, A.J.; Howard, J.P.; Bishop, M.A.; Sherding, R.G.; Parker, V.J.; Gilor, C. Dietary management of presumptive protein-losing enteropathy in Yorkshire terriers. *J. Small Anim. Pract.* **2017**, *58*, 103–108. [CrossRef]
7. Tørnqvist-Johnsen, C.; Campbell, S.; Gow, A.; Bommer, N.X.; Salavati, S.; Mellanby, R.J. Investigation of the efficacy of a dietetic food in the management of chronic enteropathies in dogs. *Vet. Rec.* **2020**, *186*, 26. [CrossRef]
8. Rossi, G.; Cerquetella, M.; Gavazza, A.; Galosi, L.; Berardi, S.; Mangiaterra, S.; Mari, S. Rapid Resolution of Large Bowel Diarrhea after the Administration of a Combination of a High-Fiber Diet and a Probiotic Mixture in 30 Dogs. *Vet. Sci.* **2020**, *7*, 21. [CrossRef]
9. Wennogle, S.A.; Martin, L.; Oleo-Popelka, F.J.; Xu, H.; Jean-Phillipe, C.; Lappin, M.R. Randomized trial to evaluate two dry therapeutic diets for shelter dogs with acute diarrhea. *Int. J. Appl. Res. Vet. Med.* **2016**, *14*, 30–37.
10. Werner, M.; Suchodolski, J.S.; Straubinger, R.K.; Wolf, G.; Steiner, J.M.; Lidbury, J.A.; Neuerer, N.; Hartmann, K.; Unterer, S. Effect of amoxicillin-clavulanic acid on clinical scores, intestinal microbiome, and amoxicillin-resistant Escherichia coli in dogs with uncomplicated acute diarrhea. *J. Vet. Intern. Med.* **2020**, *34*, 1166–1176. [CrossRef]
11. Freeman, L.M.; Abood, S.K.; Fascetti, A.J.; Fleeman, L.M.; Michel, K.E.; Laflamme, D.P.; Bauer, C.; Kemp, B.L.E.; Van Doren, J.R.; Willoughby, K.N. Disease prevalence among dogs and cats in the United States and Australia and proportions of dogs and cats that receive therapeutic diets or dietary supplements. *J. Am. Vet. Med. Assoc.* **2006**, *229*, 531–534. [CrossRef] [PubMed]
12. Becker, N.; Dillitzer, N.; Sauter-Louis, C.; Kienzle, E. Feeding of dogs and cats in Germany. *Tierarztl Prax Ausg. K Kleintiere-Heimtiere* **2012**, *40*, 391–397. [PubMed]
13. Mandigers, P.J.J.; Biourge, V.; Van Den Ingh, T.S.G.A.M.; Ankringa, N.; German, A.J. A Randomized, Open-Label, Positively-Controlled Field Trial of a Hydrolyzed Protein Diet in Dogs with Chronic Small Bowel Enteropathy. *J. Vet. Intern. Med.* **2010**, *24*, 1350–1357. [CrossRef] [PubMed]
14. Kalenyak, K.; Heilmann, R.M.; van de Lest, C.H.A.; Brouwers, J.F.; Burgener, I.A. Comparison of the systemic phospholipid profile in dogs diagnosed with idiopathic inflammatory bowel disease or food-responsive diarrhea before and after treatment. *PLoS ONE* **2019**, *14*, e0215435. [CrossRef]
15. Marks, S.; Laflamme, D.P.; McAloose, D. Dietary trial using a commercial hypoallergenic diet containing hydrolyzed protein for dogs with inflammatory bowel disease. *Vet. Ther.* **2002**, *3*, 109–118.
16. Rossi, G.; Pengo, G.; Caldin, M.; Palumbo Piccionello, A.; Steiner, J.M.; Cohen, N.D.; Jergens, A.E.; Suchodolski, J.S. Comparison of microbiological, histological, and immunomodulatory parameters in response to treatment with either combination therapy with prednisone and metronidazole or probiotic VSL#3 strains in dogs with idiopathic inflammatory bowel disease. *PLoS ONE* **2014**, *9*, e94699. [CrossRef]
17. Loeffler, A.; Lloyd, D.H.; Bond, R.; Kim, J.Y.; Pfeiffer, D.U. Dietary trials with a commercial chicken hydrolysate diet in 63 pruritic dogs. *Vet. Rec.* **2004**, *154*, 519–522. [CrossRef]
18. Oliveira, M.C.C.; Brunetto, M.A.; da Silva, F.L.; Jeremias, J.T.; Tortola, L.; Gomes, M.O.S.; Carciofi, A.C. Evaluation of the owner's perception in the use of homemade diets for the nutritional management of dogs. *J. Nutr. Sci.* **2014**, *3*, e23. [CrossRef]
19. Gaschen, F.P.; Merchant, S.R. Adverse Food Reactions in Dogs and Cats. *Vet. Clin. N. Am. Small Anim. Pract.* **2011**, *41*, 361–379. [CrossRef]
20. Busch, K.; Rade, C.; Unterer, S. Inflammatory Bowel Disease in dogs. *Kleintierpraxis* **2019**, *64*, 291–307. [CrossRef]
21. D'Angelo, S.; Fracassi, F.; Bresciani, F.; Galuppi, R.; Diana, A.; Linta, N.; Bettini, G.; Morini, M.; Pietra, M. Effect of Saccharomyces boulardii in dogs with chronic enteropathies: Double-blinded, placebo-controlled study. *Vet. Rec.* **2018**, *182*, 258. [CrossRef] [PubMed]
22. Biourge, V.C.; Fontaine, J.; Vroom, M.W. Diagnosis of Adverse Reactions to Food in Dogs: Efficacy of a Soy-Isolate Hydrolyzate-Based Diet. *J. Nutr.* **2004**, *134* (Suppl. S8), 2062S–2064S. [CrossRef] [PubMed]

MDPI
St. Alban-Anlage 66
4052 Basel
Switzerland
www.mdpi.com

Animals Editorial Office
E-mail: animals@mdpi.com
www.mdpi.com/journal/animals